ELECTROLYSIS

Edited by **Janis Kleperis** and **Vladimir Linkov**

Electrolysis
http://dx.doi.org/10.5772/2820
Edited by Janis Kleperis and Vladimir Linkov

Contributors

A.M. Fernández, U. Cano, Martins Vanags, Janis Kleperis, Gunars Bajars, Bernard Bladergroen, Huaneng Su, Sivakumar Pasupathi, Vladimir Linkov, Aleksey Nikiforov, Erik Christensen, Irina Petrushina, Jens Oluf Jensen, Niels J. Bjerrum, Shawn Gouws, Ruyao Wang, Weihua Lu, Takeshi Muranaka, Nagayoshi Shima, Yuji Imashimizu, A. H. Sulaymon, A. H. Abbar, Gustavo Stoppa Garbellini, Ehsan Ali, Zahira Yaakob, Fumio Okada, Kazunari Naya, Thomas J. Goreau

Published by InTech

Janeza Trdine 9, 51000 Rijeka, Croatia

Publishing Process Manager Vedran Greblo
Technical Editor InTech Prepress, Novi Sad
Cover InTech Design team

Additional hard copies can be obtained from orders@intechopen.com

Electrolysis, Edited by Janis Kleperis and Vladimir Linkov
p. cm.
ISBN 978-953-51-0793-4

For EU product safety concerns:
IN TECH d.o.o., Prolaz Marije Krucifikse Kozulić 3, 51000 Rijeka, Croatia,
info@intechopen.com or visit our website at intechopen.com.

Meet the editors

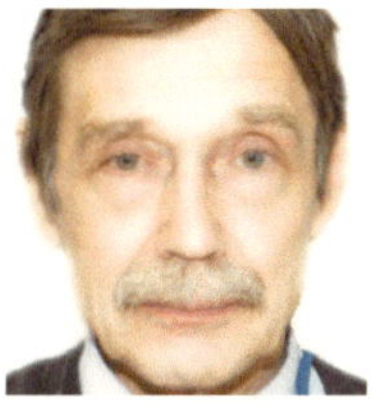

Janis Kleperis, Dr.Phys., is a senior research scientist and Head of the Laboratory of Hydrogen Energy Materials at Institute of Solid State Physics, University of Latvia. He participated in various local and international projects related to material science, energy and environment. Dr. Kleperis is an author of more than 150 scientific publications. His fields of specialization are solid state ionics, electrochromic phenomena, chemical gas sensors, instruments with artificial intellect (electronic nose), hydrogen gas sensors, materials for hydrogen storage, electrode materials for lithium and metal hydride batteries; water electrolysis, polymer membranes with mixed conductivity; differential optical absorption spectroscopy for air pollution monitoring. J. Kleperis is the member of International Society of Electrochemistry.

Vladimir Linkov, Prof Having graduated from Moscow State University and Stellenbosch University, Vladimir worked as a Senior Lecturer and Research Coordinator at SASOL Center for Separation Technology at Potchefstroom University. He was appointed Full Professor at UWC in 1997 and the same year organized Inorganic Porous Media Group at Chemistry Department. Further, he had initiated and currently leads South African Institute for Advanced Materials Chemistry (SAIAMC) at UWC. In 2003 Vladimir organized ESKOM Center for Electrocatalytic Research (ECER), which became a part of SAIAMC. Prof. Linkov started first Hydrogen and Fuel Cell research programme in South Africa (2001) that became the technical foundation of the Hydrogen and Fuel Cell National Strategy (HySA). As a part of HySA development, UWC was selected to host DST Competence Center for System Integration and Technology Validation (HySA Systems) of which Vladimir is currently Contract Director. Prof. Linkov is a recipient of many awards and has published more than 150 papers in peer-reviewed journal.

Contents

Preface

The electrolysis reaction is a process of solution splitting or solution separation in components by passing an electric current through two electrodes immersed in this solution. Water electrolysis is known for more than 200 years and is important for future hydrogen economics as clean and efficient method for hydrogen obtaining using electricity from renewables (wind, sun, water flow, geothermal heat). Hydrogen offers a significant promise as a basis for the future energy technology both for mobile and stationary applications, and is argued to be the most versatile, efficient and environmentally friendly fuel. In chemistry, metallurgy and manufacturing the electrolysis is commercially highly important because in an electrolytic cell the separation of elements from naturally occurring sources such as ores occur. The application of electrolysis to solve environmental problems became pending matter from the middle of the last century. Nowadays electrolysis, electrocoagulation, electrochemical ozonation techniques are indispensable tools to promote the trends of sustainability in the existing industrialized world. For example, in medicine, in order to solve some personal hygiene problems (e.g. removal of hairs) electrolysis is used from 1875 (Dr. Charles E. Michel (1833 - 1913), a St. Louis, Missouri ophthalmologist). It is difficult to name a single sector of human activities, where electrolysis is not required.

Most people consider that electrolysis was discovered just after Voltaic cell (1800; UK researchers A. Carlisle and W. Nicholson), when the potential from galvanic cell was applied to two electrodes in water and release of gas bubbles observed. Already in 1789 two Dutchmen A.P. van Troostwijk and R. Deiman published their researh about water splitting in burning gas (hydrogen and oxygen in a stoichiometric proportion 2:1) as a result of spark over jumping in the electrostatic generator in the presence of water. This phenomenon was given the name *electrolysis* (from the Greek ἤλεκτϱον "amber" and *λύσις* "dissolution"). Since then, discovery stimulated the scientific understanding on the behavior of electrons and ions in solution with atomic level and the role of electrode/electrolyte interface. In 1832, Michael Faraday reported that the quantity of elements separated by electric current passing through a molten or dissolved salt is proportional to the quantity of electric charge passed through the circuit. This became the basis of the first law of electrolysis, and second law also was formulated by M. Faraday: when the same amount of electricity is passed through different electrolytes, the mass of substance liberated/deposited at the electrodes is directly proportional to their equivalent weights. By using different additional stimulus (temperature, pressure, light, magnetic field, centrifugal force etc) during electrolysis, it is shown that Faraday electrolysis laws may fail.

Asking Google for different applications of electrolysis, interesting results can be noticed: electrolysis and hair removal - about 284 000 000 web pages, water electrolysis – about 7 440 000 web pages and industrial electrolysis - about 2 540 000 results. In this book three most important applications of technological electrolysis are discussed – water electrolysis (hydrogen production), industrial electrolysis and environmental electrolysis. Authors of the chapters are recognized specialists in their respective research fields and the presented material is not only from reviews and literature sources, but also original results. We hope that the reader will find useful information in the chapters of this book and are certain that the science can reveal unexpected discoveries even tomorrow, if current progress is at hand or on a shelf.

Dr. Janis Kleperis
Hydrogen Laboratory
Institute of Solid State Physics
University of Latvia,
Latvia

Vladimir Linkov
South African Institute for Advanced Materials Chemistry,
University of the Western Cape,
South Africa

Section 1

Water Electrolysis

Chapter 1

Alkaline Electrolysis with Skeletal Ni Catalysts

A.M. Fernández and U. Cano

Additional information is available at the end of the chapter

http://dx.doi.org/10.5772/50617

1. Introduction

1.1. Hydrogen as a fuel: Properties and sources

Hydrogen has been recognized as a potential energy vector for the energy future. Its high energy mass density (108,738 J/g, three times as much energy as 1 gram of gasoline), the possibility of obtaining hydrogen from many sources, the high efficiency with which its energy is extracted within fuel cells and converted into electricity, and the fact that its use does not produce any harmful emission, make of hydrogen the most attractive fuel for the new energy scenarios. These scenarios include high energy conversion efficiency technologies, zero emissions and use of sustainable and clean fuels.

The production of hydrogen needs both a hydrogen containing compound and energy to extract it. In general the processes used for its production can be divided in three: thermal, biological and electrochemical. In thermal processes, the most commercially developed, a hydrogen containing compound such as natural gas (i.e. methane) is catalytically transformed in the presence of water steam which provides thermal energy, a process known as methane steam reforming (MSR). The result of such reaction is a hydrogen rich mixture of gases which later go through an enriching (water shift reaction) and a purification stage (typically pressure swing adsorption or PSA). New processes less thermally demanded are being developed to lower hydrogen price. This type of process can also be used with different hydrogen containing compounds but the longer their molecules the more difficult is the extraction of hydrogen. In general MSR and other similar paths depend strongly on feedstock prices, i.e. natural gas prices, and are not 100% clean methods due to CO_2 generated during hydrogen production. Some approaches to this, consider carbon sequestration in conjunction with clean electricity generation to gain from hydrogen benefits. Other thermal processes include chemical cycles where a "commodity" product (intermediate chemical compound) is generated to store primary energy in order to later use it for hydrogen production. Such processes are being explored but have not reached yet competitive costs compared to commercial hydrogen.

In biological processes the activity of some living beings is either part of or completely responsible for the production of hydrogen. Some bacteria and other microorganisms, e.g. algae, are capable of generating hydrogen during their metabolism in their biological cycle. There have been hundreds of potential living systems identified that are able to generate hydrogen. Nevertheless, most of these are still under study as their production rates are low and costly. In electrochemical processes for the generation of hydrogen, photolysis is an attractive path as solar energy could be a cheap energy supply. Such systems are still in development and might arrive years ahead. The other alternative to steam reforming of methane with a commercial status is the electrolysis of water, from which two routes outstand, acid electrolysis and alkaline electrolysis. In the former, a solid acid electrolyte offers advantages as compact and less complex systems can be built. Despite acid electrolysis costs are decreasing, the maturity of alkaline electrolysis together with its generally lower costs can be advantageous as alkaline normally tend to use less noble metals as electrocatalyzer and in general kinetics tend to be better.

Once hydrogen is produced, its final use will determine whether the gas needs or not to be conditioned. In low temperature fuel cells hydrogen needs a high purity grade which gives electrolytic hydrogen an advantage over hydrogen from MSR which would need a purification stage. Electrolytic hydrogen from an alkaline process needs to scrub any traces of KOH to accomplish required purity. MSR on the other hand would not need a purification stage if produced hydrogen is going to a high temperature fuel cell, as its product can be directly fed for electricity production without much conditioning. If hydrogen is to be stored, either as a pressurized gas or in modern metallic hydride systems, hydrogen would need to be dried.

As it was mentioned above, the production of hydrogen needs a primary energy source and some type of feedstock from where hydrogen will be extracted. The combination of available primary energy sources and available feedstock give hydrogen production an almost infinite number of alternative routes. Nevertheless, electrolysis of water in combination with renewable primary energy resources is a very attractive scenario as the combination of both renewable energy and water could represent a real sustainable technology-ready alternative, especially considering that fuel cell technologies have gained credibility as a device to efficiently generate electricity from hydrogen fuel.

As an energy carrier hydrogen lends itself to build distributed generation (DG) systems to further gain efficiencies lost during traditionally centralized power plants for stationary applications like commercial or residential uses of electricity. Hydrogen is seen by many as an energy vector, similar to electricity, which can be produced in one place and consumed somewhere else. It may also be produced on-demand when economics are fulfilled. Transportation applications, for example, could have hydrogen stations where hydrogen does not need to be stored in large quantities nor needs to be transported if generated and dispatched locally.

All in all hydrogen is a very attractive path for sustainable societies, but technological and economic challenges still remain. One of these challenges includes hydrogen production at a competitive cost and those costs are directly related to the source of primary energy to produce it (up to 75% for alkaline electrolysis, according to the IEAHIA - [1]. With the blooming market of renewable energy technologies such as photovoltaic systems and wind energy turbines, electrolysis seems to have a great opportunity as a hydrogen production method. If this route accomplishes the renewable central hydrogen production cost goal of US$2.75/gge (gasoline gallon equivalent) delivered, for 2015 established by hydrogen programs such as the HFCIT of DOE [2], no doubt we will see an increase in electrolysis products in the near future.

1.2. Hydrogen storage

Despite the fact that hydrogen presents a high energy density superior to that of conventional fuels, hydrogen is a gas difficult to compress as its compression also demands high energy in order to store it for its use. Nowadays storage tanks for hydrogen gas can be commercially bought for 350 and 700bar capacities. In its last Hydrogen and Fuel Cells Program Plan [2], the U.S. Department of Energy reported a novel "cryo-compressed" tank concept that achieved a system gravimetric capacity of 5.4% by weight and a volumetric system capacity of approximately 31 g/L. Nevertheless, the energy associated with compression and liquefaction needs to be considered for compressed and liquid hydrogen technologies. Other systems based on hydrogen absorbing materials and hydrogen compounds for storing hydrogen are been studied but their energy density (both weight and volume) still needs to reach competitive values to be considered a viable paths especially for mobile applications. Energy needed to extract hydrogen from some storage systems (for example solid-state materials) as well as their life cycle (metal hydrides) also would need to be cost competitive. Certainly hydrogen storage is still a challenge for many hydrogen applications which need yet to be addressed.

2. Alkaline electrolysis

Alkaline electrolysis is considered a mature technology with many decades of commercially available products for the production of hydrogen gas. This industry grew substantially during the 1920s and 1930s. Alkaline electrolysis can be described as the use of an electrical current passing through an electrolysis cell causing the decomposition of water to generate hydrogen gas on the cell's cathode. In an electrolysis cell two electrodes containing an electrocatalyst, separated by a physical barrier, a polymeric diaphragm which only allows the passing of ions from one electrode to another, are connected to a d.c. current source. Each cell will have a high specific active area due to its porous structure and will produce hydrogen gas at the cathode while oxygen will be produced in the anode. The electrodes are in contact with an electrolyte which provides OH^- ions, from a 20%–30% solution of potassium hydroxide that completes the electrical circuit. In theory, 1kg of hydrogen would need a little less than 40kWh of electricity. Although theoretical water decomposition voltage is 1.23V (which corresponds to a theoretical dissociation energy of 286 kJ/mol or 15.9 MJ/kg at 25°C), in practice this voltage goes normally to around 2V per cell, while the total applied current depends on the cell's active area, and on the electrolyzer configuration (bipolar or unipolar). In the first case current densities (A/cm2) are higher than in unipolar configurations. There exist certain advantages in bipolar systems as they can operate under pressurized conditions, reducing or facilitating a compression stage. Higher current densities and generally a smaller foot-print are other positive characteristics of bipolar electrolyzers. As electrodes are placed closer, voltage drop from ohmic resistances are minimized saving in energy costs.

As mentioned earlier, there exist other electrolyzers that use a solid acid electrolyte based on an ionic conducting electrolyte. Those will not be treated in this document/section.

Electrode and global reactions in an alkaline electrolyzer are as follows:

$$\text{Cathode: } 2H_2O + 2e^- \rightarrow H_2 + 2OH^- \tag{1}$$

$$\text{Anode: } 2OH^- \rightarrow \tfrac{1}{2}O_2 + H_2O + 2e^- \tag{2}$$

$$\text{Global reaction: } 2H_2O \rightarrow H_2 + \tfrac{1}{2}O_2 \tag{3}$$

The operating temperature is typically around 80-90°C producing pure hydrogen (>99.8%). Higher temperatures are being also used as the electrolysis of water steam decreases electricity costs. Pressure is another characteristic operating condition which goes from 0.1 to around 3MPa. As with temperature, higher pressures help to decrease energy costs particularly as this generates high pressure hydrogen lowering compression energy when hydrogen will be stored or dispatched as a pressurized gas. Some approaches of cero pressure have been proposed generating then low pressure (about 40 bar — [3]) hydrogen without any mechanical compressor. An approach to pressurized electrolyzers sometimes comprises a first stage where hydrogen is generated at about 200-500psi (13.8-34.5 bar) followed by a mechanical compression second stage to upgrade it to 5000psi (3447 bar). Commercial units often use nickel-coated steel in their electrodes but other proprietary materials are associated too to some products. The efficiency of alkaline electrolysis is about 70% LHV (Lower Heating Value), where hydrogen cost is associated to electricity costs, one of the main technological challenges along with the wholesale manufacture which also needs to improve in order to reach technology costs goals for competitiveness. Approximately a little less than 1 liter of de-mineralized water is needed to generate 1Nm^3 of hydrogen. Once produced, hydrogen goes through a series of conditioning stages as seen in figure below (from [4]):

The main goal is to eliminate traces of KOH, water and oxygen. In the system diagram shown the transformer/rectifier plays an important role in converting A.C. current to D.C. current which is actually the type of current involved in the electrochemical process of electrolysis. As mentioned earlier, the actual cost of hydrogen will be dependent on the source of electricity. Therefore renewable energy, especially as D.C. electricity, can greatly contribute to a more competitive hydrogen cost. It is reported also that electrolysis is better respondent to a load-following condition compared to other hydrogen production methods (e.g. MSR), making renewable energy related hydrogen a more suitable energy integrated system.

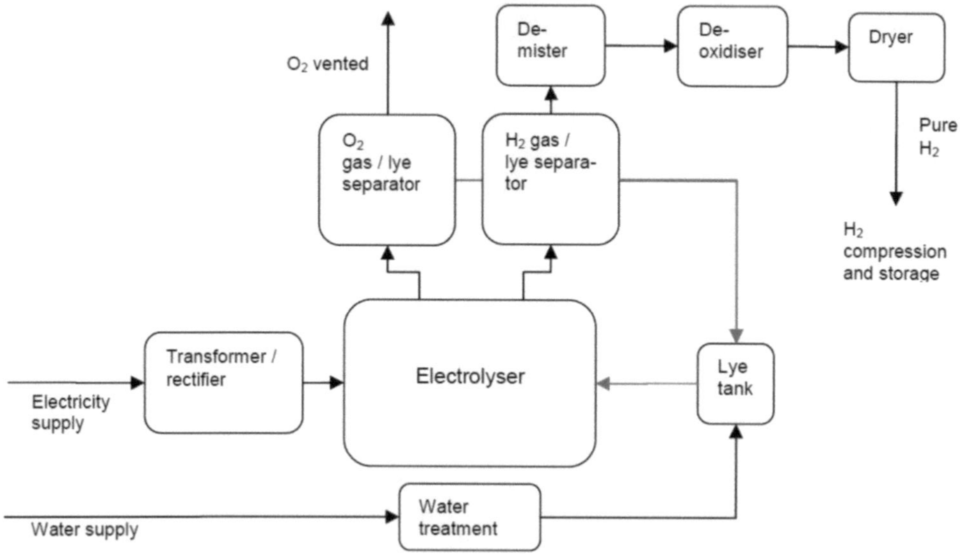

Figure 1. Schematic diagram of an alkaline electrolyzer [4]

2.1. Alkaline electrolysis components:

2.1.1. Electrodes

One of the materials most used in the hydrogen industry for alkaline electrolysis is nickel. This is due to its excellent catalysts properties and its corrosion resistance at the high pH values of KOH electrolyte in particular at the highly imposed anodic potential for the oxidation reaction. Often Pt alone or together with Ni, is used in commercial products improving electrode kinetics performance but this increases the overall costs of the unit. Electrodes are manufactured so that they show a large electroactive area, therefore porous electrodes are normally made to produce large density currents and better hydrogen production rates. One type of nickel electrode material used is Ni Raney or skeletal Ni, a material developed almost 100 years ago as a catalyst for hydrogenation of oils. Skeletal Ni is prepared from a Ni-Al alloy at specific concentration depending on the desired properties. This alloy is then leached in an alkaline solution to dissolve an aluminum containing phase leaving behind a porous structure. During the leaching process there is production of hydrogen which is said to remain in the porous making the resulting material pyrophoric and difficult to handle. Also the remaining aluminum can act as a trap for hydrogen as this metal is known to adsorb hydrogen molecules, which dissociate to produce hydrogen atoms within the metal- [6]. This hydrogen is believed to serve as a hydrogen-"prepared" surface for the further hydrogen evolution process at the cathode regardless of the reaction mechanism [5]. The following figures show a microstructure of the Ni-Al alloy and its typical phases formed after alloying (a) and after the leaching process (b and c).

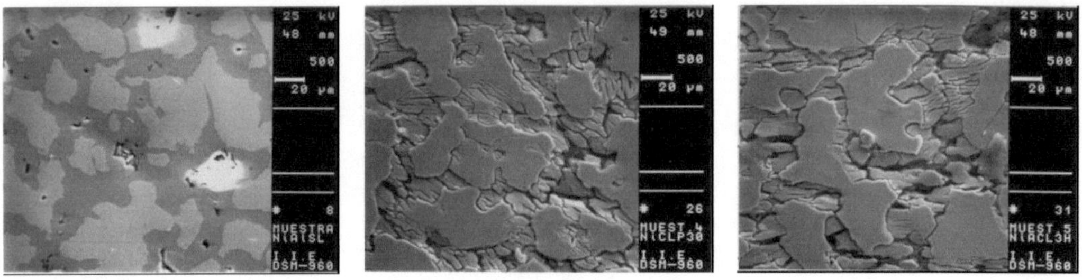

a) before leaching b) leached 30 minutes c) leached 3 hours

Figure 2. SEM micrographs (500X) showing microstructure of Nickel Raney electrodes and the leaching time effect in alkaline solution [5].

As nickel oxidizes in an alkaline solution forming oxides and hydroxides of the metal, it is also believed that most of the catalyst properties in Ni electrodes come from those phases. For example, Subbaraman et al. [7] combined nickel oxide and platinum to produce a more effective catalyst than either component alone. The authors propose a mechanism in which nickel helps to cleave the O-H bond while platinum directs the separated H intermediates to form H_2. Section 3 reviews some literature on different electrode systems researched in the last few years in order to make more effective and economically attractive electrodes.

2.1.2. Electrolyte

As mentioned earlier KOH solutions are used in alkaline electrolysis. This electrolyte is contained in a closed circuit to avoid contact with CO_2 from the atmosphere that could impose an adverse technical challenge due to the precipitation of carbonates formed. The KOH concentration is very important as it determines the ionic conductivity that should remain high in order to avoid the use of higher voltages from ohmic losses. These losses could increase the energy use and therefore the hydrogen cost.

2.1.3. Diaphragm

In alkaline electrolysis a diaphragm is a separator which should keep a good ionic conductivity while effectively separating hydrogen and oxygen gases generated at the cathode and anode sides respectively. It typically consists of a matrix porous material which should chemically stand (and contain) a corrosive environment (~25% KOH solutions) at relatively aggressive temperatures (~85°C). It should also be mechanically strong to withstand changes in dimensions (compression) due to stresses associated to structural design and due to temperature changes. Although asbestos have been used for many decades, alternative polymeric matrixes are preferred by main developers due to the potential asbestos exposure health risks. Along with the economics in the selection of materials, efficient operation needs to be ensured by materials with low gas permeability, low ionic resistance (but high electronic resistance). The effective thickness, i.e. the product of MacMullin number and thickness of the matrix material, is normally determined while selecting diaphragms as it better represents its effectiveness. The MacMullin number, defined as the ratio of the specific resistivity of the electrolyte-saturated matrix material to the resistance of the same volume of electrolyte, can be obtained using Electrochemical Impedance Spectroscopy (EIS) measurements.

From the technical point of view there have been corrosion problems by the use of alkaline solutions in recent international projects, which turned into down time of the equipment. Nevertheless, this could be more a good engineering practice issue (for example materials selection) than a technological problem.

3. Raney nickel and its alloys, as a catalytic material

During alkaline electrolysis, there are three important reactions that may take place during the Hydrogen Evolution Reaction (HER). Depending on the mechanism, the HER in alkaline media is typically treated as a combination of two steps: the Volmer step-water dissociation (reaction 4) followed by either the formation of the reactive intermediate H_{ad}, Heyrovsky step (path in reaction 5) or the Tafel recombination step (path in reaction 6).

$$H_2O + M + e^- \rightleftarrows M\text{-}H_{ad} + OH^- \tag{4}$$

$$\left.\begin{aligned} M + H_2O + e^- &\leftrightarrow MH_{ads} + OH^- \\ MH_{ads} + H_2O + e^- &\leftrightarrow H_2 + M + OH^- \end{aligned}\right\} \tag{5}$$

$$\left.\begin{aligned} M + H_2O + e^- &\leftrightarrow MH_{ads} + OH^- \\ MH_{ads} + MH_{ads} &\leftrightarrow H_2 + 2M \end{aligned}\right\} \tag{6}$$

Nickel Raney electrodes have as main characteristic a high surface area but along its preparation its catalytic performance is enhanced to promote the hydrogen evolution reaction in alkaline media. Many element and compounds have been explored to further improve this feature. For example, compounds materials based on $NiCo_2O_4$ have been used to prepare, via electrodeposition, electrolysis electrodes [12]. This material was compared with Ni electrodes finding an exchange current density (i_o) seven times larger, obtaining a good chemical stability in alkaline solutions. On the other hand, studies on Ni-Co alloys [13], also prepared by electrodeposition, have shown that the overpotential for the hydrogen evolution reaction is smaller when the material has Cobalt content between 41%-65% weight percent. Additional work [14], where Ni was alloyed with 47.7% weight percent Co, showed good electrocatalyst activity and electrochemical studies revealed that the Volmer reaction is the controlling step of the hydrogen evolution reaction, promoting the Ni-H_{ads} species formation. The roughness factor of Ni-Co/Zn [15] structures, prepared by electrodeposition followed by an alkaline KOH 1M leaching process, was evaluated using electrochemical impedance. It was observed that at high current densities the effective area and the active sites promoting the hydrogen evolution reaction are reduced. In this work another finding made was the fact that the lesser the Co content on the external surface, the less catalytic activity.

In a different work using bronze substrates, Ni and Ni-Zn were electrodeposited [16] to obtain bronze/Ni, bronze/NiZn and bronze/Ni/NiZn configurations. NiZn electrodes were leached in alkaline media in order to obtain a porous electrode structure. Their major finding was the fact that the bronze/Ni/NiZn material showed a better corrosion resistance.

Nanocrystals of electrodeposited CoNiFe [17] were used to study the hydrogen evolution reaction by electrochemical impedance finding that the controlling reaction mechanism is via Volmer and then Tafel paths, where a Heyrosky stages was favored. The electrocatalytic activity of these nanocrystals depended on operating temperature.

Electrodes based on electrodeposited NiCoZn [18] showed better stability during studies. These electrodes were synthesized by electrodeposition of Ni, Co, Zn, on a Cu substrate followed by a leaching stage in NaOH, producing a compact porous structure. By electrochemical impedance and linear polarization resistance studies the authors suggest the formation of a passive layer on the electrodes surface allowing a better corrosion resistance but decreasing catalytic activity for the hydrogen evolution reaction. The addition of small quantities of noble metals such as Ag, Pd and Pt to NiCoZn layers after the leaching process allows for a better catalytic activity. These studies showed NiCoZn-Pt electrodes with better catalytic activity. The use of noble metals limits an extensive use for these compounds as some of the components are not abundant in nature and their extraction and refining processes are expensive for hydrogen production.

For that reason metals like Fe to form Ni-Fe are used to prepare electrodes by electrodeposition [19]. For these electrodes the alloy's Fe content has shown to determine the catalytic activity, while the hydrogen evolution reaction is controlled by the Volmer route. NiFe [20] electrodes were added with carbon loads in order to control the grain size and with this, the current density for hydrogen evolution in alkaline media. A carbon 1.59% weight percent produced a 3.4 nm grain size and the lowest overpotential for the HER at polarization current density 0.12 A/cm2. In a layer by layer preparation of Ni-Fe-C electrodes on Cu [21], Tafel slope of 116 mv dec^{-1} was measured.

Carbon substrates were coated with nickel [22] by electroless and electrodeposition methods showing that the latter is more catalytically active for hydrogen evolution in alkaline media.

Mixing rare earth metal powders of Cerium or Lanthanum with Ni, electrodes were prepared by generation of a solid solution [23]. The synthesis was realized at 500°C forming CeO_2 and NiO while with La, $LaNiO_3$ particles transformed to $La_4Ni_3O_{10}$ allowing a particle size increment. Also the formation of the intermetallic $LaNiO_3$ favors current densities for the hydrogen evolution reaction. Ni-Mo alloys [24] are also materials that have been studied in order to increase the chemical stability of Ni-based electrodes. Studies on these materials have shown the formation of intermetallic Ni_4Mo compounds presenting an adequate stability for the hydrogen evolution reaction.

Electrodes made with Ni-Ti alloys [25], prepared by thermal arc spraying, presented catalytic activity that improves with Ti content. Previously Ni-Ti was added with aluminum for a leaching process, allowing a better activation. For this system, it was found that using electrochemical techniques the HER takes place via the Volmer-Heyrosky mechanism.

Coating layers of NiCu [26], prepared by electrodeposition of Cu-Ni, present a compact non homogenous structure, which gives the material a good stability, i.e. corrosion resistance during electrolysis studies. These electrodes also presented good electrical conductivity also giving advantageous features during the electro-oxidation process.

The EIS analysis confirms that Ni-Bi [27] alloys show that charge transfer resistances decrease and the double layer capacitance values increase for binary coatings. An improvement in catalytic activity for the alloys in the range of 0.22% to 0.49% in weight of Bi of composition was shown compared to nickel electrodes.

The formation of amorphous alloys, contributes to develop larger specific area materials compared to nickel alloys. For example, Ni-S-Mn based alloy showed an improved performance and better stability for the Hydrogen evolution reaction [28]. This was achieved by the co-deposition of Ni, S and Mn, using a galvanostatic method. It was found that the catalytic activity of Ni-S compound decays due to leaching induced by the presence of sulfur. Recently, it was reported work on nickel amorphous alloys using boron, niobium and tantalum [29], and other similar Ni-Mo-B based alloys. From that work it was concluded that the $Ni_{66.5}Mo_{28.5}B_5$ and $Ni_{63}Mo_{27}B_{10}$ alloys have better catalytic activity. The authors also tested Nb, Ta and Si based materials and reported hydrogen absorption.

Some studies [30] have reported the making of electrodes using PTFE as binder for nickel Raney, as well as some additives to study the charge/discharge properties during ageing and polarity changes. From these studies it was found that nickel Raney added with PTFE increased its electroactive area in 102-103 times compared to the geometric area. Using Ti, Cr and Fe as additives a stable electrochemical performance was seen at 60°C.

Changing growth conditions during electrodeposition process showed an effect on catalytic properties of nickel alloys, in particular Ni-W, which grown under super gravity conditions [31], tends to develop a finer grain size therefore, increasing its electroactive and considerably reducing microcraks, showing a better corrosion performance.

4. Preparation and characterization of Raney nickel

Many features of Raney nickel that favor the use of electrodes for the generation of hydrogen derive from its original microstructural properties. In figure 3, the Ni-AI equilibrium phase diagram [32] shows the different phases that are obtained when Ni and Al are combined in different proportions. This diagram is characterized by a liquid phase, a face-cubic-center (fcc) phase at both the Al and Ni rich ends, a non-congruently melting compound Al_3Ni and three intermediate phases with a variable range of solubilities, AI_3Ni_2, AlNi and $AlNi_3$. When the Ni–Al alloys (Ni between of 40–50 wt.%) are melted at 1300° C, the alloys are located in the area of β(Ni_2Al_3 phase) + liquid. After melting and quenching, the system may either keep the phase compositions at high temperatures (β) or rich the equilibrium phases reach in δ, depending on the quenching rate, being transferred to the area of lower temperatures, namely the area of δ(Ni_2Al_3 phase) + liquid ($NiAl_3$ + δ), if the cooling rate is not too high. So the content of the Ni_2Al_3 phase is more abundant and the $NiAl_3$ phase becomes less abundant. Ni–Al alloys (Ni/Al 50/50 w/w) are in equilibrium in the phase area of $NiAl_3$ + δ(Ni_2Al_3 phase) when they are cooled by ambient cooling.

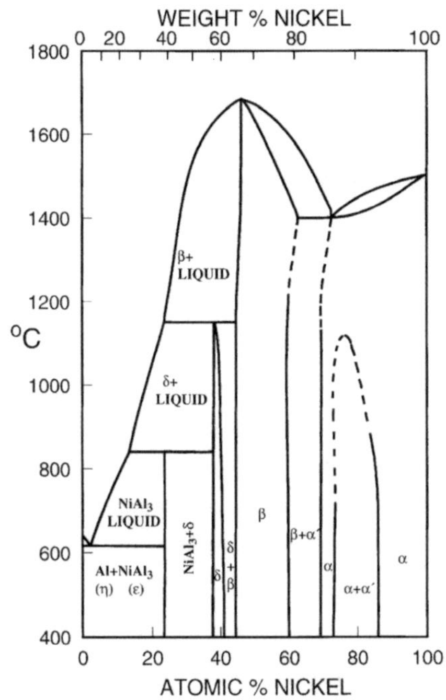

Figure 3. Phase diagram of Ni-Al alloy

As mentioned earlier nickel Raney was first elaborated by Murray Raney, who registered its invention in 1927 (US patent No. 1628190). The main use of the material at that time was for hydrogenation of oils, fats and waxes. At present this material is still used in that industry; its use as a catalyzer for alkaline electrolysis was later proposed. To make skeletal nickel, this metal and aluminum in the form of powder, granules or solid are used. Nickel Raney is Ni-Al alloy with high catalytic features besides being a low cost material as its preparation does not require noble metals. It presents a spongy skeletal morphological structure. Its generic name derives from that. Its grain size reaches a nanometric range.

Due to its activity and selectivity, it can be used to promote various reactions. It is used in the partial hydrogenation of acetylene compounds, reduction of 1-(nitrometil)-cyclohexane. In alkaline solution reduces –CO- to –CHOH- and ArCO- to ArCH3-. In its skeletal form, nickel can be used for hydrogenolysis of sulfur compounds, desulfurization of thiophene, desulfurization of ethylene-thioacetals. It also catalyzes the reduction of carbonyl groups, either double or triple activated links.

In the conventional technique used for making Ni Raney, this metal and aluminum are melted together in the desired proportion (typically 50% weight each) and let cooled down at ambient conditions. After that, a leaching process follows using a soda solution to dissolve most of the aluminum and leaving behind a highly porous structure in the remaining nickel, that act as active sites. Other techniques for the preparation of Raney nickel electrodes include the co-electrodeposition of both metals to create the wanted alloy. Sintering of nickel and aluminum polysilicates is another method to make nickel Raney. In this process, several are grown on a substrate and then an interdifusion process is promoted. Mechanical alloying is another technique recently used for the making of skeletal nickel.

During the leaching process, the alloy is immersed in a sodium hydroxide solution (NaOH) which action produces a high surface area porous material. In general aluminum is alloyed to nickel as aluminum dissolves in alkaline solutions, but other metals can be used too. With aluminum a proportion of 50% weight is used but using other metals may require other compositions to attain the desired properties. The leaching process is also time dependent, the longer the process, the more aluminum is dissolved and more porous will form. Lesser times can be used to allow some aluminum to remain in the final electrodes but less skeletal features will be produced reducing the active surface area.

$$2(\text{Ni-Al})_s + 2\text{OH}^- + 6\text{H}_2\text{O} \rightarrow 2\text{Ni-Raney} + 2\text{Al}(\text{OH})_4(\text{aq}) + 3\text{H}_2(\text{g}) \quad (7)$$

During leaching hydrogen gas is also released considering this as an activating process as hydrogen is easily adsorbed mainly by aluminum. As early as 1947 it was reported [33] that aluminum absorbs very large quantities of hydrogen and although electrochemical processes occur at the surface level, hydrogen has been reported to move around in metal lattices, especially defects such as grain boundaries allowing hydrogen reach the surfaces of the metal. These features have allowed the development of metal-hydrogen systems such as NiMH batteries as well as the Al containing Mischmetal alloy used in hydrogen storage [34] where high hydrogen desorption rates have allowed the development of both technologies at the commercial level. This ease in hydrogen availability can be seen as a catalytic feature that gives Ni Raney its high activity for many hydrogen-involving reactions. On the other hand, hydrogen absorption on metals is known to be a fast process [35], with hydrogen interacting in three principal ways: (i) by dissociative chemisorption at the surface; (ii) by physical adsorption as molecules at very low temperatures; and (iii) by dissolution or occlusion. Besides these interactions there have been added many intermediate states of various features of importance in catalysis but the reader is encouraged to go directly to specialized literature.

Once the leaching process has been completed, a process to eliminate the alloy's pyrophoric properties is often applied by immersion in water of the alloy which promotes the formation of an oxide layer on top of the material stabilizing it. There have not been studies to establish the effect (adverse or not) of the stabilization process on the catalytic activity of Ni Raney as it has been used only to facilitate its handling. Once fabricated, from the electrocatalytic point of view, skeletal nickel will be more active and will have more surface area than pure solid nickel.

Parameters such as temperature and reaction time, etc. affect the resulting properties of the catalyzer. The complex liquid-solid reactions can be described by the following stages: a) liquid hydroxide diffusion around solid particles b) penetration and diffusion of solution into the unreacted material's structure, c) reaction of solution with aluminum to form aluminate and hydrogen, d) diffusion of products to the outside of the forming structure and e) diffusion of products to the bulk of the solution [36]. A similar model, the shrink core model, can be used to determine the kinetics of the leaching process. This model describes the irreversible desorption of hydrogen followed by diffusion in the porous solid through the pores. When mass transfer rate of the solute in the no extracted inner part is much slower than that in the outer part, where most of the solute has been extracted, or the solute concentration is much higher than the solubility of the solute in the solvent phase, a sharp boundary may exist between the outer and the inner region. A core of inner region shrinks with the progress of the extraction. These situations can be modeled by the shrinking-core model [36]. The model is adequate to describe adsorption process, ionic interchange and solid-fluid reactions.

The effect of temperature has been reported for example for treatments at 100°C and 50°C [37]. In that work skeletal nickel showed crystal sizes from 1 to 20 nm. Also, those values can be obtained if the temperature and caustic solution reduces.

The presence of additives during the leaching process allows better activity, either by using metals or their oxides. The purpose of these additives is to increase the surface area of the resulting material. When the $CuAl_2$ phase is formed, the addition of Zn to form $Cu(Zn)Al_2(s)$ [38] improves the leaching process by allowing a higher surface area of skeletal copper by the dissolution of these additives.

In nickel Raney the phases identified are a mixture of $NiAl_3$, Ni_2Al_3 and the eutectic phases of Al-$NiAl_3$. Such phases are seen with Ni:Al 50%-50% and 31.5%-68.5% in weight percent [39]. $NiAl_3$ and the eutectic phases are very reactive in sodium hydroxide solution so that aluminum is dissolved, forming the skeletal configuration of nickel. Although the Ni_2Al_3 phase reacts at a slower rate, rising the temperature above 50°C it reacts faster, decomposing at boiling temperature.

It has been recently reported [40] that using nickel powder with Ni-14%Al, in a mechanical mill the inter-metallic $NiAl_3$ compound is formed. Microstructure studies using Synchrotron X-ray micro tomography [41] on nickel Raney (50%-50%) synthesized by a gas atomization process, revealed the presence of several grain sizes. The authors report an improvement in its hydrogenation capacity when synthesized with grain sizes of 106 to 150 microns. This was achieved by the addition of metals to the intermetallic $NiAl_3$. Other studies [42] using B, Zr, Cr and Mo showed that the intermetallic compound $NiAl_3(B,Zr)$ presents the best corrosion resistance, in alkaline medium.

By the spraying technique nickel Raney was prepared using different Ni:Al proportions going from 65% to 75% Al weight, founding that by cooling the metal during the spraying its rugosity is improved, obtaining more $NiAl_3$ phase allowing by that a better electrocatalytic performance.

5. Final remarks

Skeletal or nickel Raney has proved to be a very interesting and costly effective system for the catalysis of both chemical and electrochemical reactions. Having said that, it is also recognized that the several factors involved in its preparation affect the performance of the final product, i.e. type of metals in the alloy, their proportion (i.e. composition), used additives, solution concentration and time and temperature during the leaching stage. Possibly its final conditioning to eliminate its pyrophoricity, could affect its catalyzing properties but little has been investigated in this respect. As with any electrocatalysts, conditions at the surface have a great influence on the product's properties, e.g. crystalinity, crystal planes, defects (grain boundaries, stepped and kinked surfaces, etc.), rugosity/porosity, i.e. surface area, microstructure such as the type of phases (metallic, intermetallic, other compounds), grain/particle sizes, and other structural features like the presence of clusters and other agglomerates. In this respect, it should also be mentioned the efforts in the preparation of amorphous materials to increase corrosion resistance and surface area. Among combined materials for improving nickel properties there have been efforts in testing metals such as Pt, Pd, Ti, Zr, Cr, Bi, La, Co, Zn, Ag, Cu, Fe, Mn, Nb, Mo, and others, as well as the addition of C, S, B, etc.

Corrosion resistance is also a property sought in new electrode systems. Nevertheless, many metals passivated in alkaline media which may change their surface properties including catalytic performance. Both features, corrosion resistance and catalytic performance, should be sought simultaneously and it should include the evaluation of the effect of the presence of oxides and hydroxides on the electrode surface in alkaline media.

Different manufacturing routes have also been tested such as conventional melt alloying, mechanical alloying, arc gas spraying, electrodeposition, electroless coating, laser alloying of aluminum with electrodeposited nickel [43], interdiffusion of more than one electrodeposited layer, etc.

During the leaching stage, additives or alloying components have been used to promote several effects on the final product. Materials tested for better leaching process results include metals like Cu, Zn and also oxides. Certainly, other parameters such as temperature and reaction time during leaching are tested especially for new systems as these can change compared to conventional Ni alloys.

Although there have been reports on improvement catalyzers for the HER, many proposed systems still include platinum, while the economics is still recognized as one of the main challenges to overcome. For this reason, the most used composition in alkaline electrolysis is still the 50/50 weight percent nickel with aluminum, but many laboratory results show new systems with great potential, including higher exchange current densities and corrosion resistance, to be incorporated in economical commercial hydrogen production systems.

Author details

A.M. Fernández
Universidad Nacional Autónoma de México, México

U. Cano
Instituto de Investigaciones Eléctricas, México

6. References

[1] Presentation on alkaline electrolysis from IEA/HIA Task 25, High temperature Hydrogen Production Process, http://www.ieahia.org
[2] Hydrogen and Fuel Cells Program Plan, the U.S. Department of Energy, September 2011
[3] www.acagen.com
[4] Andrew D Moore, Barbara Nebel, Simon Whitehouse, Final Report, Hydrogen Production from Renewable Energy by Electrolysis, PE Australasia Ltd., CREST Project 1.2.3, Centre for Research into Energy for Sustainable Transport (CREST), Perth, Australia, March 2010.
[5] Martínez-Millán Wenceslao, Ph.D. Thesis, CIE-UNAM, México, September 2006
[6] W. Martínez, A. Fernández, U. Cano-Castillo, "Synthesis of nickel-based skeletal catalyst for an alkaline electrolyzer", International Journal of Hydrogen Energy, Volume 35, Issue 16, August 2010, 8457-8462, 2010.
[7] Ram Subbaraman, Dusan Tripkovic, Dusan Strmcnik, Kee-Chul Chang, Masanobu Uchimura, Arvydas P. Paulikas, Vojislav Stamenkovic, Nenad M. Markovic, Enhancing Hydrogen Evolution Activity in Water Splitting by Tailoring Li+-Ni(OH)2-Pt Interfaces, SCIENCE, vol 334, 2 December 2011, www.sciencemag.org.
[8] Jerzy Chlistunoff, Final Technical Report, Advanced Chlor-Alkali Technology, LAUR 05-2444, DOE Award 03EE-2F/ED190403, Project Period 10:01 – 09:04, Los Alamos National Laboratory, Los Alamos, NM 87544
[9] B. Kroposki, J. Levene, and K. Harrison, P.K. Sen, F. Novachek, Technical Report, NREL/TP-581-40605, September 2006, Electrolysis: Information and Opportunities for Electric Power Utilities.
[10] Richard Bourgeois, Alkaline Electrolysis, Final Technical Report, Award DE-FC36-04GO14223, General Electric Global Research Center, 31 March 2006
[11] Michael Stichter, Final Technical Report, Reporting Period:02/01/2004 to 09/30/2007, Hydrogen Generation from Electrolysis, DOE Award # DE-FC36-04GO13028; Amendment No. A002.
[12] Bao Jin-Zhen, Wang Sen-Lin, Acta Phys-Chim. Sin. 2011, 27 (12), 2849-2856
[13] C. Lupi, A. Dell'Era, M. Pasquali, Int. J. of Hydrogen Energy 34(2009)2101-2106.
[14] I. Herraiz-Cardona, E. Ortega, J. García Antón, V. Pérez-Herranz, Int. J. of Hydrogen Energy 36(2011)9428-9438.
[15] I. Herraiz,-Cardona, E. Ortega, V. Perez-Herranz, Electrochimica Acta 56(2011)1308-1315

[16] Ramazan Solmaz, Gu lfeza Kardas, Energy Conversion and Management 48 (2007) 583–591

[17] M. Jafarian, O. Azizi, F. Gobal, M.G. Mahjani, Int. J. of Hydrogen Energy 32(2007)1686-1693.

[18] R. Solmaz, A. Döner, I. Sxahin, A.O. Yü ce , G. Kardasx, B. Yazıcı, M. Erbil, Int. J. of Hydrogen Energy 34(2009)7910-7918.

[19] Ramazan Solmaz, Gülfeza Kardas, Electrochimica Acta 54 (2009) 3726–3734.

[20] L.J. Song, H.M. Meng, Int. J. of hydrogen Energy 35(2010)10060-10066.

[21] Reza Karimi Shervedani, Ali Reza Madram, Electrochimica Acta 53 (2007) 426–433.

[22] : Boguslaw Pierozynski and Lech Smoczynski, J. of The Electrochemical Society, 156(9)(2009)B1045-B1050

[23] M.A. Domínguez-Crespo, A.M. Torres-Huerta, B. Brachetti-Sibaja, A. Flores-Vela, Int. J. of Hydrogen Energy 36(2011)135-151

[24] N.V. Krstajic′, V.D. Jovic′,, Lj. Gajic′-Krstajic′, B.M. Jovic′, A.L. Antozzi, G.N. Martelli, Int. J. of Hydrogen Energy 33(2008)3676-3687.

[25] Andrea Kellenberg, Nicolae Vaszilcsin, Waltraut Brandl, and Narcis Duteanu, Int. J. of Hydrogen Energy 32(2007)3258-3265.

[26] Ramazan Solmaz, Ali Döner, Gülfeza Kardas, Int. J. of Hydrogen Energy 34(2009)2089-2094.

[27] Mehmet Erman Mert, Gülfeza Kardas, J. of Alloys and Compounds 509(2011)9190-9194.

[28] Zhongqiang Shan,Yanjie Liua, Zheng Chen, GarryWarrender, Jianhua Tian, Int. J. of Hydrogen Energy 33(2008)28-33.

[29] L. Mihailov, T. Spassov, I. Kanazirski, I. Tsvetanov, J Mater Sci 46 (2011)7068–7073.

[30] Paolo Salvi, Paolo Nelli, Marco Villa, Yohannes Kiros, Giovanni Zangari, Gianna Bruni, Amedeo Marini, Chiara Milanese. Int. J. of Hydrogen Energy 36(2011)7816-7821.

[31] Mingyong Wanga, Zhi Wang, Zhancheng Guo, Zhaojun Li, Inter J. of Hydrogen Energy 36(2011)3305-3312.

[32] A Pasturel, C Colinet, AT Paxton: and M van Schilfgaarde, J. Phys: Condens. Matter 4 (1992) 945-959.

[33] C. E. Ransley & H. Neufeld, Absorption of Hydrogen by Aluminium Attacked in Caustic Soda Solution, Nature, 159, 709-710 (24 May 1947)

[34] Y Osumi, A Kato, H Suzuki, M Nakane, Hydrogen absorption-desorption characteristics of mischmetal-nickel-aluminum alloys, Journal of the Less Common Metals, Volume 66, Issue 1, July 1979, Pages 67–75 1979

[35] Geoffrey C. Bond, Chapter 3. Chemisorption and Reactions of Hydrogen, in Metal-Catalysed Reactions of Hydrocarbons, Fundamental and Applied Catalysis book series, 2005, pp93-152, DOI 10.1007/0-387-26111-7_3,

[36] O. Levenspiel, Chemical Reaction Eng., 2nd edition, pag 371, (Wiley International Ed., 1972).

[37] M.S. Wainwright, in G. Erl, H. Knozinger, J. Weitkamp (Eds.) Handbook of Heterogeneous Catalysis, vol. 1, VCH, New York, 1997, pp. 64-67.

[38] A.J. Smith, D.L. Trimm, Ann. Rev. Mat Res. 2005, 35:127-42.

[39] F. Devred, A.H. Gieske, N. Adkins, U. Dahlborg, C.M. Bao, M. Calvo-Dahlborg, J.W. Bakker, B.E. Nieuwenhuys, Applied Catalysis A:General 356(2009)154-161.
[40] Hongxing Dong, Ting Lei, Yuehui He, Nanping Xu, Baiyun Huang, C.T. Liu, : International Journal of Hydrogen Energy 36(2011)12112-12120.
[41] F. Devred, G. Reinhart, G.N. Iles, B. van der Klugt, N.J. Adkings, J.W Bakker, B.E. Nieuwenhuys, Catalysis Today 183(2011)13-19.
[42] Grzegorz D. Sulka, Pawel Jozwik, : Intermetallics 19(2011), 974-981.
[43] J.Senthil Selvan, G Soundararajan, K Subramanian, Surface and Coatings Technology, Volume 124, Issues 2–3, 21 February (2000), Pages 117–127.

Chapter 2

Water Electrolysis with Inductive Voltage Pulses

Martins Vanags, Janis Kleperis and Gunars Bajars

Additional information is available at the end of the chapter

http://dx.doi.org/10.5772/52453

1. Introduction

The main idea of Hydrogen Economy is to create a bridge between the energy resources, energy producers and consumers. If hydrogen is produced from renewable energy sources (wind, solar, hydro, biomass, etc.), and used for energy production in the catalytic combustion process, then the energy life cycle does not pollute nature longer. With transition to Hydrogen Economy the Society will live accordingly to the sustainable development model, defined in the 1987 (Our Common Future, 1987).

Hydrogen is not available on Earth in free form; therefore the production process is representing a major part of the final price of hydrogen (Hydrogen Pathway, 2011). This is the main reason while research for effective electrolysis methods is very urgent. On our Planet the hydrogen is mainly located in compounds such as hydrocarbons, water, etc. and appropriate energy is needed to release hydrogen from them. In principle the amount of consumed energy is always greater than that which can be extracted from the hydrogen, and in the real operating conditions, the cycle efficiency does not exceed 50% (The Hydrogen Economy, 2004). The current problem is motivated to seek improvements to existing and discovering new technologies to produce hydrogen from the water – widely available and renewable resource on the Earth.

Water electrolysis is known more than 130 years already, and different technologies are developed giving power consumption around 3.6 kWh/m^3 - high temperature electrolysis, and 4.1 kWh/m^3 - room temperature alkaline electrolysers and proton exchange membrane electrolysers (The Hydrogen Economy, 2004). Lower hydrogen production costs is for technologies using closed thermo-chemical cycles, but only in places where huge amount of waste heat is available (for example, nuclear power stations (The Hydrogen Economy, 2004). Nevertheless what will be the hydrogen price today, in future only hydrogen obtained from renewable resources using electricity from renewable energy sources will save the World, as it was stated in 2nd World's Hydrogen Congress in Turkey (Selected Articles, 2009). For Latvia the hydrogen obtained in electrolysis using electricity from renewables (wind, Sun, water) also would be the best solution to move to Hydrogen Economics (Dimants et all, 2011). That is because all renewables available in Latvia's geographical situation are giving non- stable and interrupt power, for which the storage solutions are necessary. Usage of hydrogen as energy carrier to be produced from electricity generated by renewables, stored and after used in fuel cell stack to generate electricity is the best solution (Zoulias, 2002). Efficient and stable electrolysers are required for such purposes. Smaller electrolysis units are necessary also for technical solutions were hydrogen is produced and used directly on demand, for example, hydrogen welding devices, hydrogen powered internal combustion cars (Kreuter and Hofmann, 1998).

DC power typically is used for electrolysis, nevertheless pulse powering also is proposed (see, for example, Gutmman and Murphy, 1983). Using a mechanically interrupted DC power supply (Brockris and Potter, 1952; Bockris et all, 1957) next interesting phenomena was noticed: immediately upon application of voltage to an electrochemical system, a high but short-lived current spike was observed. When the applied voltage was disconnect, significant current continues to flow for a short time. In 1984 Ghoroghchian and Bockris designed a homopolar generator to drive an electrolyser on pulsed DC voltage. They concluded that the rate of hydrogen production would be nearly twice as much as the rate for DC.

The Latvian Hydrogen Research Team is developing inductive pulse power circuits for water electrolysis cell (Vanags et all, 2009, 2011a, 2011b). The studies revealed a few significant differences compared to conventional DC electrolysis of water. New model is established and described, as well as the hypothesis is set that water molecule can split into hydrogen and oxygen on a single electrode (Vanags et all, 2011a). There has been found and explained the principle of high efficiency electrolysis. A new type of power supply scheme based on inductive voltage pulse generator is designed for water electrolysis. Gases released in electrolysis process from electrodes for the first time are analyzed quantitatively and qualitatively using microsensors (dissolved gases in electrolyte solution nearby electrode) and masspectrometer (in atmosphere evolved gases). The hypothesis of hydrogen and oxygen evolution on a cathode during the process of pulse electrolysis is original, as well as interpretation of the process with relaxation mechanisms of electrons emitted by cathode and solvated in electrolyte (Vanags, 2011b).

2. Literature review

2.1. A brief history of the electrolysis of water

Adriaan Paets van Troostwijk, 1752.–1837., and Johan Rudolph Deiman, 1743.–1808., while using Leyden jar and a powerful electrostatic generator noticed a gas evolution on the electrodes of water electrolysis cell as a result of spark over jumping in the electrostatic generator. The evolved gases displaced water out of the Leyden jar during the experiment and spark jumped into the collected gas mixture creating an explosion. The researchers decided that they have decomposed water into hydrogen and oxygen in a stoichiometric proportion 2:1; they published the results in 1789, which is considered to be the year of discovering the water electrolysis (Zoulias et all, 2002; De Levie, 1999). In more that hundred years, in 1902 there were more than 400 industrial electrolysers used all over the world, but in 1939 the first large water electrolysis plant was commissioned with the hydrogen production capacity of 10000 Nm^3/h (Zoulias et all, 2002). The high-pressure electrolysers were produced in 1948 for the first time; in 1966 General Electric built the first electrolysis system with solid electrolyte and in 1972 the first solid oxide high-temperature electrolyser was built. However the development of electrolysis devices is in progress nowadays as well along with the development of proton exchange membrane, which can be used in the water electrolysers and fuel cells, along with the development of high-temperature solid oxide electrolysers likewise the optimization of alkaline electrolysers (Kreuter W, and Hofmann H (1998).

2.2. Direct current water electrolysis

When dissolving acid in the water (e.g. sulfuric acid), molecules of water and acid dissociate into ions. The same happens if alkali (e.g. KOH) is dissolved in the water, the solution dissociates into ions, creating ionic conductor or electrolyte. There has been formed an ionic conductor, where a direct current will be passed through. The processes taking place on electrodes are in the case of sulfuric acid - the positive hydroxonium ions H_3O^+ (cation) move to the side of negative electrode. When cations reach the electrode, they receive missing electrons (Zoulias et all, 2002):

$$2H_3O^+ + 2e \Rightarrow H_2 + 2H_2O \tag{1}$$

Hydrogen is produced as gas from the medium, in its turn, water dissociates into ions again. The reaction on an anode or positive electrode in an alkaline medium is (Zoulias et all, 2002):

$$2OH^- - 2e \Rightarrow \frac{1}{2}O_2 + H_2O \tag{2}$$

Oxygen evolves as gas, but water dissociates into ions again. There are produced three parts of volume of gaseous substance in the process described - two parts of hydrogen and one - oxygen. In the case of alkaline electrolyte there are polarized water molecules, which have their hydrogen atoms oriented toward an electrode, near the cathode and dissociation reaction takes place:

$$2H_2O + e \Rightarrow 2OH^- + H_2 \tag{3}$$

The first law of thermodynamics for the open system states that:

$$Q - W_s = \Delta H \tag{4}$$

where Q is the amount of heat supplied to the system, W_s the amount of appropriate work performed by the system and ΔH is the change in system's enthalpy. The work done is electricity used up in electrolyser, therefore W_s is:

$$W_s = -nFE \tag{5}$$

where:
n - The amount of transferred electrons;
F – Faraday constant: = 23,074 cal/volts·g-equivalent;
E - Electric potential of the cell in volts.

Using equation (5), transform the expression (4), resulting:

$$E = \frac{\Delta H - Q}{nF} \tag{6}$$

In an isothermal reversible process (without loss) Q is:

$$Q_{atg} = T\Delta S \tag{7}$$

where T is absolute temperature and ΔS is system entropy change. From (6) and (7) the value of reversible reaction potential is obtained, where it is impossible to decompose water into hydrogen and oxygen in real time:

$$E_{rev} = \frac{\Delta H - T\Delta S}{nF} \tag{8}$$

$(\Delta H - \Delta S)$ is the change in Gibbs free energy ΔG. At normal temperature and pressure ((25 ^{o}C temperature and 1 atm pressure) ΔH equals 68 320 cal/gmol and ΔG equals 56 690 cal/gmol. Therefore the reversible potential of the cell is:

$$E_{rev} = \frac{\Delta G}{nF} = \frac{56{,}690}{2(23{,}074)} = 1{,}23V \tag{9}$$

Potential where Q equals zero and supplied energy transforms into chemical energy, is called thermo-neutral voltage (Oldham and Myland, 1993; Bockris and Potter, 1952):

$$E_{thermo} = \frac{\Delta H}{nF} = 1{,}48V \tag{10}$$

The voltage to split water in practical electrolysis devices is higher than thermo-neutral cell voltage due transformation into heat, which heats up the cell. Therefore industrial electrolyser requires additional cooling and the value of DC voltage is defined:

$$E = E_{rev} + loss \tag{11}$$

where the

$$loss = E_{anode} + E_{cathode} + E_{mt} + IR \tag{12}$$

In equation (12) E_{anode} – activation overvoltage of the anode; $E_{cathode}$ – activation overvoltage of the cathode; E_{mt} – overvoltage of the mass transfer and IR – ohmic overvoltage (includes resistance in an electrolyte, on electrodes, leads). Current density must be higher than 100

mA/cm^2 in industrial electrolysers, therefore voltage applied to individual cell partly transforms into the heat, becoming typical loss in DC water electrolysis.

It is possible to write an expression for the efficiency factor of the water electrolysis, calculated versus the thermo-neutral voltage, using relations above (Bockris and Potter, (1952):

$$\eta = \frac{\Delta H}{\Delta G + loss} = \frac{E_{thermo}}{E} \tag{13}$$

When ΔG is negative, the reactions are spontaneous and work has been done by releasing the energy. When ΔG is positive, for reaction to happen external work must be used. As for ensuring the reaction, work must be done, water electrolysis cell does not operate spontaneously. The reaction in fuel cells is spontaneous because of the catalyst and during reaction the energy is released (Salem, 2004).

Hydrogen **E**volution **R**eaction (HER) is one of the most widely researched reactions in the electrochemistry. The studies of HER are carried out in different kind of systems and following each other processes is divided (Salem, 2004; Heyrovsky, 2006; Murphy, 1983; Bockris, 1957; El-Meligi, 2009; Sasaki and Matsuda, 1981; Noel and Vasu, 1990; Kristalik, 1965): Volmer electrochemical discharge step, Heyrovsky electrochemical desorption step, Tafel catalytic recombination step. Each of steps can be a reaction limiting step in a certain system during the whole reaction. This means, that each step can have different reaction rate, and the slowest step will determine the speed of reaction. Charge transfer may begin when the reagent is next to the electrode. Two most typical steps are charge transfer ending with the adsorption of hydrogen atom, and recombination of the adsorbed atoms with next desorption of H_2 molecule.

The general equation of the electrochemical reaction links current with potential (Noel and Vasu, 1990):

$$i_c = \vec{i} - \overleftarrow{i} = i_0 \left[\exp\left(-\frac{\beta \lambda \eta F}{RT} \right) - \exp\left(\frac{(1-\beta)\lambda \eta F}{RT} \right) \right] \tag{14}$$

β is symmetry factor (0, ½, 1 for process without activation, normal process and barrier-free process accordingly).

2.3. Interface between electrode and electrolyte: double layer

When two equal electrodes (conductors) are immersed in an electrolyte, initially there is no measurable voltage between them. But when the current is caused to flow from one rod to the other by a battery, charge separation is naturally created at each liquid/solid interface and two electrochemical capacitors connected in series are created. Typical capacitors store electrical charge physically, with no chemical or phase changes taking place, and this process is highly reversible; the discharge-charge cycle can be repeated over and over again, virtually without limit. In electrochemical capacitor at electrode/electrolyte interface solvated ions in the electrolyte are attracted to the electrode surface by an equal but opposite charge in it. These two parallel regions of charges at interface form the "double layer" where charge separation is measured in molecular dimensions (i.e., few angstroms), and the surface area is measured in thousands of square meters per gram of electrode material (Miller and Simon, 2008).

Double-layer phenomena and electro-kinetic processes are the main elements of electrochemistry. It is considered that the behavior of the interface is and should be described in terms of a capacitor. It is a consequence of the "free charge" approach that in the case of a continuous current flow through the interface a strict distinction should be made between the so-called non-Faraday and Faraday currents. The former is responsible for charging of the double-layer capacitor, while the latter is the charge flow connected with the charge transfer processes occurring at the interface (Horányi, Láng, 2006). The state of an interface at constant pressure and temperature can be changed by changing the concentration of the components in the bulk phases, and by constructing an electrical circuit with the aid of a counter electrode and forcing an electric current through the circuit, which can be expressed as:

$$i = i' + i'', \tag{15}$$

Were i` is charging current of double layer, and i`` – charge transfer or Faraday current. Double layer charging current can be viewed as an ideal capacitor charging current equal to $C\frac{\partial \eta}{\partial t}$, were η is overvoltage and C – double layer capacitance. Rewriting equation (15), next current equation is obtained:

$$i = C\frac{\partial \eta}{\partial t} + i'' \tag{16}$$

which has caused much debates. Paul Delahais wrote in 1966 that this equation permits the decoupling of the non-Faraday from Faraday processes, but at the same time concludes that Faraday charge transfer and charging processes cannot be separated a priori in non-steady-state electrode processes because of the phenomenon of charge separation or recombination at the electrode-electrolyte interface without flow of external current. Charging behaviors as ideal polarized or reversible electrode represent only two limiting cases of a more general case (Delahay, 1966). Nisancioglu and Newman (2012) in their article even without going into the assumptions and basing only on the mass balance equation, obtained next current equation: $i=dq/dt + i''$ and showed that a priori separation of double-layer charging and Faraday processes in electrode reactions is the component mass balance for the electrode surface. Equation (1) is valid if the rate of change of concentration of the species, which take part in the electrode reaction, can be neglected at the electrode surface.

2.4. Water splitting with the pulse electrolysis

There are different ways of water splitting first reviewed by Bockris et all (1985), that sharply differs from conventional DC water electrolysis. The most common could be:

thermo chemical, sonochemical, photocatalytic, biological water splitting; water splitting under the magnetic field and centrifugal force of rotation; pulse electrolysis and plasma electrolysis. Regarding pulse electrolysis, Ghoroghchian and Bockris in 1956 already defined that the pulse electrolysis is more effective than conventional electrolysis. Many new patents on pulse electrolysis appeared in 1970-1990 (Horvath, 1976; Spirig, 1978; Themu, 1980; Puharich, 1983; Meyer, 1986; Meyer, 1989; Meyer, 1992a, 1992b; Santilli, 2001; Chambers, 2002) stating to be invented over-effective electrolysis (i.e. the current efficiency is higher than 100%). The water splitting scheme described in these patents initiated a huge interest, but nobody has succeeded in interpreting these schemes and their performance mechanisms up to now, and what is more important, nobody has succeeded to repeated patented devices experimentally as well.

In interrupted DC electrolysis the diffusion layer at the electrode can be divided into two parts: one part, which is located at the electrode surface is characterized with pulsed concentration of active ions, and another part is fixed, similar to the diffusion layer in case of DC. The concentration of the active ions in pulsing diffusion layer changes from defined initial value when the pulse is imposed, to a next value when it expired. The concentration of active ions in pulse may fall or can not fall to zero. Time, which is necessary for active ion concentration would fall to zero, is called the transition time τ. The transition time is depending from pulse current i_p and pulse duration T. If depletion in stationary diffusion layer is small, i.e. $c'_e \approx c_0$, were c'_e is concentration of ions in pulsed-layer outer edge, and c_0 is bulk concentration, the transition time can be found from Sand equation (Bott, 2000):

$$\tau = \frac{\pi D c_0^2 (zF)^2}{4 i_p^2} \tag{17}$$

were F is Faraday's constant, z – charge number and D is its diffusion coefficient. As can be seen, transition time τ depends on the ion concentration in bulk volume c_0 and pulse current density i_p. Thickness of pulsed layer δ_p at the end of the pulse depends only from the density of pulse current:

$$\delta_p = 2\left(\frac{DT}{\pi}\right)^{1/2} \tag{18}$$

With very short pulses extremely thin pulsating layer can be reached. This thin layer would allow temporary to impose a very high current densities during metal plating (more than 250A/cm^2, which is 10,000 times higher than currents in conventional electrolysis), which accelerates the process of metal electroplating (Ibl et all, 1978). The rough and porous surface is formed during metal plating with direct current, when the current value reached the mass transport limit. When plating is done with pulse current, pulsating diffusion layer always will be much thinner than the surface roughness, what means that in case the mass transfer limit is reached, the plated surface is homogeneous still and copy the roughness of substrate. This feature gives preference to pulse current in metal plating processes, comparing with conventional DC plating, because the highest possible power can used (current above mass transfer limit) to obtain homogeneous coatings in shortest times (Ibl et all, 1978).

Pulse electrolysis is widely investigated using various technologies (Hirato et all, 2003; Kuroda et all, 2007; Chandrasekar et all, 2008). In all of these technologies rectangular pulses are mostly used which have to be active in nature. Shimizu et all (2006) applied inductive voltage pulses to water electrolysis and showed significant differences with conventional DC electrolysis of water. The conclusion of this research is that this kind of water electrolysis efficiency is not dependant on the electrolysis power, thus being in contradiction to the conventional opinion of electrolysis.

We studied inductive voltage electrolysis and promoted the hypothesis that pulse process separates the cell geometric capacitance and double layer charging current from the electrochemical reaction current with charge transfer (Vanags (2009), Vanags (2011a, 2011b). To prove this we have done plenty of experiments proving double layer charging process separation from the electrochemical water splitting reaction. There are no studies about the usage of reactive short voltage pulse in the aqueous solution electrolysis; also no microelectrodes are used to determine the presence of the dissolved hydrogen and oxygen near the cathode in electrolysis process.

3. Experimental

3.1. Materials and equipment

Materials, instruments and equipment used in this work are collected in Tables 1 and 2.

3.2. Inductive reverse voltage pulse generator

The inductive voltage pulses were generated in the electric circuit (Fig. 1) consisting of a pulse generator, a DC power source, a field transistor BUZ350, and a blocking diode (Shaaban, 1994; Smimizu et all, 2006). A special broad-band transformer was bifilarly wound using two wires twisted together. Square pulses from the generator were applied to the field transistor connected in series with the DC power source. The filling factor of pulses was kept constant (50%). To obtain inductive reverse voltage pulses, the primary winding of the transformer is powered with low amplitude square voltage pulses. In the secondary winding (winding ratio1:1) due to collapse of the magnetic field induced in the coil very sharp inductive pulse with high amplitude and opposite polarity with respect to applied voltage appears. Pulse of induced reverse voltage is passed through the blocking diode, and the resulting ~1 μs wide high-voltage pulse is applied to the electrolytic cell. A two-beam oscilloscope GWinstek GDS-2204 was employed to record the voltage (i.e. its drop on a reference resistance) and current in the circuit.

MOSFET (IRF840) is used as semiconductor switch between DC power supply and ground circuit. Pulse transformer is a solenoid type with bifilar windings; length is 20 cm and a coil diameter of 2.3 cm and ferrite rod core. Number of turns in both the primary and the secondary winding is 75, so ratio is 1:1. Inductance of solenoid is approximately 250 μH. Super-fast blocking diode with the closing time of 10 ns is included in the secondary circuit, to pass on electrolysis cell only the pulses induced in transformer with opposite polarity. Direct pulses are blocked by diode.

	No	Name	Parameters	Producer
Chemicals	1.	KOH	99%	Aldrich
	2.	NaOH	99%	Aldrich
	3.	LiOH	99.9%	Aldrich
	4.	K_2CO_3	99.8%	Aldrich
	5.	H_2SO_4	95%	Aldrich
	6.	$(NH_2)_2CO$	98%	Aldrich
	7.	H_2O	0.1 μS	Deionised
Metals	1.	Stainless Steel (parameters Table 2)		316L
	2.	Tungsten	95%	Aldrich
	3.	Platinum	99.9%	Aldrich
Equipments and Instruments	1.	DC power supply Agilent N5751A	300V; 2.5A	Aligent echnologies
	2.	Frequency Generator GFG-3015	0 – 150 MHz	GW-Instek
	3.	Oscilloscope GDS-2204	4 beams, resolution 10 ns	GW-Instek
	4.	Power Meter HM8115-2	16A, 300V	Hameg Instruments
	5.	Water Deionization Crystal – 5	Water - 0.1 μS	Adrona Lab.Systems
	6.	Masspectrometer RGAPro 100	0 – 100 m/z units	Hy-Energy
	7.	X-ray fluorescence spectrometer EDAX/Ametek, Eagle III		Ametek
	8.	Microsensors for dissolved gases H_2 and O_2	Resolution 0.1 μmol/l	Unisense, Denmark

Table 1. Materials and equipment used in this work.

Element	C	Si	P	S	Ti	Cr	Mn	Fe	Ni	Cu
Quantity, wt%	0.12	0.83	0.04	0.02	0.67	17.88	2.02	68.36	9.77	0.29

Table 2. Composition of Stainless steel 316L used for electrodes (wt%).

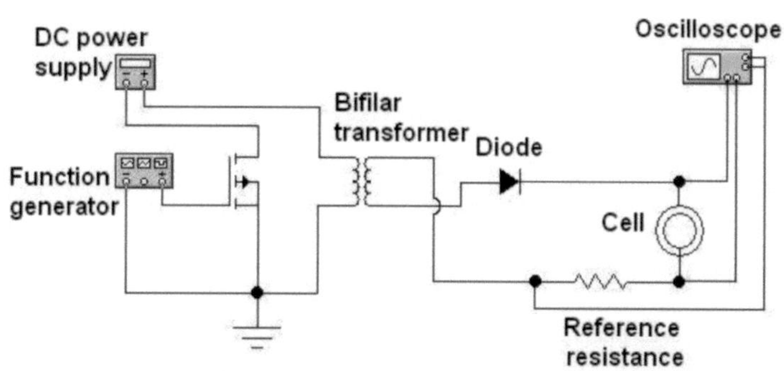

Figure 1. Experimental circuit for generation of inductive reverse voltage pulses.

3.3. Construction of electrolyses cells

Experiments in this chapter are divided into five parts. In the first part the gas evolution rate is explained and performance efficiency coefficients defined (current efficiency and energy efficiency). The second part examines kinetics of the inductive voltage pulse applied to electrolysis cell were electrolyte concentration and the distance between electrodes are changing. The third part describes the application of respiration microsensors to measure concentration of dissolved hydrogen gas directly to the cathode surface in an electrolysis cell, powered with inductive voltage pulses. The fourth experiment studied inductive voltage pulse kinetics in very dilute electrolyte solutions. The fourth experiment also noticed interesting feature in current pulse kinetics, therefore additional experiment is performed, to measure concentration of evolved hydrogen at the cathode with oxygen microsensor. This experiment devoted to fifth.

Amount of released gases during electrolysis was determined with volume displacement method (Fig. 2).

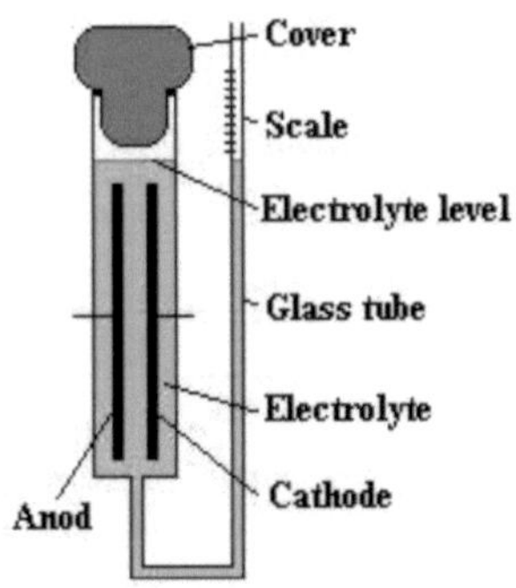

Figure 2. Principal scheme to determine the volume of released gases.

Electrolysis cell is in a separate chamber closed with a sealing cap. Glass tube bent in 180 degrees is attached to the bottom of the electrolysis chamber. The tube is graduated in units of volume above the level of the electrolyte. Gases arising in electrolysis process are pressing on electrolyte and the level in calibrated tube is increasing giving approximate volume of gases produced. In measured value of volume the 5% relative error is from different reasons; the biggest uncertainty is determined by the pressured gas generated during electrolysis - higher than atmospheric pressure. Gases are produced in electrolysis by volume 2/3 hydrogen and 1/3 oxygen. Knowing the mass of hydrogen generated in period t_{exp}, the charge necessary to produce such amount can be calculated and compared with consumed energy – result is current efficiency of particular electrolysis cell. Energy efficiency is calculated from consumed energy compared to what can be obtained from burning the produced amount of hydrogen at highest calorific value - 140 MJ/kg.

Self-made water electrolysis cell with movable electrode was used in experiments. It consists of a polyethylene shell with built in micro-screw from one side. Using stainless steel wire the micro-screw is connected to the movable electrode, situated perpendicular to the electrolyte cavity (diameter 40 mm). Stainless steel stationary electrode with same area is situated against a moving electrode. SUS316L stainless steel plate electrodes with equal area (2 cm^2) were used in experiments. Before experiments the electrodes were mechanically polished and washed with acetone and deionized water. As an electrolyte KOH solution in water was used in different concentrations. At each electrolyte concentration the distance between electrodes was changed with micro-screw from 1 mm to 5 mm. During experiment the appropriate concentration of the electrolyte solution was filled and cell attached to an inductive voltage pulse generator. At each electrolyte concentration the current and voltage oscillograms were taken for 1 to 5 mm distance between electrodes (step 1 mm). Oscillograms were further analyzed calculating consumed charge, pulse energy, and in some cases - energetic factors.

To measure the concentration of dissolved hydrogen at the cathode during electrolysis, self-made cell was used (Fig. 3). Cell consists from three cameras connected with ion conducting bridges.

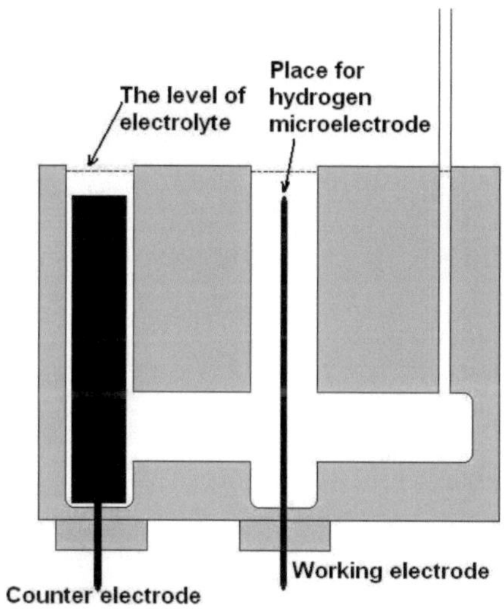

Figure 3. Three-camera electrolysis cell to measure concentration of dissolved hydrogen.

The first camera is for nickel plate counter electrode, second – for working electrode - smooth wires (diameter 0.5 mm, length 100 mm) of tungsten and platinum, but third camera was used for reference electrode in some specific experiments. Pt and W electrodes were cleaned before experiments, etching them 24 hours in concentrated alkali solution and rinsing with deionized water. The concentration of dissolved hydrogen was determined with respiration microsensor used typically in biological experiments (Unisense, 2011). The Unisense hydrogen microsensor is a miniaturized Clark-type hydrogen sensor with an internal reference electrode and a sensing anode. The sensor must be connected to a high-sensitivity picoammeter where the anode is polarized against the internal reference. Driven by the external partial pressure, hydrogen from the environment will pass through the sensor tip membrane and will be oxidized at the platinum anode surface. The picoammeter converts the resulting oxidation current to a signal. In our experiments sensor H2100 having the tip with diameter 110 μm was placed as closely as possible to cathode (<1 mm distance). Before experiments the microsensor was graduated in two points – zero H_2 concentration (Ar gas is bubbled through deionized water) and 100% or 816 mmol/l at 20 °C (H_2 gas is bubbled through deionized water – from Unisense, 2011 user manual). The experiment was carried out as follows: separate inductive voltage pulses was delivered to the cell and voltage and current oscillograms recorded. At the same time the concentration of the dissolved hydrogen was measured using microsensor.

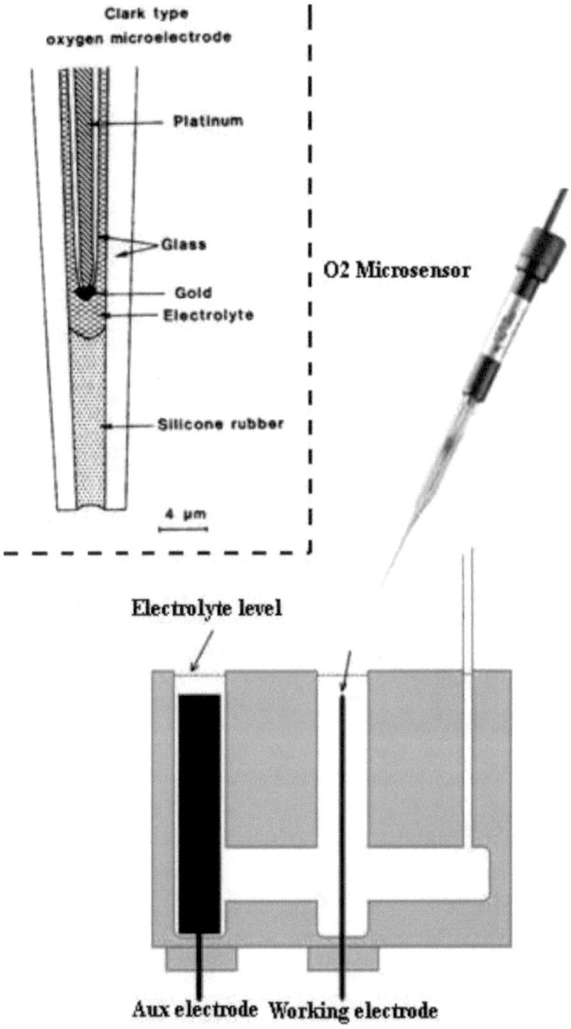

Figure 4. Water electrolysis cell to measure the concentration of dissolved oxygen.

The oxygen microsensor (also from Unisence, 2011) was used to measure the concentration of dissolved hydrogen close to cathode during inductive pulse electrolysis (Figure 4). The oxygen micro-sensor is all Clark-type sensor based on diffusion of oxygen through a silicone membrane to an oxygen reducing cathode which is polarized versus an internal Ag/AgCl anode. The flow of electrons from the anode to the oxygen reducing cathode reflects linearly the oxygen partial pressure around the sensor tip (diameter 100 μ) and is in the pA range. The current is measured by a picoammeter. To generate short inductive voltage pulses, the same circuit (chapter 3.2) is used. Cell is filled with deionized water, and the generator set in mode when expressed negative current peak is observed in oscillograms.

Concentration of oxygen is measured simultaneously with microsensor, previously calibrated in deionized water bubbled with oxygen.

Specific electrolysis cell was made for study the kinetics of inductive pulse electrolysis in diluted electrolytes (Figure 5). It was made from glass bowl with two separate electrode holders equipped with screws for electrodes from stainless steel 316L wires (diameter 2 mm, length 100 mm). Steel electrodes were cleaned before experiments, etching them 24 hours in concentrated alkali solution and rinsing with deionized water together with glass bowl of electrolysis cell. Very diluted electrolyte was prepared pouring in the cell 350 ml deionized water and adding drops of 5 M electrolyte from calibrated volume dropper (0.05±10% ml). Four electrolytes (KOH, NaOH, LiOH, H_2SO_4) were used in experiments and measurements were registered after each drop.

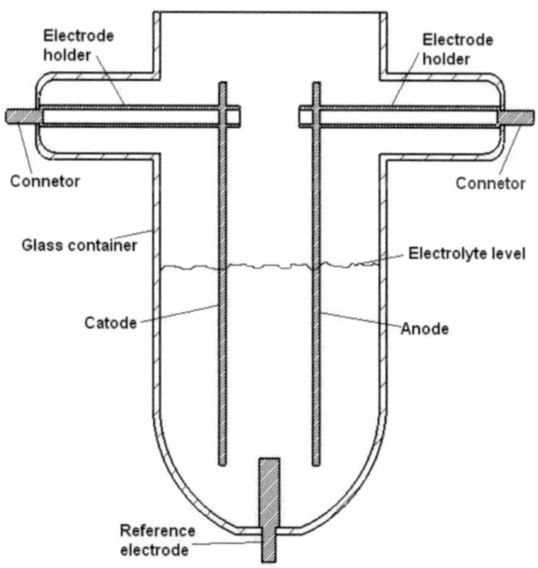

Figure 5. Water electrolysis cell to measure the kinetics of pulse electrolysis.

4. Results and analysis

4.1. Current and energy efficiencies

Average values of voltage and current, as well as flow of generated hydrogen gas depending on KOH concentrations are shown in table 3. Theoretically maximal current is calculated knowing the hydrogen gas flow by assumption that 2 electrons generate one molecule of hydrogen. Using data from Table 3, current and energy efficiency coefficients are calculated for pulse electrolysis process (see Table 4) on the assumption that the pulse transformer primary side and secondary side are two separate systems, which are only bind by average value of current flowing in the cell.

KOH concentration [mol/kg]	Average value for current pulse [mA]	Average voltage value [V]	Current value calculated from mass of generated hydrogen [mA]	Hydrogen flow [cm^3/min]
0.1	6.5	2.1	3.2	0.043
1	8	2.1	3.7	0.054
2	8.3	2.1	4	0.057
3	8.6	2.1	4.2	0.059

Table 3. Parameters of registered voltage and current pulses on an electrolysis cell.

This assumption is not entirely correct, but acceptable. When viewed from the primary circuit side, the pulse generator is with a reactive element included in its scheme - an induction coil (the primary winding of pulse transformer).

KOH concentration [mol/kg]	Current efficiency coefficient [%]	Energy efficiency coefficient [%]
0.1	49	66
1	46	64
2	48	68
3	49	68

Table 4. Current and energy efficiency coefficients

By disconnecting secondary side, primary side does not consume anything (except the power that is distributed on elements with active resistances included in the primary circuit). When connecting the secondary side, the active 1 V amplitude of the voltage pulse in the primary side is unable to consume more, because it is necessary to exceed electrolysis overvoltage – at least 1.23 V (ratio of windings in coil is 1:1).

Therefore, the average current values in Table 3 are replaced with the current consumed in the power supply system. Voltage value is read from the oscilloscope by measuring the voltage pulse on primary coil. Thus, equipment errors associated with variations in voltage values are excluded. Then, the resulting pulse is averaged over time and the resulting voltage values are shown in second column of Table 5.

It should be mentioned that adjusted energy efficiency coefficients were calculated without any reference to the circuit elements and the quantity of generated gas flow. As it is seen from Table 4.3., it is necessary to determine current and voltage values with oscilloscope within scheme of this experiment, which eliminates the pulse schemes for analogue measuring errors.

KOH concentration	Power supply voltage [V]	Average current value on the cell [mA]	Hydrogen flow [cm^3/min]	Energy efficiency coefficient [%]
0.1	1.43	6.5	0.043	97
1	1.48	8	0.054	96
2	1.53	8.3	0.057	94
3	1.49	8.6	0.059	97

Table 5. Adjusted parameters of voltage, current and efficiency.

4.2. Pulse kinetics at different concentration solutions and electrode distances

In Figure 6 the voltage and current pulse oscillograms are shown for steel electrode plates in 0.1 M KOH solution, where the maximum voltage pulse value is approximately 5.5 V when distance between the electrodes is 5 mm and it drops to about 3 V when the distance between the electrodes is 1 mm. In 0.3 M KOH solution (curves similar to 0.1 M solution) the maximum voltage pulse value is 3.5 V, when distance between the electrodes is 5 mm and it drops to 2.6 V when distance between the electrodes is reduced to 1 mm. In even more concentrated solution, ie., 0.5 M KOH, the maximum voltage pulse value at the electrode distance of 5 mm is approximately 2.9 V and when the electrode distance is 1 mm, it drops to 2.4 V. Current peak value does not change significantly depending on the electrode distance, or the concentration, but there are observed changes in the discharge tail length, suggesting that higher charge in electrolysis cell flows at more concentrated electrolyte solution.

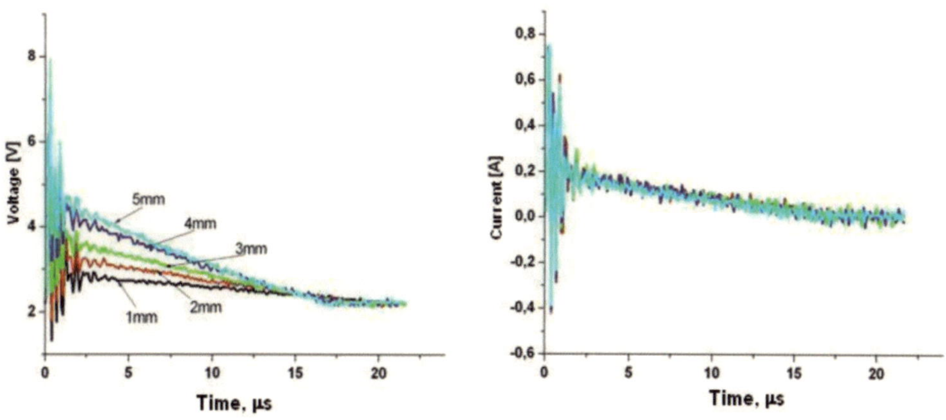

Figure 6. Current and voltage pulses registered with oscilloscope in 0.1 M KOH.

When looking at pulse generation scheme in the experimental method section (Figure 1), it is clear that high-voltage pulse generated in the transformer is reactive in nature. Reactive pulse amplitude will depend on the quality factor of capacitive element. Capacitor with a large leak (concentrated electrolyte solution) will not be able to hold the reactive pulse with large amplitude, though in the previous figures it is shown that the amplitude of those achieved in the secondary circuit on the electrolysis cell is greater than the direct pulse amplitude. This means that at the first moment when short inductive pulse is applied, the water electrolysis cell behaves as good capacitor, also at the voltage region, in which water electrolysis can occur. But after starting the discharge tail, the energy stored in the capacity transforms into the chemical energy in the process of water electrolysis.

4.3. The concentration of dissolved hydrogen at cathode

Current and voltage pulses registered with oscilloscope (Fig. 7) show that changing the electrode material, the rising front and relaxation of voltage pulses does not change, while voltage pulse amplitude decreases with increasing current pulse amplitude when solution concentration increases. Current pulses also are not different on the platinum and tungsten electrodes with identical concentration of KOH solutions (Fig. 7). To evaluate pulse energy supplied to the cell, pulse voltage and current values were multiplied and resulting curves was integrated with time (Table 6). Each row in the Table 6 shows electrode material and the concentration of the solution, and in the next column – calculated supplied energy to the system during the pulse.

Figure 8 presents each electrode's voltage and current oscillograms in the same time scale in order to better evaluate the phase shift angle between current and voltage. There are not noticeably significant differences observable between the phase shift angles depending on the electrode material.

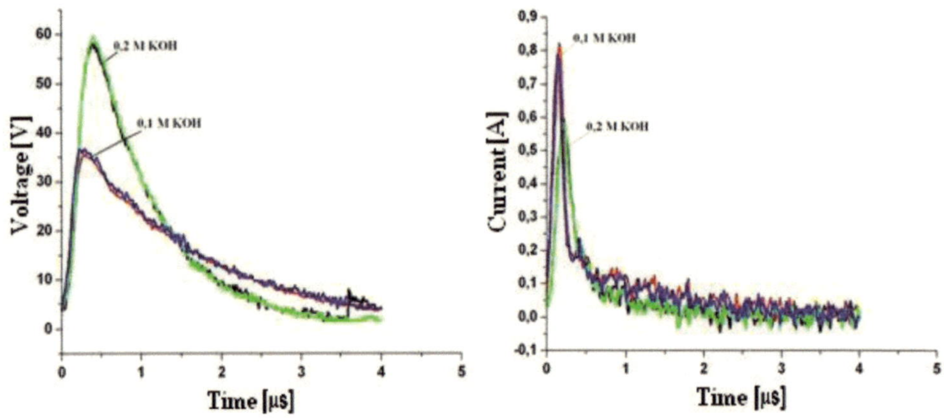

Figure 7. Current and voltage pulse oscillograms of Pt and W electrodes (Pt – black and blue, W – green and red accordingly).

Electrode material and solution concentration	Energy, mJ
Pt in 0.1M KOH solution	8.5
Pt in 0.2M KOH solution	7.7
W in 0.1M KOH solution	8.2
W in 0.2M KOH solution	7.6

Table 6. Energy supplied to the cell during the pulse, calculated from voltage and current oscillograms.

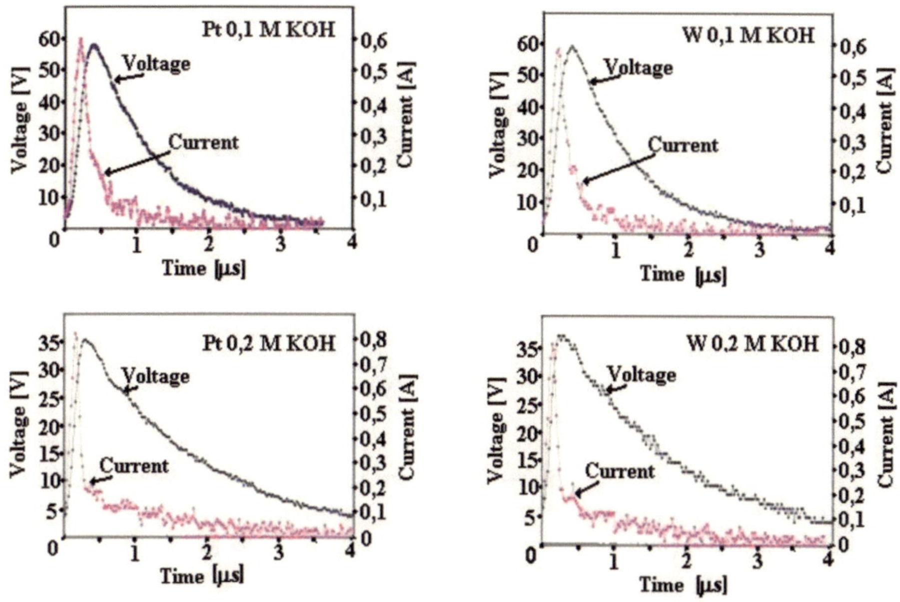

Figure 8. Current and voltage pulse oscillograms of Pt and W electrodes in 0.1 M and 0.2 M KOH

In each experiment with microsensor to measure the concentration of dissolved hydrogen, the measuring time lasted 100 s (curves at Fig. 9). As it is seen, the curves with largest slope are an electrolyte with a higher concentration, and the tungsten electrode, rather than platinum.

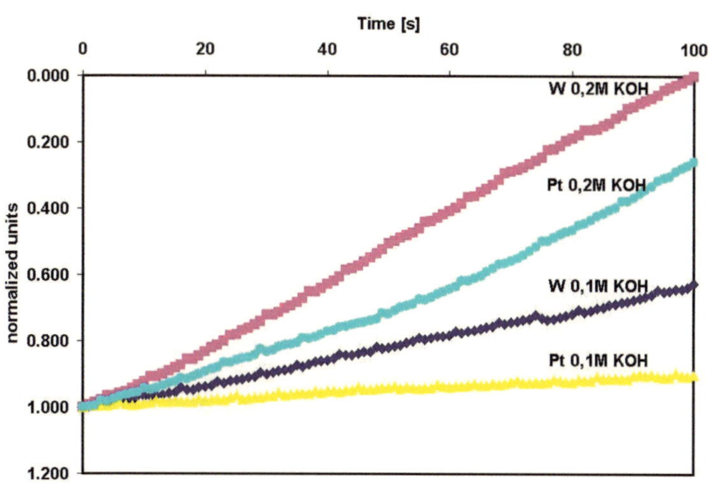

Figure 9. Changes in concentration of dissolved hydrogen gas during pulse electrolysis.

It means that on the tungsten electrodes the concentration of dissolved hydrogen increases faster than on the platinum electrodes. As it is seen from cathodic region of voltamperic curves (Fig. 10), for platinum electrode the characteristic hydrogen adsorption/absorption peak at negative currents appears at potential -0.5 V, but not for tungsten electrode.

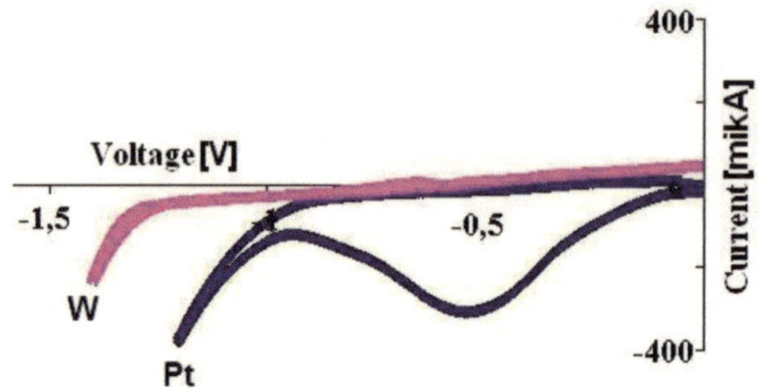

Figure 10. Voltamperic characteristics for platinum and tungsten electrodes in 0.1 M KOH solution measured in two electrode configuration at scan rate 10 mV/s.

The pulse energy of inductive reverse voltage pulses is limited. Voltage and current during the pulse reacts in such way that their multiplication and next integration in time would be equal in the same concentration of electrolyte without reference to the material of electrodes that are used. Pulse energy decreases whilst solution concentration increases, suggesting that reactive energy component has decreased. Therefore it is observable that phase shift angle between current and voltage is smaller in a more concentrated solution. Since the inductive voltage pulse energy is limited, on platinum electrode it is consumed in the adsorption area, thus structuring hydrogen adsorption monolayer on the platinum electrode. There is no hydrogen adsorption/absorption peak for tungsten electrode and during a very short voltage pulse, electrons from the metal discharge directly on hydrogen ions at interface electrode/electrolyte and hydrogen molecules are formed intensively which are detected with dissolved hydrogen microsensor.

4.4. Pulse kinetic measurements in highly dilute solutions

Voltage pulse growth at various concentrations of KOH solution (Figure 11) is equal in all concentrations, while the discharge tile after voltage pulse at various concentrations is different. Amplitude of voltage pulse is maximal in deionized water, but pulse dynamics in cell with slightly diluted electrolyte is exactly what is in the case of the open circuit, only the amplitude is smaller. Continuing to increase the concentration of electrolyte in cell, the value of voltage pulse amplitude continues decrease, while the discharge tail will increase.

Current changes the direction from negative to positive with increasing concentration of electrolyte passing through the point where the current pulse has not descending a long tail (Fig. 12). Current pulse in deionized water most of the momentum is negative. By increasing the concentration of solution up to 1 mM, current pulse appears in both positive and immediately following a negative pulse, while discharge tail almost disappears. Continuing to increase the concentration of electrolyte, the negative values of current pulse disappear and the discharge tail remains positive and increasing, which indicates that the charge injected in the cell during pulse increases. More increase of concentration does not change the view of current pulse and it remains like from the previous concentrations (Fig. 12).

When electrolyte concentration increases, voltage peak drop is observed (Fig. 13). The peak value of voltage pulse is decreasing exponentially, and it stabilizes around the value of 9 V for solutions, while in deionized water that value is over 600 V. These curves almost coincide in different alkali solutions, while in the sulfuric acid the peak values are falling faster.

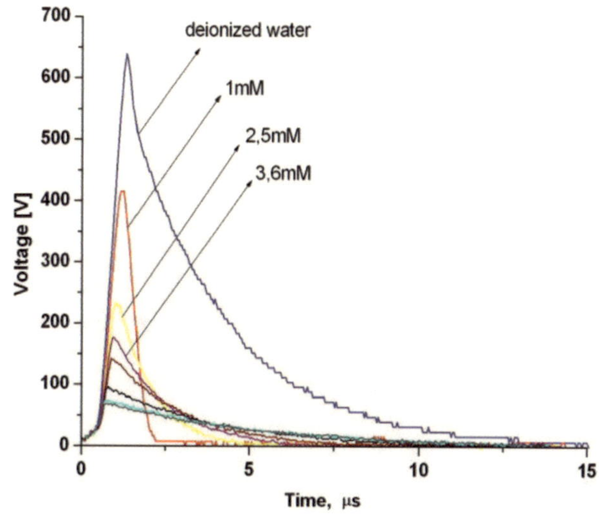

Figure 11. Inductive discharge voltage pulses with different concentrations of KOH.

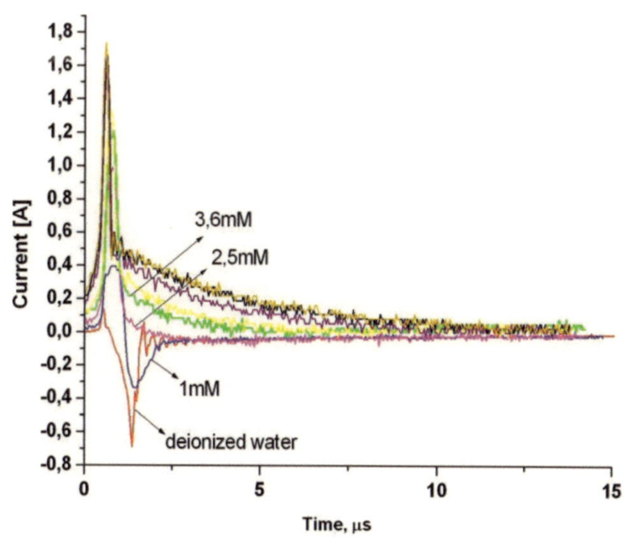

Figure 12. Current pulses initiated by inductive voltage pulses on cell with different concentrations of KOH solution

The pulse charge (integral of the current pulse) increases with increase of electrolyte concentration and tend saturate at some value (Figure 14).

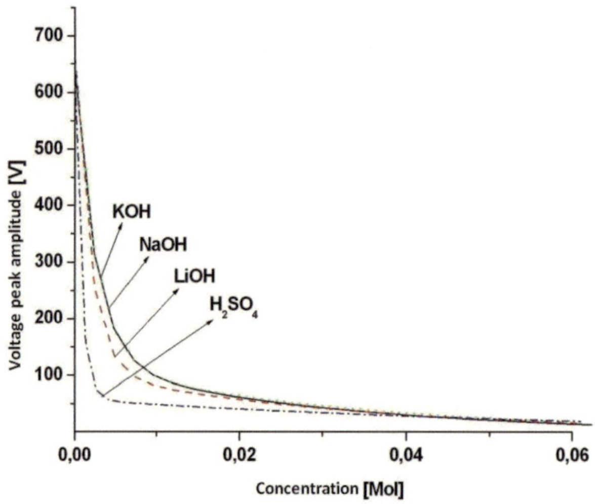

Figure 13. Decrease of voltage pulse amplitude with increasing concentration of electrolyte.

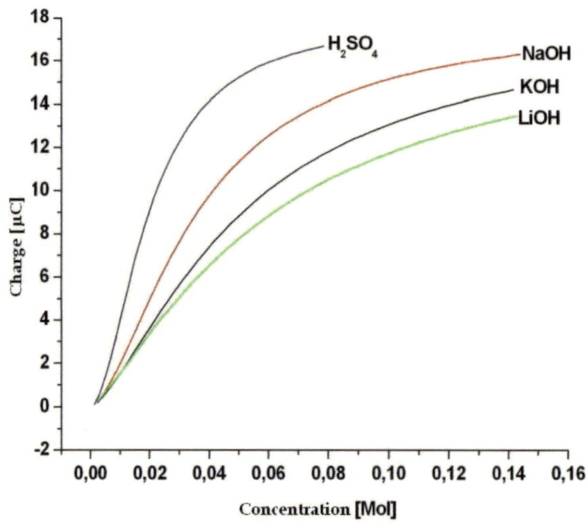

Figure 14. Changes of pulse charge (integrated current pulse) in various solutions with increasing concentration of electrolyte.

In alkali solutions, the charge behavior is nearly identical for alkali tested, while in the cell with sulfuric acid solution, the increase of charge is more rapid. Concerning occurrence of the negative currents following hypothesis is proposed. Voltage pulse kinetics demonstrates that around the electrode spatial charge density appears, i.e., when voltage rapidly grows in two-electrode system, electrons are emitted from the cathode environment. Since water ion concentrations in deionized water is low (H_3O^+ molar concentration is in the order of 10^{-7} M), then, most likely, the emitted electrons are solvated between polar water molecules and than will attach a neutral water molecule, which is described by following hydration reaction:

$$H_2O + e^- \leftrightarrow H + OH^- \quad (19)$$

If OH^- ions and solvated electrons don't manage to discharge at the cathode, then around the cathode a spatial charge appears. In case of arising the spatial charge around electrode, it is more likely of electrons to move back into the metal. If the electron donor is the OH^- ion, then oxygen evolution should appear at the cathode. In principle, according to the experimental circuit, such electron returning back in metal in large amounts what results from the negative current pulse value presented in Fig. 12, is not possible since this current component is blocked by the diode incorporated in the circuit. Therefore behind the diode, parasitic element with the inductive nature must exist in the measurement circuit (Fig. 1) which becomes comparatively small and solvated electrons are discharged by ions in electrolyte, therefore decreasing negative current. To confirm this hypothesis, it is necessary to determine if oxygen does not appear near the cathode (the solvated electrons OH^- form allows a reverse reaction (19) on the cathode).

4.5. Measurement of dissolved oxygen near the cathode

The concentration of dissolved oxygen in a solution near the cathode during pulse electrolysis in dependence of time is presented in Fig. 15. During the first 60 seconds current pulses has an explicit negative peak in the cell. After 60 seconds, the generator is set to manage the negative current peak disappeared. From Fig. 15 it is seen that when a negative current peak occurs, the oxygen evolves at the cathode, but when the negative current during voltage pulses is prevented, the oxygen at cathode is no longer released and concentration decreases.

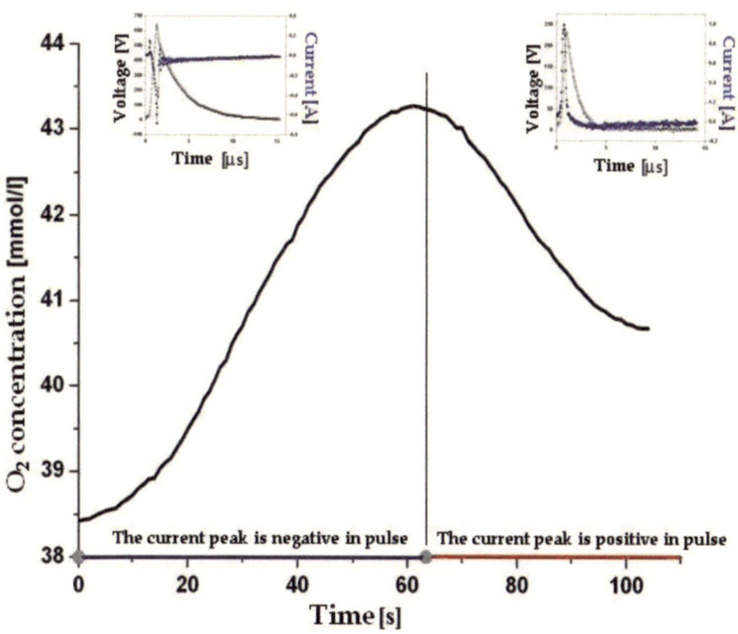

Figure 15. Dissolved oxygen concentration near the cathode in two regimes of pulse generator – when current peak is negative (left, dissolved oxygen concentration is increasing) and positive (right, dissolved oxygen concentration is decreasing).

5. Conclusions

Reactive short voltage pulse generator is designed to power water electrolysis cells of different constructions, both with spatially separated and with variable distance electrodes. Required value of electrolysis voltage in the primary circuit of power supply can be reduced by inserting the electrolysis cell in secondary circuit of power supply together with inductive element and reverse diode. For example, in this work electrolysis is provided with direct pulse amplitude 1 V, which induces a short high voltage pulse (tens, hundreds of volts, depending on the conductivity of electrolyte) in the secondary circuit. For studying the process of electrolysis the microelectrode sensors are used to measure concentration of dissolved hydrogen and oxygen gas in the direct vicinity of cathode for the first time.

By changing the distance between the electrodes and concentration of electrolyte, it is experimentally proved that the electrolysis cell is capacitor with high Q factor when short voltage pulse (width below 1 μs) is applied. During this short time the capacitor (electrolysis cell) is charged, which can be interpreted as charging of double-layer on interface cathode/electrolyte. After the interruption of short voltage pulse the energy accumulated in double-layer capacitor slowly discharges (pulse discharge tail), thus activating the process of electrolysis. Consequently, it is shown that with short voltage pulse electrolysis the charging of cell can be separated from the electrochemical reactions in electrolysis process. Kinetics of charging the electrolysis cell with reactive high voltage short pulses does not depend on concentration of the electrolyte, whereas kinetics of the subsequent long-discharge process depends on the electrolyte concentration (faster in dilute solutions, slower in more concentrated solutions). If concentration of electrolyte in the electrolysis cell is above 3 mM, reactive voltage pulse energy does not depend on concentration of the solution. The current polarity switch from cathodic to anodic and back is observed in oscillograms, when de-ionized water electrolysis is performed with short reactive voltage pulses. Polarity changes only during short time when reactive voltage pulse is applied; at the beginning of the discharge tail the current is cathodic again. Measurements of dissolved oxygen concentration by microsensor in the immediate vicinity of the cathode show that oxygen concentration increases (reverse process to hydrogen evolution) in the presence of anodic current. The hypothesis is proposed that high-voltage pulse causes the emission of electrons from the cathode metal into electrolyte, where at first the electrons are solvated, then dissociating the water molecules forms H atoms and OH^- ions; next generated OH^- ions can discharge on cathode at the moment when applied voltage pulse reduces, giving the release of oxygen detected by microsensor. Current efficiency of 50% is registered in high-voltage reactive short pulse electrolysis, while the energetic efficiency is in range 70-100%.

Platinum and tungsten electrodes are studied to find the impact of electrode material on the process of short reactive voltage pulse electrolysis. Experimental results show that the voltage and current characteristics of inductive voltage short pulse electrolysis are the same for both metals, but the concentration of dissolved hydrogen grows faster at the tungsten electrode. Delay in the release of hydrogen on a platinum electrode is explained by the tendency of platinum to adsorb hydrogen on the surface.

Author details

Martins Vanags, Janis Kleperis and Gunars Bajars
Institute of Solid State Physics, University of Latvia, Riga, Latvia

Acknowledgements

The authors acknowledge laboratory colleagues Liga Grinberga and Andrejs Lusis for stimulating discussions, and Vladimirs Nemcevs for technical assistance. Authors thank Professor Robert Salem for helpful guidance and advices, and deep compassion to his family on the death of Professor in 2009. Financial support from the European Social Fund project "Support for doctoral studies at the University of Latvia" is acknowledged by M.Vanags. All authors thank the National Research Program in Energy supporting the development of hydrogen infrastructure in Latvia.

6. References

Bockris O`M and Potter E.C. (1952) The mechanism of the Cathodic Hydrogen Evolution reaction, Journal of the Electrochem. Society, Vol. 99:169 – 186.

Bockris J.O'M., Ammar I.A., Huq A.K.S. (1957). The Mechanism of the Hydrogen Evolution Reaction on Platinum, Silver and Tungsten surfaces in Acid Solutions. *J Chem Phys,* 61:879–886.

Bockris J.O`M., Dandapani B., Cocke D. and Ghoroghchian J. (1985) On the Splitting of Water. *International J. Hydrogen Energy,* Vol. 10:179-201, 1985.

Bott A.W. (2000) Controlled Current Techniques. *Current Separations* 18/4:125-127.

Chambers, S.B. (2002) Method for producing orthohydrogen and/or parahydrogen. *US Patent 6419815* (2002).

Chandrasekar M.S., Pushpavanam M. (2008) Pulse and pulse reverse plating—Conceptual, advantages and applications. *Electrochimica Acta,* 53:3313–3322.

Delahay P. (1966) Electrode Processes without a Priori Separation of Double-Layer Charging, *The Journal of Physical Chemistry,* 70: 2373–2379.

De Levie R (1999). The electrolyses of water. *Journal of Electroanalytical Chemistry* 476:92-93.

Dimants J, Sloka B, Kleperis J, Klepere I (2011) Renewable Hydrogen Market Development: Forecasts and Opportunities for Latvia. Paper No 319GOV, Proceedings of *International Conference on Hydrogen Production ICH2P-11,* June 19-22, 2011, Thessaloniki, Greece, p.1-6

El-Meligi A.A., Ismail N. (2009) Hydrogen evolution reaction of low carbon steel electrode in hydrochloric acid as a sourse for hydrogen production, *International Journal of Hydrogen Energy* 34:91 – 97.

Ghoroghchian J., Bockris J.O`M. (1985) Use Of A Homopolar Generator In Hydrogen Production From Water, *International Journal of Hydrogen energy*, vol. 10:101-112.

Gutmann F, Murphy O.J (1983) The Electrochemical Splitting of Water, In boock: R.E. White, J.O'M. Bockris, B.E. Conway (Eds.), *Modern aspects of electrochemistry*, vol. 15, p. 1, Plenum, New York (1983) pp. 5–13

Heyrovsky M (2006) Research Topic – Catalysis of Hydrogen Evolution on Mercury Electrodes, *Croatica Chemica Acta* 79 (1) (2006) 1-4

Hirato T., Yamamoto Y., Awakura Y. (2003) A new surface modification process of steel by pulse electrolysis with asymmetric alternating potential. *Surface and Coatings Technology* 169/170:135–138.

Horányi G., Láng G.G. (2006) Double-layer phenomena in electrochemistry: Controversial views on some fundamental notions related to electrified interfaces, *Journal of Colloid and Interface Science* 296:1–8.

Horvath, St. (1976) Electrolysis apparatus. *US Patent 3954592* (1976).

Hydrogen Pathway (2011). Welcome to the Roads2HyCom Hydrogen and Fuel Cell Wiki: Cost Analysis; 18.11.2011; Available from: www.ika.rwth-aachen.de/r2h/index

Ibl N., Puippe J.Cl. and Angerer H. (1978) *Electrocrystallization in pulse electrolysis*. Surface Technology, 6:287 – 300.

Kisis G., Zeps M., Vanags M. (2009) Parameters of an efficient electrolysis cell, *Latvian Journal of Physics and Technical Sciences*. Riga, 2009, N3. 6 p.

Kreuter W, and Hofmann H (1998) Electrolysis: the important energy transformer in a world of sustainable energy, *Int. J. Hydrogen Energy* 23(8): 661-666.

Kristalik L.I. (1965) Barrierless Electrode Process, *Russian Chemical Reviews*, Vol.34:785 – 793.

Kuroda K., Shidu H., Ichino R. and Okido M. (2007) Osteoinductivity of Titania/Hydroxyapatite Composite Films Formed Using Pulse Electrolysis, *Materials Transactions*, 48:328-331.

Meyer, S.A. (1986) Electric pulse generator. *US Patent 4613779* (1986).

Meyer, S.A. (1989) Gas generator voltage control circuit. *US Patent 4798661* (1989).

Meyer, S.A. (1992) Process and apparatus for the production of fuel gas and the enhanced release of thermal energy from such gas. *US Patent 5149407* (1992).

Miller J.R. and Simon P. (2008) Fundamentals of Electrochemical Capacitor Design and Operation, *The Electrochemical Society Journal Interface*, Spring:31-32.

Murphy G.F. (1983). In: White RE, Bockris JO'M, Conway BE, editors. *Modern aspects of electrochemistry*, vol. 15. NewYork: Plenum Press; 1983. p. 5-13

Nisancioglu K. and Newman J. (2012) Separation of Double-Layer Charging and Faradaic Processes at Electrodes, *Journal of The Electrochemical Society*, 159:E59-E61.

Noel M, Vasu K.I. (1990) *Cyclic Voltammetry and the Frontiers of Electrochemistry*, Oxford and IBH Publishing Co. Pvt. Ltd., New Delhi, 1990., 695 p., ISBN 81-204-0478-5.

Oldham K.B., Myland J. C. (1993) *Fundamentals of Electrochemical Science*, United Kingdom: Academic Press Limited, 1993, 474 p.

Our Common Future (1987) Book, Oxford: Oxford University Press. ISBN 0-19-282080-X

Puharich, H.K. (1983) Method & Apparatus for Splitting Water Molecules. *US Patent 4,394,230* (1983).

Salem R. R.(2004) *Chemical Thermodynamics*, Fizmatlit, 2004., 352 p, ISBN 5-9221-0078-5

Santilli, R.M. (2001) Durable and efficient equipment for the production of a combustible and non-pollutant gas from underwater arcs and method therefore. *US Patent 6183604* (2001).

Sasaki T., Matsuda A. (1981) Mechanism of Hydrogen Evolution Reaction on Gold in Aqueous Sulfuric Acid and Sodium Hydroxide, *J. Res. Inst. Catalysis*, Hokkaido Univ., Vol. 29:113 – 132.

Selected Articles of Hydrogen Phenomena The Book, Editors: T. Nejat Vezģroğlu, M. Oktay Alniak, Ġenay Yalçin; 1.Basım: Ekim 2009, ISBN: 978-605-5936-23-5, p. 39-45.

Shaaban A.H. (1994) Pulsed DC and Anode Depolarisation in Water Electrolysis for Hydrogen Generation, HQ Air Force Civil Engineering Support Agency Final Report, August, 1994. 21.11.2011. Available from: http://www.free-energy-info.co.uk/P1.pdf

Shimizu N., Hotta S., Sekiya T. and Oda O. (2006) A novel method of hydrogen generation by water electrolysis using an ultra-short-pulse power supply. *Journal of Applied Electrochemistry*, 36:419–423.

Spirig, E. (1978) Water decomposing apparatus. *US Patent 4113601* (1978).

The Hydrogen Economy (2004): Opportunities, Costs, Barriers, and R&D Needs. 21.11.2011; Available from: http://www.nap.edu/openbook.php?record_id=10922&page=R1

Themu, C.D. (1980) High voltage electrolytic cell. *US Patent 4316787* (1980).

Unisense Microsensors (2011). Tool Guide 15.11.2011. From: UNISCIENCE Science: Available from: http://www.unisense.com/Default.aspx?ID=458); http://www.unisense.com/Default.aspx?ID=443

Vanags M, Shipkovs P, Kleperis J, Bajars G, Lusis A (2009) Water Electrolyses – Unconventional Aspects" In Book: *Selected Articles Of Hydrogen Phenomena.* Editors: T. Nejat Vezģroğlu, M. Oktay Alniak, Ġenay Yalçin; 1.Basım: Ekim 2009, ISBN: 978-605-5936-23-5, p. 39-45.

Vanags M, Kleperis J and Bajars G (2011a) Electrolyses Model Development for Metal/Electrolyte Interface: Testing with Microrespiration Sensors. *International Journal of Hydrogen Energy*, vol. 36:1316-1320.

Vanags M, Kleperis J and Bajars G (2011b) Separation of Charging and Charge Transition Currents with Inductive Voltage Pulses. *Latvian Journal of Physics and Technical Sciences*, No 3, p. 34-40.

Zoulias E., Varkaraki E., Lymberopoulos N., Christodoulou C.N. and Karagiorgis G.N. (2002) A Review On Water Electrolysis. Centre for Renewable Energy Sources (CRES), Pikermi, Greece. 21.11.2011. Available from: http://www.cres.gr/kape/publications/papers/dimosieyseis/ydrogen/A%20REVIEW%20ON%20WATER%20ELECTROLYSIS.pdf

Chapter 3

Overview of Membrane Electrode Assembly Preparation Methods for Solid Polymer Electrolyte Electrolyzer

Bernard Bladergroen, Huaneng Su, Sivakumar Pasupathi and Vladimir Linkov

Additional information is available at the end of the chapter

http://dx.doi.org/10.5772/52947

1. Introduction

In search for improved overall efficiency, higher current density, lower membrane cost and maximized utilization of relatively expensive catalytic materials, the scientific community has produced hundreds of publications, assisting those looking for the opportunity to turn electrolyzer prototypes into commercially viable products [1]. While other chapters of this book focus on the development of catalysts and proton conductive membranes, this chapter describes the principle functions of "membrane electrode assemblies" (MEAs), followed by an overview of methods designed and developed to produce effective MEAs. The process conditions chosen during the preparation/production of MEAs have a great impact on the logistics of proton-, electron-, reagent- and product-transport. Each different method is aimed at achieving an architecture yielding optimal accessibility, stability and numbers of "three phase boundaries" (TPB) that contribute to the productivity and efficiency of the electrolyzer system. Proper design of the MEA is essential since the true potential of the most appropriate membrane and the most active catalysts will only be revealed in a successful MEA configuration.

List of abbreviations:

MEA - membrane electrode assembly
TPB - three phase boundary
SPE - solid polymer electrolyte
PEMFC - polymer electrolyte membrane fuel cell
Pt/C - carbon supported platinum
CL - catalyst layer
CCG - catalyst coated GDL

CCM - catalyst coated membrane
GDL - gas diffusion layer
PFSA - perfluorosulphonic acid
PTFE - polytetrafluoroethylene
PVD - physical vapour deposition
CVD - chemical vapour deposition
ES - electro spraying
EPD - electrophoretic deposition
MWNT - multi walled carbon nano tube
TEM - transmission electron microscope
SEM- scanning electron microscope
NML - noble metal loading
EASA - electrochemical active surface area
MT - mass transfer

2. The principle functions of the MEA

In both Solid Polymer Electrolyte (SPE) and Fuel Cell MEAs the principle function of the MEA is to efficiently control the flow of electrons liberated at the electron donating reaction (the anode) to the electron accepting reaction (cathode). This is typically achieved by separating the cathodic reaction from the anodic reaction by using a membrane that conducts protons (H^+) only. Electrons are channelled through an external circuit from the anode to the cathode. By controlling the flow and direction of electrons, H_2O can either be used to produce H_2 and O_2 (Electrolysis) or can be produced from H_2 and O_2 (fuel cell). The anode and the cathode reaction for an SPE electrolyzer are shown in equation 1 and 2 respectively;

$$2H_2O \rightarrow 4H^+ + O_2 + 4e^- \quad \text{(anode, electron donating reactions)} \quad (1)$$

$$4H^+ + 4e^- \rightarrow 2H_2 \quad \text{(cathode, electron accepting reactions)} \quad (2)$$

Since the enthalpy of H_2 and O_2 is higher than the enthalpy of H_2O, a sufficiently high electrical potential has to be applied between the two electrodes to force a flow of electrons from the anode to the cathode. One of the aims in SPE MEA design is to minimize power losses during the electrolysis process. Hereafter, SPE MEAs will be referred as MEAs in this chapter. Total power losses are related both to the material properties of MEA components and the MEA production methods (the way the components are put together). The various components in a typical MEA are listed in Table 1;

Different groups of researchers have been optimizing the properties of each component. Components with the most promising properties have been assembled into MEAs. The most commonly used membrane is Nafion® 115 with mainly commercial IrO_2 as anode catalyst, and commercially available carbon supported Pt as cathode. Ti-fibre as anode backing and E-TEK / Toray carbon paper/cloth as commercial backing on the cathode. Typically, MEA performance is given as the current density as function of applied potential. Table 2 gives an overview of various published MEA performances with listed components and production methods used.

No	MEA COMPONENT	PROPERTIES
1	Membrane	Conductance (1/S), gas permeance, electroosmotic properties (all as function of temperature)
2	Cathode catalyst	Catalytic activity of the specific material, electrical resistance of the cathode layer, particle size
3	Anode catalyst	Catalytic activity of the material, electrical resistance of the anode layer, particle size
4	Cathode backing layer	Mass transfer of reactants and reaction products
5	Anode backing layer	Mass transfer of reactants and reaction products

Table 1. MEA components and their main properties

The results reveal that the true potential of the MEA is not only determined by the properties of the individual components but mainly by the MEA production method. Before describing various MEA preparation methods, the concept of TPB is explained in conjunction with the logistics around electrons, protons, reactants and products.

3. Logistics at three phase boundaries

Since MEAs have no significant capacitance, accumulation of protons or electrons at either surface of the MEA is negligible. Therefore, both half cell reactions will happen at exactly the same rate. In order for an MEA to function, the following conditions must be met;

a. The combined electrode overpotential should be low enough for the reactions to occur at a reasonable rate, typically <2V.
b. Each electron liberated at the anode catalyst in accordance with equation 1 has to find its way to a cathode catalyst particle in order to recombine in accordance with equation 2. For the electrons to flow, a path of sufficient electro conductivity running from the anode catalyst particle to the cathode catalyst particle must be available. Concurrently, a path with sufficient proton conductivity must be available to transport the H^+ from the anode catalyst, through the membrane, to the cathode catalyst.
c. For the reactions to continue, reactants (water) should be supplied continuously to the anode catalyst site (that holds both required protonic and electronic connection with the cathode catalyst). Products such as O_2 and H_2, need to be removed on a continuous bases from the anode and cathode respectively.

The three dimensional interfaces where catalyst, ionomer and reactant convene to meet all three conditions, are called the three phase boundaries (TPBs). From conditions B and C it can be concluded that not only the number of TPBs are important but also the logistics of proton-, electron-, reagent- and product-transport. Both factors determine the rate and efficiency at which the MEA can produce H_2 and O_2 from water. A schematic representation of the TPB is shown in Figure 1.

Ref	MEA	Production Methode		Hot pressing conditions					Electrolyte	Thick-	Catalyst type		Loading	Backing		Test Cell
	Size	Anode	Cathode	Pmin	Pmax	Time	Tmin	Tmax		ness	Anode	Cathode	(A/C/IA /IC*)	Anode	Cathode	clamping
	(cm^2)			Mpa	Mpa	Sec	oC	oC		(mm)						force
2	100	CCM (spray)	CCS (paste spreading)						Nafion 115	127	IrO2 (inhouse)	30%Pt/C ETEK	3/0.6/33/33	Ti-grid 260	GDL ELAT	15kg/cm2
2	100	CCM (spray)	CCS (paste spreading)						Nafion 115	127	IrO2 (inhouse)	30%Pt/C ETEK	3/0.6/33/33	Ti-grid 500	GDL ELAT	15kg/cm2
2	100	CCM (spray)	CCS (paste spreading)						Nafion 115	127	IrO2 (inhouse)	30%Pt/C ETEK	3/0.6/33/33	Ti-grid 500	GDL ELAT	25kg/cm2
2	100	CCM (spray)	CCS (paste spreading)						Nafion 115	127	IrO2 (inhouse)	30%Pt/C ETEK	3/0.6/33/33	Ti-grid 500	GDL ELAT	25kg/cm2
3	100	CCM (spray)	CCS (spread)						Nafion 115	125	IrO2 (inhouse)	30%Pt/C ETEK	2.5/0.5/33/33	Ti-Mesh	GDL ELAT	25kg/cm2
4	100	CCM (spray)	GDE Etek						Nafion 115	125	IrO2 (inhouse)	GDE Etek	2.5/0.6/?/33	Ti-Mesh	Carbon Cloth	7Nm
4	5	CCM (spray)	CCS (paste spreading)						Nafion 115	125	IrO2 (inhouse)	30%Pt/C ETEK	3/0.6/33/33	Ti-Mesh	GDL ELAT	9Nm
4	5	CCM (spray)	CCS (paste spreading)						Nafion 115	125	IrO2 (inhouse)	30%Pt/C ETEK	3/0.6/33/33	Ti-Mesh	GDL ELAT	9Nm
5	X	CCM (spray)	CCS (paste spreading)						Nafion 115	125	IrO2/TiOx	30%Pt/C ETEK	1/ 1 /33 /33	Ti-Grid - mod)	GDL ELAT	75kg/cm2
5	X	CCM (spray)	CCS (paste spreading)						Nafion 115	125	IrO2/TiOx	30%Pt/C ETEK	1/ 1 /33 /33	Ti-Grid - mod)	GDL ELAT	75kg/cm2
6	160	prototype Giner GS-10 electrolyzer, no details provided														
6	160	prototype Giner GS-10 electrolyzer, no details provided														
7	27	CCS (x)							Nafion 115	127	IrO2	Pt	x /x / x/ x	Ti Felt (Pt mod)	Toray TGP-H-090)	
7	27	CCS (x)							Nafion 115	127	IrO2	Pt	x /x / x/ x	Ti Felt (Pt mod)	Toray TGP-H-090)	
8	5	CCM (spray)	CCS (paste spreading)						Nafion 115	127	IrO2 (inhouse)	30%Pt/C ETEK	2.5/0.5/33/33	Ti-Mesh	GDL ELAT	
9	2500	No details							Nafion 112	50	No details					
10	4	CCS (spray on silicon rubber)		4	17	90	168	198	30PES SPEEK	127	Ir Black	20% Pt/C (JM)	5/1.5/15/x	Ti mesh 40	Toray CP	
10	4	CCS (spray on silicon rubber)			8	90	130		Nafion 115	127	Ir Black	20% Pt/C (JM)	5/1.5/15/x	Ti mesh 40	Toray CP	
11	5	CCM (spray)			10	x	140		Nafion 1035	89	Ir0.4Ru0.6O2	RuO2/C (25%)	1/0.5/25/25	Toray TGP-H-060	Toray TGP-H-060)	
11	5	CCM (spray)			10	x	140		Nafion 1035	89	Ir0.4Ru0.6O2	Pt/C (28%)	1/0.5/25/25	Toray TGP-H-060	Toray TGP-H-060)	
12	4	CCS (spray on silicon rubber)			8	90	130		Nafion 112	50	Ir black	Pt Black (JM)	3.8/2.3/13/11	Ti mesh 40	Toray CP	
13	50	CCS brush coating	CCS SPD		x	x	130		Nafion 115	127	Ir black	20%Pt/C	0.3/x/x/5	Ti-Mesh (Pt mod)	Porous CP	
14	100	CCS brush coating	CCS SPD		x	x	130		Nafion 115	127	Ir black	20%Pt/C	0.3/0.4/x/5	Ti-Mesh (Pt mod)	Porous CP	
14	9	CCS brush coating	CCS SPD		x	x	130		Nafion 112	50	Ir black	20%Pt/C	0.3/0.4/x/5	Ti-Mesh (Pt mod)	Porous CP	
15	x	CCS (ink "applied" to GDL)			5.5	180	140		Nafion 112	50	Pt & Ir black 85:15		4/4/20/20	LT140W Elat	LT140W Elat	
16	27	Hot pressed electrodes to membrane							Nafion 112	50	Pt & IrO2 mixed			Ti felt (Bekinit)	Toray 090	
17	4	CCM (spray)	CCM (spray)						Nafion 115	127	Ir0.6Ru0.4O2	20% Pt/C (JM)	2/0.4/5/10	Ti sinter	Ti sinter	
18	4	CCM (spray)	CCM (spray)						Nafion 115	127	IrO2	20% Pt/C (JM)	2/0.4/5/10	Ti sinter	Ti sinter	
19	4	CCM (spray)	CCM (spray)						Nafion 115	127	IrO2	20% Pt/C (JM)	2/0.4/5/10	Ti sinter	Ti sinter	
20	5	CCM (spray)	CCGDL						Nafion 115	127	IrO2	30% Pt/C (JM)	3/0.6/33/33	Carbon cloth	Carbon cloth	
21	5	CCM (spray)	CCM (spray)						Nafion 115	127	IrO2	40% Pt/C (JM)	2/0.5/33/33	Carbon cloth	Carbon cloth	
22	1	CCGDL	CCGDL						Nafion 212	51	IrO2/SnO2	40% Pt/C (JM)	1.5/0.3/30/30	Carbon cloth	Carbon cloth	
23	100	CCM	CCM	hot press					B2-type	51	IrO2	40% Pt/C (JM)	4/2.4/25/25	GP CP	GP CP	
24	5	CCM (spraying)	CCM (spraying)	catalyst on sprayed nafion layer					Nafion 115	127	IrO2	Pt black	2.5/2/30/30	GP CP	CP	

X=missing details; GP=Gold plated; CP= Carbon paper; GDL= Gas diffusion layer; CCM= Catalyst Coated Membrane
CCS= Catalyst Coated Substrate; SPD=Screen Print Diffusion

Table 2. MEA performance, components and production methods (Ref [2-24])

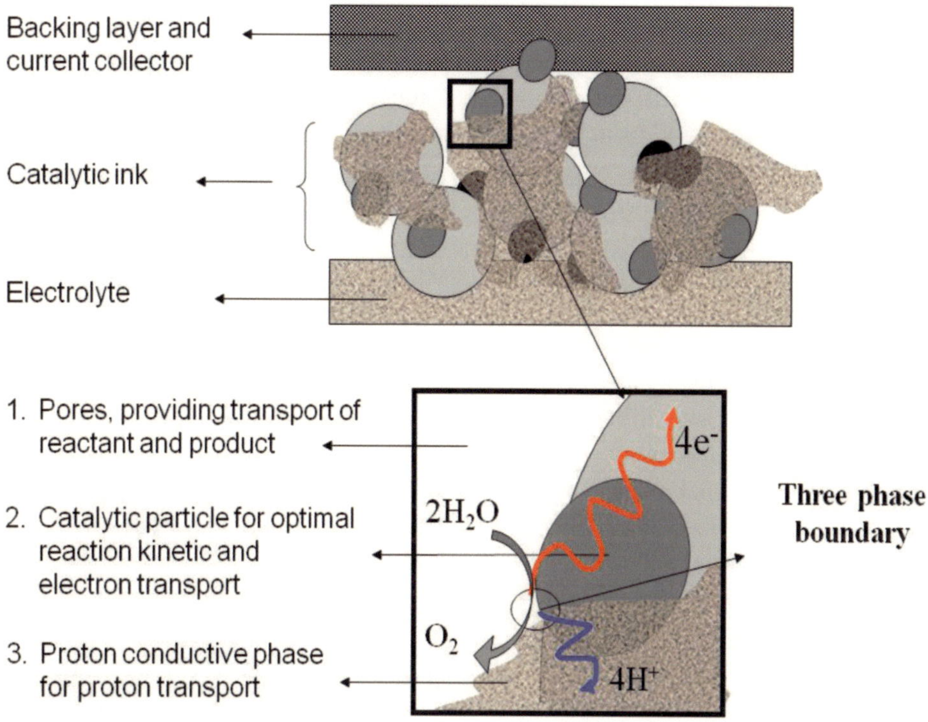

Figure 1. Schematic overview of the three phase boundary

For the cathode of a SPE electrolyzer, the catalyst layer (CL) structure is similar with that for PEMFC. Normally, carbon supported platinum (Pt/C) catalysts are used in these CLs. For the anode of a SPE electrolyzer, the CL differs somewhat from that for PEMFC due to the unavailability of carbon support in anode catalyst. For example, oxides such as IrO2, Ir–Ru, Ir–Sn, Ir–Ta, Ir–Ru–Sn and Ir–Ru–Ta etc were widely used as oxygen evolution electrocatalysts in anode CL. Particles of supported catalysts are well dispersed and extend deep into layers of carbon particles, which requires more ionomer to extend the TPB whereas in unsupported catalysts, the catalyst layer is only one to two layers thick and therefore requires less ionomer for optimum TPB.

The principle functions of the MEA are summarized in the left column of Table 3. The right column provide parameters that influences these functions

4. Membrane electrode assembly production methods

As mentioned in the previous section, each MEA preparation method is aimed at achieving a MEA architecture with an optimal accessibility, stability and number of TPB. Generally the approaches taken to produce MEA can be divided into A and B as depicted in Figure 2. Catalysts are either applied directly onto the proton conductive membrane or onto the electrically conductive supports. The different ways catalysts are applied on either surface is discussed in the following paragraphs.

No	FUNCTIONS OF THE MEA	
1	Separate cathode and anode reactions	Membrane properties (gas permeance)
2	Channelling electrodes from the anode to the current collector and from the current collector to the cathode	Catalyst type, electrical resistivity of the membrane material
3	Provide optimal electrochemical activity	Catalyst type
4	Provide optimal number of TPBs	MEA preparation method, catalyst type
5	Provide structure that facilitate electron transport from current collector to TPBs at the minimal overall electrical resistance	MEA preparation method, catalyst type, support type, ionomer type and loading
6	Provide structure that facilitates proton transport from TPBs to PEM with the optimal proton conductivity	MEA preparation method, ionomer content, ionomer type
7	Provide structure that allows for products to be transported away from TPBs, as accumulation of products will have a negative impact on the rate at which reactants can reach the TPBs	MEA preparation method, ink composition including hydrophobic / hydrophilic components
8	Provide structure that allows for reactants to reach TPBs	MEA preparation method, ink composition including hydrophobic / hydrophilic components

Table 3. MEA components and their main properties

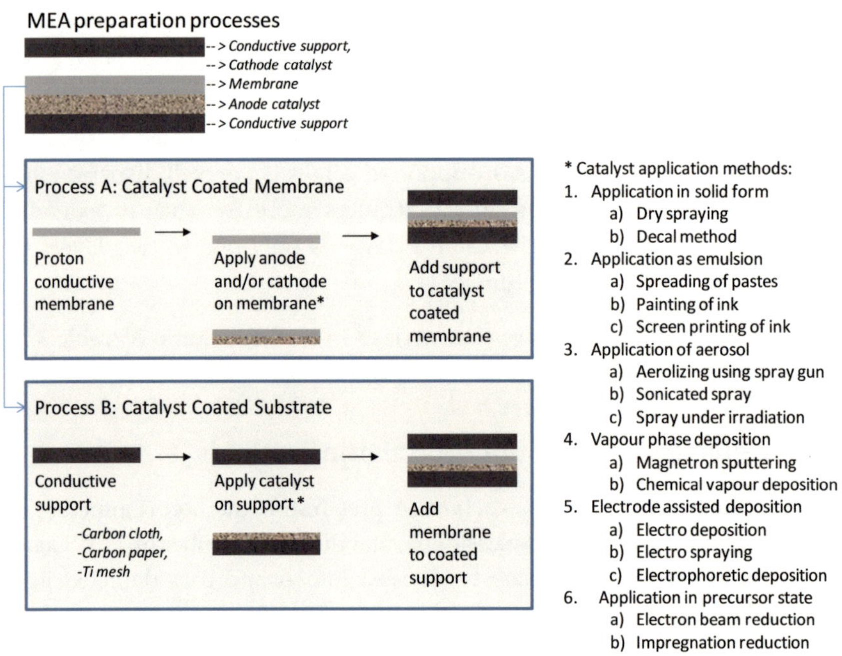

Figure 2. Schematic of catalyst-coated GDL (CCG) and catalyst-coated membrane (CCM) structures.

5. Application of the catalyst in solid form

a. Dry spraying

Dry spraying is a method of depositing CL from dry powder electrode material, based on the adaptation of a rolling process. After mixing the reactive materials (catalysts, PTFE, PFSA powder and/or filler materials) in a knife mill, it is then atomized and sprayed in a nitrogen stream through a slit nozzle directly onto the GDL or membrane. This reactive layer is fixed and thoroughly connected to the membrane by passing them through a calender. Although adhesion of the catalytic material on the surface is strong, the layer is further fixed by hot rolling or pressing to improve the electric and ionic contact. Depending upon the degree of atomization, a completely, uniformly covered reactive layer with thickness down to 5 μm can be prepared with this technique [25]. A consistent particle size distribution is a reason for using dry spraying. The procedure is simple and, as a dry process, avoids the use of any solvents and drying steps.

b. Decal method

Decal method is the most commonly used route for fabricating CL on either side of SPE membrane [26-27]. In this method, the catalyst ink (electrolyte ionomer, catalysts and solvent mixture) is coated on a decal that has precisely the same dimensions as that of the active area in the water electrolyzer. The widely used decal material for this purpose is fiberglass reinforced Teflon (200-300 mm thick). Prior to catalyst coating, the decal surface must be cleaned with a solvent, then treated with a Teflon release agent and dried at room temperature. Desired catalyst loading can be achieved by repeated painting and drying of the decal. Drying the slurry-coated decals at high temperature (100-150 °C) in a vacuum oven for 30-60 minutes between each coating ensures complete solvent evaporation and ionomer redistribution. The pretreated membrane is sandwiched between two catalyst-coated decals, with the catalyst-coated sides facing each other. The decal/membrane assembly is then enveloped between two PTFE blanks, followed by sandwiching two stainless-steel sheets and compressed using a preheated hot press at a compression pressure for few minutes. The next step is transfer of CL from the decal to the PEM to produce a CCM. This is done again at high temperature and pressure using a hot press. After cooling the hot-pressed assembly to room temperature, the decals can be peeled away from the membrane, then two thin casting layers of catalysts are left on the membrane. The GDLs can then be added to the CCM either by hot-pressing or without the need for hot-pressing, if a uniform compression force and low contact resistance can be ensured during the cell assembly.

6. Application of the catalyst as emulsion

Various techniques have been categorized as application of the catalyst as emulsion including spreading, painting and screen printing. These techniques involve one of the following routes as discussed earlier: coating the CL on the membrane (CCM) or coating the CL on the GDL (CCG). In either route, the first step is the preparation of catalyst-electrolyte ionomer-solvent emulsion, which will be coated as slurry or paste on the SPE membrane or GDL.

a. Spreading of pastes

In this technique, the previously prepared catalyst emulsion is spread onto a membrane or wet-proofed GDL using a heavy stainless-steel cylinder (spreading mill) on a flat surface or by rolling-in between two rotating cylinders [28]. The thickness of the prepared CL can be controlled by adjusting the distance between the rolling cylinders. The roller-pressed CLs are uniformly thin and the catalyst loading is directly proportional to the CL thickness. The CCM or CCG thus prepared must be hot-pressed with the GDL or SPE membrane to fabricate the MEA.

b. Painting of ink

In the painting method, the catalyst ink is brush-painted directly onto a dry ion-exchanged membrane in the Na^+ form and then baked in an oven to evaporate the solvent in the ink. A more uniform CL might be difficult to achieve through this method as there could be a significant distortion on the membrane during painting and drying. This can be overcome by drying the membrane in a special vacuum table heating fixture. Also the bulk of the solvent is removed at a lower temperature to alleviate cracking, and later the final traces of solvent are rapidly removed at higher temperature (> 80 °C). In the last step, the catalyzed membranes are rehydrated and ion-exchanged to H^+ form by immersing them in slightly boiling sulphuric acid followed by rinsing in deionized water.

c. Screen printing of ink

The screen printing method has not been widely used for MEA preparation as much as spreading or painting for SPE water electrolyzer. Four items are essential for screen printing: the printing medium (catalyst ink), a substrate onto which the print will be made (GDL or membrane), a screen to define the required patterns, and a squeegee to force the ink through the screen [29]. In this method, a screen sieve is held above the substrate, while the pre-prepared catalyst slurry or ink is applied over it. As the squeegee travels over the screen, it presses it down into contact with the substrate pushing the paste through the screen thus depositing the catalyst ink onto the substrate surface. The pore size in the screen must be optimized to be about the same size as that of catalyst particles to get an optimum print quality. The limitation of this method is that larger particles tend to get clogged and could produce irregularly printed patterns on the substrate surface.

7. Application of the catalyst as vapour

The vapour deposition methods also yield thin CLs, but unlike the slurry (ink)-based preparation methods, vapour deposition does not yield a uniform layer of electrolyte matrix in the CL. Here the unsupported catalyst is deposited as a metal on the SPE membrane or GDL from its vapour phase. The need for an electrolyte matrix (ionomer) inside the CL is altogether eliminated in vapour deposition processes owing to their ability to fabricate ultra-thin CLs (as low as 1 μm). The most common vapour deposition methods are physical vapour deposition (PVD; e.g., magnetron sputtering) and chemical vapour deposition (CVD).

a. Magnetron sputtering

Sputtering is a PVD process in which the atoms from the source material (target) erode and get deposited on the substrate. It is performed in a vacuum chamber or in a controlled environment chamber that uses argon plasma. Despite the ability to produce CLs as thin as 1 mm, the performance of MEA prepared using sputter-deposited CLs varies by several orders of magnitude primarily due to the variation in CL thickness and particle size (<10 nm). Sputtering thick CLs (>10 um) is disadvantageous because of the absence of ionomer inside the CL [30]. In such cases, impregnation of PTFE and carbon power into the porous substrate is crucial to enhance ionic transport. Methods to create three dimensional reaction zones from two-dimensional thin film structure have demonstrated improved performance. Although the sputtering technique provides an easier way of direct deposition of CL, the main drawback of this technique is the poor adherence of platinum to the substrate. As a result, the catalyst is prone to dissolution and sintering under variable operating conditions and is often not durable enough to meet the long-term requirements of the MEA application. So far, published material related to magnetron sputtered MEA were related to PEMFC MEAs.

b. Chemical vapour deposition

Chemical vapour deposition is similar to a PVD process in many aspects, but instead of using solid precursors (targets) to deposit a thin film of solid material, it uses gas-phase precursors. This process chemically transforms the gaseous precursor molecules into a solid material in the form of thin film or powder on the substrate surface. In essence, the CVD method enables the platinization of the dispersed carbon particles and does not directly produce a CL. Platinum particles selectively deposit on the surface defects produced by the acid pretreatment of the carbon particles; hence, the particles are small (<5 nm) and highly dispersed [31].

8. Electrically assisted catalyst deposition

The electrically assisted catalyst deposition is a novel technique for electrode fabrication under the influence of electric field (electrochemical processes), including electro deposition, electro spraying (ES) and electrophoretic deposition (EPD). Similar to vacuum deposition methods, the electrochemical methods provide the feasibility of fabricating ultra-thin CLs with superior properties.

a. Electro deposition

Electro deposition process for MEA preparation involves several steps including impregnation of the porous GDL with ionomer, exchange of cations in the ionomer with a cationic complex, followed by electro deposition of catalyst from this complex onto the support [32]. This results in the deposition of catalyst only at sites that are accessible by both support and ionomer, thus providing good utilization. This technique is capable of producing loadings as low as 0.05 mg cm^{-2}. Another method of electro deposition process involves deposition of catalyst from the electrolyte through the membrane and the catalyst is deposited where it encounters the electrically conducting carbon. This process deposits the metal catalyst only at locations where both protonic and electronic conduction are possible, thus yielding a loading as low as 10 μg cm^{-2}.

b. Electro spraying

Electro spraying is as the name indicates, spraying a jet of catalyst ink from a capillary tube under the influence of high electric field [33]. The ES apparatus consists of a capillary tube (similar to a spray gun) in which the catalyst ink is forced to flow by using a pressurized inert gas (nitrogen or argon) toward the substrate. Very high electric field (3–4 kV) is applied between this capillary tube and the substrate. A jet of catalyst ink emerging out of the capillary tube is converted into a jet of highly charged particles under the influence of electric field. Owing to solvent evaporation and coulomb expansion (droplet division resulting from high charge density), the droplets of electro sprayed ink reduce in size before reaching the substrate. Thus a thin layer of catalyst-ionomer is deposited onto the GDL, which can then be hot-pressed with the membrane. Both morphological and structural improvements have been observed that contribute to a better catalyst utilization compared to more conventional methods.

c. Electrophoretic deposition

Electrophoretic deposition is a process in which charged particles in a colloidal suspension move toward oppositely charged electrodes under the influence of a high electric field [34]. The particles coagulate into a dense mass during deposition, which can produce complex geometries and functionally graded materials, suitable for preparing graded CLs. The catalyst suspension must possess good electrochemical stability to avoid any parasitic faradaic reactions even at very high cell voltages. For preparing a MEA using electrophoretic process, the catalyst-ionomer CL can be directly coated onto the membrane (CCM) without the need for hot-pressing or decal transfer.

9. Application of catalyst in precursor state

a. Electron beam reduction

The electron-beam reduction for electrode preparation is a novel method that utilizes the idea of reducing ions of the catalyst species right on the carbon cloth fibers and the multi walled carbon nano tubes (MWNTs) via direct electron-beam bombardment [35]. The basic procedure of the method involved an ionic solution, such as $PtCl_4$, then wetted onto the TEM specimen copper grid and dried in the air. The air-dried copper grid on the specimen holder inserted the whole module into the TEM column. With an accelerating voltage of 80 kV and a 2×10^{-6} Torr vacuum, the Pt^{4+} ions were struck by 80 keV electrons (e^-) and reduced to Pt^0 nanoparticles on the copper grid. Meanwhile, the $Cl_2(aq)^-$ and $H_2O(l)$ were transformed into Cl_2(g) and H_2O(g) and were immediately removed by the TEM vacuum pump. In this process, high energy electron-beams emit from transmission electron microscopes (TEM), scanning electron microscopes (SEM), or electron-beam writers, so basically this method transforms the TEM or SEM into a device that fabricates catalyst particles for electrodes.

b. Impregnation reduction

This process is also known as electroless deposition [36]. In this method, the membrane is ion exchanged with NaOH to the Na^+ form. It is then equilibrated with an aqueous mixture of hydrazine $(NH_3)_4PtCl_2$ and a co-solvent of H_2O/CH_3OH, in an impregnation step. Following impregnation, the vacuum-dried PFSA ionomer membrane in the H^+ form is exposed to air on one face and to aqueous reductants such as hydrazine N_2H_4 or $NaBH_4$ on the other face, during which the platinum ions are reduced to form metallic platinum, in the reduction step. This method has been found to produce catalyst loadings on the order of 2–6 mgPt cm^{-2}. Following the reduction step, the fabricated CCM is equilibrated with 0.5 mol L^{-1} sulphuric acid prior to hot-pressing step with the GDL. Note that a carbon-coated membrane is an essential precursor for this process in order to provide support to the catalyst particles.

10. Application of the catalyst as aerosol

a. Aerosolizing using spraying gun

Similar to spreading and tape casting, the prerequisite for this technique is the catalyst ink. Spraying is one of the most popular methods for CL fabrication [37]. In this method, the catalyst ink is sprayed using an air-brush onto the membrane or GDL using a pressurized stream of inert gas such as argon or nitrogen. Unlike the spreading method, spraying is done in multiple steps in order to achieve the desired loading or thickness. The sprayed catalyst ink is then evaporated, and sometimes sintered at 80-120 °C, before spraying the next coating of thin layer. Manually sprayed CLs are not as uniform as the roller-pressed ones, but computer-controlled industrial sprayers are known to produce uniform layers.

b. Sonicated spraying

Sonicated spraying is a novel technique for electrodes preparation based on ultrasonic and sonoelectrochemical devices [38-40]. In the process, the catalyst inks are first inserted in a sonicated syringe prior to atomisation in a nozzle and sprayed at a flow rate up to 2.4 ml min^{-1}. Various passes are performed in view of obtaining the appropriate loading. Here, the ultrasonic spray incorporate an ultrasonic atomizing nozzle, vibrating at high frequency ultrasound (120 kHz) created by piezoelectric transducers inside the nozzle's titanium housing. The catalyst inks are pumped through the nozzle and are atomized into a fine mist at the nozzle tip to produce highly repeatable thin films of micron-sized droplets, with coating thicknesses from 200 nm to 500 μm. The ultrasonic-spray method distribute the catalyst ink more evenly leading to better catalyst utilisation compared to the hand-painted method and this is further evident at lower catalyst loadings.

11. Factors influencing the MEA performance

In addition to the MEA preparation methods, the MEA performance also depends on key physical properties such as the membrane glass transition temperature, pressure distribution across and within the MEA, feed water quality, gas crossover (especially at high pressure), gas departure, contact resistance between the sublayers, catalyst loading, ionomer content, and the long-term stability of the sublayers.

Pressure Distribution

Uniform and optimal pressure distribution is essential during the design, engineering, and assembly of an electrolyzer or stack. Nonuniform pressure distribution, especially for MEA with big-size, could cause performance issues such as gas leakage, high contact resistance, malfunctioning of cells, or even physical damage to stack components. Uneven pressure distribution may also result in localized hot spots creating pinholes in the membrane, which could have a detrimental effect on the electrolyzer performance.

Water Quality

The conductivity of water supply for SPE electrolyzer has very important effect on the long-term performance of MEA. Due to the use of titanium materials and IrO_2 catalyst at the anode, the corrosion or aging of the MEA was not likely to happen in the primary operation. Therefore, the performance degradation in the primary running is mainly caused by a contamination from the feed water. There is considerable accumulation of ionic species in the feed water with the process of water electrolysis, the main reason leading to short period performance decline of the MEA. These ionic species mainly originate from the water tank, piping and other components of the test stand, which can be dissolved in water in trace amounts , and not only concentrated with the water being electrolyzed, but also added from time to time with refilling the water tank. To maintain good performance, it is therefore crucial to use high quality feed water.

Electrochemical Active Surface Area (EASA)

The EASA is the important parameter that determines the MEA performance than the catalyst loading itself. Higher EASA would result in better catalyst utilization although it is not always guaranteed. The performance of water electrolysis also depends on the type of materials used for the MEA components, bipolar plates, and other stack components. Although the operating conditions also play an important role in determining the performance of electrolyzer, it can be conclusively said that the MEA has a more dominant role. A water electrolyzer polarization curve consists of three regions, namely the activation (kinetic), ohmic, and mass transfer regions. The myriad of material, chemical, and electrochemical properties of each MEA sublayer have a profound influence on the performance in each of these regions. This is conceptually described in the polarization curve shown in the Figure 3.

Activation Polarization

The activation polarization loss (kinetic loss) occurs because of the sluggishness in oxygen evolution kinetics at the anode and can be minimized by using a high active catalyst. The CL properties such as catalyst loading, type of catalyst, utilization, electrochemical active area, and the stability of the catalyst support are some properties that play a role in determining the water electrolysis performance in the activation region of the polarization curve.

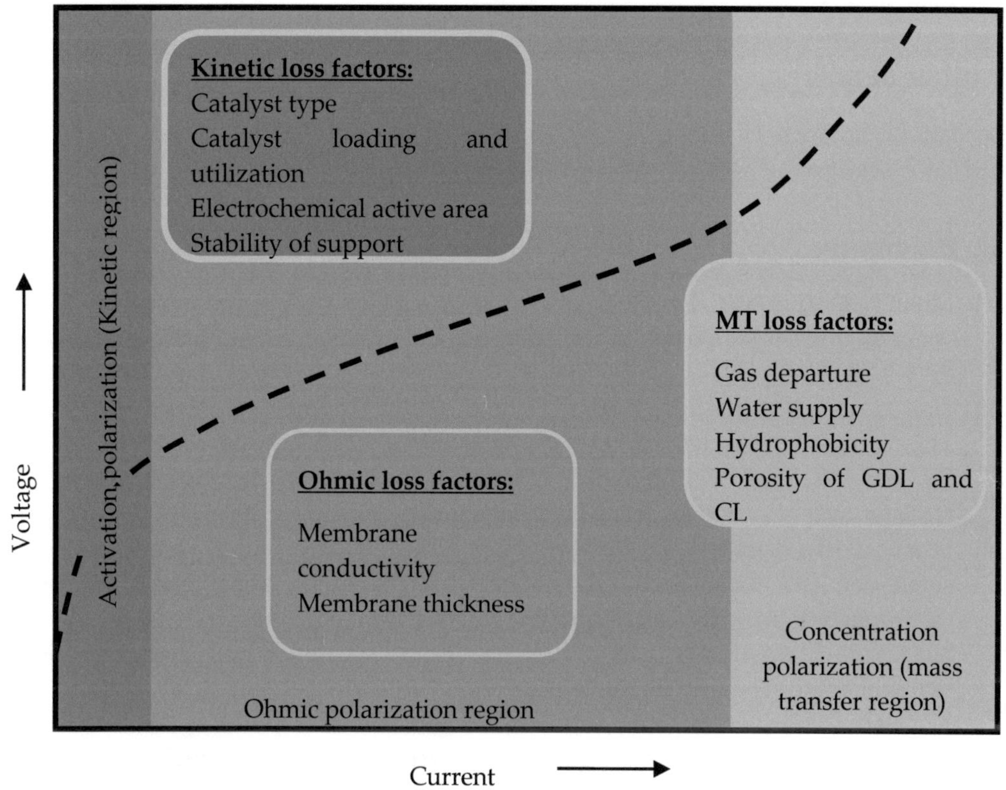

Figure 3. Conceptual representation of performance influencing MEA properties in the activation, ohmic, and mass transfer (MT) regions of an electrolyzer polarization curve.

Ohmic Polarization

The performance of MEA in the ohmic region is largely dependent on the electrical conductivity of stack components especially the bipolar plates. In the case of a single cell, the membrane ionic conductivity is the single most dominant factor to influence the ohmic polarization behavior. Properties such as membrane ionic conductivity, thickness, contact resistance between the MEA sublayers, compression pressure, and the electronic conductivity of the GDL are some key factors that play a role in determining the performance in the ohmic region of the polarization curve.

Concentration Polarization

At high current densities, the MEA performance is affected by mass transfer limitations for the diffusion of gaseous products and water transport inside the pores of MEA sublayers. One must ensure effective two-phase transport inside the pores of bilayer GDL and CL in order to keep the MEA dry under wet operating conditions. Properties such as PTFE content (hydrophobicity), porosity, pore size, and compression pressure (which determines the interfacial gaps between the MEA sublayers during different compression load cycles during operation) determine the MEA performance in the mass transfer region of the polarization curve.

Author details

Bernard Bladergroen, Huaneng Su, Sivakumar Pasupathi and Vladimir Linkov
SAIAMC, University of the Western Cape, South Africa

12. References

[1] Millet P., Dragoe D., Grigoriev S., et al. (2009) GenHyPEM: A research program on PEM water electrolysis supported by the European Commission. Int. J. Hydrogen Energy 34: 4974-4982

[2] Siracusano S., Di Blasi A., Baglio V., et al. (2011) Optimization of components and assembling in a PEM electrolyzer stack. Int. J. Hydrogen Energy 36: 3333-3339

[3] Siracusano S., Baglio V., Briguglio N., et al. (2012) An electrochemical study of a PEM stack for water electrolysis. Int. J. Hydrogen Energy 37: 1939-1946

[4] Siracusano S., Baglio V., Di Blasi A., et al. (2010) Electrochemical characterization of single cell and short stack PEM electrolyzers based on a nanosized IrO_2 anode electrocatalyst. Int. J. Hydrogen Energy 35: 5558-5568

[5] Siracusano S., Baglio V., D'Urso C., et al. (2009) Preparation and characterization of titanium suboxides as conductive supports of IrO_2 electrocatalysts for application in SPE electrolysers. Electrochim. Acta 54: 6292-6299

[6] Medina P., Santarelli M. (2010) Analysis of water transport in a high pressure PEM electrolyzer. Int. J. Hydrogen Energy 35: 5173-5186

[7] Ito H., Maeda T., Nakano A., et al. (2010) Effect of flow regime of circulating water on a proton exchange membrane electrolyzer. Int. J. Hydrogen Energy 35: 9550-9560

[8] Siracusano S., Baglio V., Stassi A., et al. (2011) Investigation of IrO_2 electrocatalysts prepared by a sulfite-couplex route for the O_2 evolution reaction in solid polymer electrolyte water electrolyzers. Int. J. Hydrogen Energy 36: 7822-7831

[9] Oi T., Sakaki Y. (2004) Optimum hydrogen generation capacity and current density of the PEM-type water electrolyzer operated only during the off-peak period of electricity demand. J. Power Sources 129: 229-237

[10] Wei G., Xu L., Huang C., et al. (2010) SPE water electrolysis with SPEEK/PES blend membrane. Int. J. Hydrogen Energy 35: 7778-7783

[11] Cheng J., Zhang H., Ma H., et al. (2010) Study of carbon-supported IrO_2 and RuO_2 for use in the hydrogen evolution reaction in a solid polymer electrolyte electrolyzer. Electrochim. Acta 55: 1855-1861

[12] Wei G., Wang Y., Huang C., et al. (2010) The stability of MEA in SPE water electrolysis for hydrogen production. Int. J. Hydrogen Energy 35: 3951-3957

[13] Giddey S., Ciacchi F., Badwal S. (2010) High purity oxygen production with a polymer electrolyte membrane electrolyser. J. Membr. Sci. 346: 227-232

[14] Badwal S.P.S., Giddey S., Ciacchi F. (2006) Hydrogen and oxygen generation with polymer electrolyte membrane (PEM)-based electrolytic technology. Ionics 12: 7-14

[15] Jung H.Y., Park S., Popov B.N. (2009) Electrochemical studies of an unsupported PtIr electrocatalyst as a bifunctional oxygen electrode in a unitized regenerative fuel cell. J. Power Sources 191: 357-361

[16] Hwang C.M., Ishida M., Ito H., et al. (2011) Influence of properties of gas diffusion layers on the performance of polymer electrolyte-based unitized reversible fuel cells. Int. J. Hydrogen Energy 36: 1740-1753

[17] Marshall A.T., Sunde S., Tsypkin M., et al. (2007) Performance of a PEM water electrolysis cell using $Ir_xRu_yTa_zO_2$ electrocatalysts for the oxygen evolution electrode. Int. J. Hydrogen Energy 32: 2320-2324

[18] Marshall A., Børresen B., Hagen G., et al. (2006) Electrochemical characterisation of $Ir_xSn_{1-x}O_2$ powders as oxygen evolution electrocatalysts. Electrochim. Acta 51: 3161-3167

[19] Marshall A., Børresen B., Hagen G., et al. (2006) Iridium oxide-based nanocrystalline particles as oxygen evolution electrocatalysts. Russ. J. Electrochem. 42: 1134-1140

[20] Cruz J., Baglio V., Siracusano S., et al. (2011) Nanosized IrO_2 electrocatalysts for oxygen evolution reaction in an SPE electrolyzer. J. Nanopart. Res. 13: 1639-1646

[21] Ma L., Sui S., Zhai Y. (2009) Investigations on high performance proton exchange membrane water electrolyzer. Int. J. Hydrogen Energy 34: 678-684

[22] Xu J., Miao R., Zhao T., et al. (2011) A novel catalyst layer with hydrophilic-hydrophobic meshwork and pore structure for solid polymer electrolyte water electrolysis. Electrochem. Commun. 13: 437-439

[23] Yamaguch R., Development of High Performance Solid Polymer Electrolyte Water Electrolyzer in WE-NET. In *Proceedings of the 32nd intersociety energy conversion engineering conference*, 1997; Vol. 3, pp 1958-1965.

[24] Zhang Y.J., Wang C., Wan N.F., et al. (2007) Study on a novel manufacturing process of membrane electrode assemblies for solid polymer electrolyte water electrolysis. Electrochem. Commun. 9: 667-670

[25] Schulze M., Schneider A., Gülzow E. (2004) Alteration of the distribution of the platinum catalyst in membrane-electrode assemblies during PEFC operation. J. Power Sources 127: 213-221

[26] Wilson M.S., Gottesfeld S. (1992) High Performance Catalyzed Membranes of Ultra-low Pt Loadings for Polymer Electrolyte Fuel Cells. J. Electrochem. Soc. 139: L28

[27] Wilson M.S., Gottesfeld S. (1992) Thin-film catalyst layers for polymer electrolyte fuel cell electrodes. J. Appl. Electrochem. 22: 1-7

[28] Mehta V., Cooper J.S. (2003) Review and analysis of PEM fuel cell design and manufacturing. J. Power Sources 114: 32-53

[29] Sebastian P., Solorza O. (1998) Mo-Ru-W chalcogenide electrodes prepared by chemical synthesis and screen printing for fuel cell applications. Int. J. Hydrogen Energy 23: 1031-1035

[30] Hirano S., Kim J., Srinivasan S. (1997) High performance proton exchange membrane fuel cells with sputter-deposited Pt layer electrodes. Electrochim. Acta 42: 1587-1593

[31] Morse J.D., Jankowski A.F., Graff R.T., et al. (2000) Novel proton exchange membrane thin-film fuel cell for microscale energy conversion. Journal of Vacuum Science & Technology A: Vacuum, Surfaces, and Films 18: 2003-2005

[32] Thompson S.D., Jordan L.R., Forsyth M. (2001) Platinum electrodeposition for polymer electrolyte membrane fuel cells. Electrochim. Acta 46: 1657-1663

[33] Baturina O.A., Wnek G.E. (2005) Characterization of proton exchange membrane fuel cells with catalyst layers obtained by electrospraying. Electrochem. Solid-State Lett. 8: A267

[34] Morikawa H., Tsuihiji N., Mitsui T., et al. (2004) Preparation of membrane electrode assembly for fuel cell by using electrophoretic deposition process. J. Electrochem. Soc. 151: A1733

[35] Pai Y.H., Huang H.F., Chang Y.C., et al. (2006) Electron-beam reduction method for preparing electrocatalytic particles for membrane electrode assemblies (MEA). J. Power Sources 159: 878-884

[36] Hwang B.J., Liu Y.C., Hsu W.C. (1998) Nafion-based solid-state gas sensors: Pt/Nafion electrodes prepared by an impregnation-reduction method in sensing oxygen. J. Solid State Electrochem. 2: 378-385

[37] Mosdale R., Wakizoe M., Srinivasan S. In *Fabrication of electrodes for proton exchange membrane fuel cells using a spraying method and their performance evaluation*, The Electrochemical Society, Pennington, NJ: 1994; p 179.

[38] Millington B., Whipple V., Pollet B.G. (2011) A novel method for preparing proton exchange membrane fuel cell electrodes by the ultrasonic-spray technique. J. Power Sources 196: 8500-8508

[39] Pollet B.G. (2009) A novel method for preparing PEMFC electrodes by the ultrasonic and sonoelectrochemical techniques. Electrochem. Commun. 11: 1445-1448

[40] Pollet B.G., Valzer E.F., Curnick O.J. (2011) Platinum sonoelectrodeposition on glassy carbon and gas diffusion layer electrodes. Int. J. Hydrogen Energy 36: 6248-6258

Chapter 4

Advanced Construction Materials for High Temperature Steam PEM Electrolysers

Aleksey Nikiforov, Erik Christensen, Irina Petrushina, Jens Oluf Jensen and Niels J. Bjerrum

Additional information is available at the end of the chapter

http://dx.doi.org/10.5772/51928

1. Introduction

1.1. Principles of polymer electrolyte membrane (PEM) water electrolysis

There are main 3 types of electrolyzers: alkaline, acidic and solid oxide electrolyzer cell (SOEC). This chapter will be concentrated on acidic electrolyzers, where reactions follow the pathes (reactions 1, 2).

$$\textit{cathode} : 2\,\mathrm{H}^{+} + 2\,\mathrm{e}^{-} \rightarrow H_2 \tag{1}$$

$$\textit{anode} : \mathrm{H_2O} \rightarrow \frac{1}{2}\mathrm{O_2} + 2\,\mathrm{H}^{+} + 2\,\mathrm{e}^{-} \tag{2}$$

The efficiency of water splitting by electrolysis is rather low for conventional electrolyzers and there is hence a large potential for improvement.

The modern acidic electrolysers use polymeric proton conducting membranes, e.g. Nafion® or PBI, doped with phosphoric acid.

One of the potential advantages of PEM cells over more abundant alkaline electrolyzers is that they were shown to be reversible [21, 31, 69]. The type of an electrochemical cell working both as a fuel cell and a water electrolyzer is called a unitized regenerative fuel cells (URFC) [13, 38, 67, 70]. These devices produce hydrogen from water in the electrolysis mode, while electricity can be inversely produced in the fuel cell mode. This mode of working is beneficial when the lack of electricity changes with the excess energy available (periods of low consumption) [61].

PEM water electrolysis technology is frequently presented in literature as a potentially very effective alternative to more conventional alkaline water electrolysis [43, 46, 47]. Among advantages are higher production rates and energy efficiency [60]. In a future "hydrogen society" this method is envisioned as a part of the "energy cycle", where hydrogen acts as an energy carrier. In this cycle, electricity from renewable energy sources is used in electrolysis for electrochemical splitting of water [26]. PEM cells usually use perfluorinated ion-exchange membranes as an electrolyte (known under the trademark Nafion®).

In a PEM cell the electrolyte is a solid ion-exchange membrane, which does not involve compulsory circulation of electrolyte through it. Different types of ionic membranes will be discussed in Section 2. The electrodes are usually directly sprayed or pressed on the opposite sites of the solid polymer electrolyte (SPE), thus being the origin of a membrane electrode assembly (MEA)[1, 5, 67, 78]. Also, electrodes can be sprayed on the gas diffusion layer (GDL), and then put together leaving the SPE between them [56, 79]. A PEM electrolyzer stack consists of a combination of several cells (as many as 100), electrically connected in series [54]. The cells are separated from each other by a metal plate (also called a bipolar plate), which serves both as a current collector and as an interconnect to the next cell in the stack. Flows of evolved hydrogen and oxygen are usually swept out through the bipolar plate by water flow and further separated from it outside the cell.

Several commercial types of PEM electrolyzers are available today on the market [27–29]. Some units have power up to 44 kW and claimed lifetime up to 40,000 hours. Still, the main drawbacks of such systems are the price of materials and complex system components, which ensures save and reliable function.

1.2. High temperature PEM electrolysis. Advantages and drawbacks

At temperatures above the boiling point of water, the energy efficiency of water splitting can be significantly improved because of decreased thermodynamic energy requirements, which is one of the major advantages of these systems.

Since water electrolysis becomes increasingly heat consuming with temperature (Figure 1), larger portion of the total energy demand can be provided as heat at elevated temperatures. This provides an opportunity to utilize the Joule heat, that is inevitably produced due to the passage of electrical current through the cell. In this way, the overall electricity consumption and, thereby, the H_2 production price can be reduced.

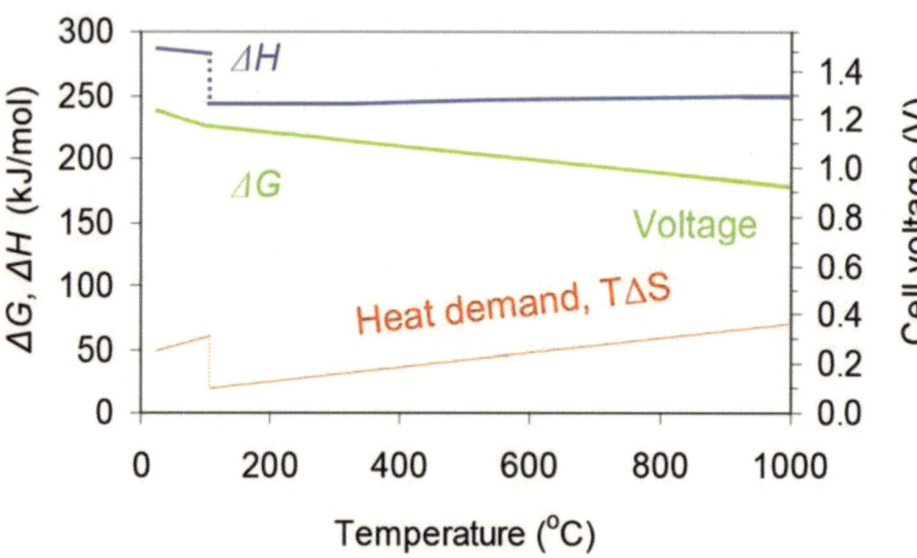

Figure 1. The theoretical cell voltage as a function of temperature [35]

Since the reaction of water splitting is not spontaneous and the Gibbs free energy is positive, a positive change of a factor $T \cdot \Delta S^{\ominus}$ in the equation 3 means less energy needs to be applied. As the reaction has a positive entropy, the equilibrium will be displaced towards the products for high temperatures. The term $T \cdot \Delta S^{\ominus}$ increases with increasing temperature, thus increasing the contribution of thermal energy to the total needs for the water splitting reaction [66]. Therefore, the part of heat, which can be used for the reaction is higher, meaning that the production costs of hydrogen are decreased [58]. It was noticed much earlier that econonomic reasons force to move towards high temperature electrolysis (120-150 °C) [73].

$$\begin{aligned} \Delta G^{\ominus} &= \Delta H^{\ominus} - T \cdot \Delta S^{\ominus} \\ \Delta G^{\ominus} &= -nF\Delta E^{\ominus} \\ \Delta S^{\ominus} > 0 \rightarrow \frac{dG^{\ominus}}{dT} &< 0 \\ \frac{d\Delta E^{\ominus}}{dT} &= -\frac{d\Delta G^{\ominus}}{dT} > 0 \end{aligned} \tag{3}$$

According to the Arrhenius' equation 4, the kinetics of the electrode reactions is enhanced at elevated temperatures. It is associated with lower overpotentials at the electrodes, giving higher efficiency for electrolysis.

$$k_f = E_{A_f} exp\left(\frac{\Delta G_c^{\ddagger}}{RT}\right) \tag{4}$$

Another positive opportunity provided while operating at temperatures above 100 °C is that water is not in a liquid phase (at ambient pressure) and this fact significantly simplifies water/gas management. In this case all the reactant/product flows are in a steam phase and the transport of them is easier, which provides simplified stack construction and operation [44].

The heat management is also easier for high temperature systems, as the heat flow out of the system is proportional to the temperature gradient between the cell and the ambient. This means that cooling is more effective in elevated-temperature systems, as there are less efficiency losses associated with the forced cooling of the cell [36].

Another benefit of high temperature systems is attributed to the decreased sensibility of catalysts towards poisoning by adsorption of inhibiting agents. This effect is acknowledged to be a considerable advantage of high temperature PEM fuel cells (HTPEMFCs) [44]. The inhibition mechanism usually involves chemisorption of species on the catalyst surface, covering and screening it from interacting with the reactants. This adsorption is weaker at higher temperatures, giving higher tolerance to impurities [16].

However, with increasing temperature, the probability and rate of side processes, such as dissolution of the electrodes and components corrosion, is higher. This decreases the lifetime of the whole system and increases demands to all materials used with respect to corrosion and thermal stability [19, 46]. The corrosion issue for construction materials in such cells will be addressed in the following Section 3.

2. Acidic electrolytes and polymer electrolyte membranes technology

The electrolyte is one of the main components of any type of electrolyzer. The importance of the electrolyte is emphasized by the fact that the type of electrolyzer, as well as of fuel cell, is named after the type of electrolyte used. In the classical view it is traditionally a solution of acid, base or water soluble salt in the water. When added, those dissociate into ions in the solution, which increases conductivity of pure water.

The acidic water electrolysis traditionally uses sulphuric or phosphoric acid as they are stable within the potential window of water. The acid increases the conductivity of water through the donating protons.

The polymer electrolyte membrane (PEM) is a membrane which acts as a proton conductor in electrolyzer cell. Usually, the ionic membrane consists of a solid polytetrafluoroethylene (PTFE) backbone, which is chemically altered and contains sulfonic ionic functional groups thus the pendant side chains terminated with $-SO^{3-}$. The acid dissociates and release protons by the following mechanism (5):

$$SO_3H^- => SO_3^- + H^+ \tag{5}$$

The membranes of this type allow water molecules to penetrate into its structure, while remaining not permeable to molecular H_2 and O_2. The sulfonic groups are responsible for the transfer of protons during electrolysis, where a hydrated proton H_3O^+ can freely move within the polymer matrix, while a sulfonate ion SO^{3-} is fixed to the side chain of polymer. When electric current flows across the membrane, the hydrated protons are attracted to the cathode, where they are combined into hydrogen. Nafion® is the most known trademark among ionic membranes and is patented by Du Pont Company in 1966 [15]. A typical membrane has a thickness in the range of 50-100 μmeters and this type of membrane is commonly used as an electrolyte for conventional PEM water electrolyzers [23, 56, 79]. These membranes have excellent chemical stability, high ionic conductivity and excellent mechanical strength [56]. Water electrolysis using Nafion® as an electrolyte is a promising technology for large-scale hydrogen production [8, 75].

The conductivity of such membranes decreases significantly at temperatures above 80 °C, which is associated with the ion of water content [42, 45]. Sufficient efficiency is achieved using Poly[2,2'-(m-phenylene)-5,5'-bibenzimidazole (PBI) membranes doped with phosphoric acid in PEM fuel cells at temperatures up to 200 °C under ambient pressure [36, 44, 45]. The structure of PBI is shown in Figure 2. Doped PBI membranes are a potential electrolyte for use in PEM steam electrolyzer systems. The ionic conductivity of membranes increases with temperature [45], which means the higher the working temperature is, the lower ohmic losses through electrolyte are. In spite of that, the conditions of extremely low pH combined with high overpotentials at the anodic compartment of the oxygen evolution electrode (OEE) impose serious limitations on materials which can be used in these cells.

In the laboratory conditions commonly 0.5M sulphuric acid is used for screening electrode materials in a 3-electrode electrochemical cells, simulating conditions of the Nafion®-based systems, which work at temperatures below 100 °C [51, 52]. Since high temperature PEM cells are working at temperatures around 150 °C, H_2SO_4 cannot be used to simulate conditions in the 3-electrode cell, even at high concentration. Instead, concentrated H_3PO_4 can ben used, which permits to work at temperatures as high as 150-160 °C, depending on the composition [14].

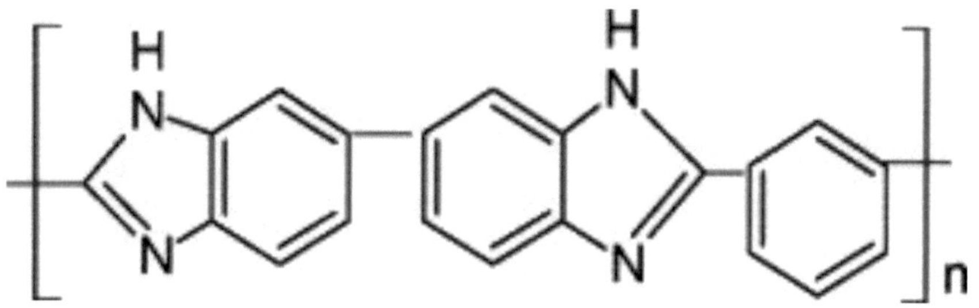

Figure 2. General structure of Poly[2,2′-(m-phenylene)-5,5′-bibenzimidazole (PBI) [45].

However, it was noticed by Appleby and Van Drunen that the Tafel slopes for noble electrocatalysts are significantly higher in concentrated phosphoric acid than those for more diluted solutions, being the apparent cause of adsorption of electrolyte on the electrode surface [7].

As it was stated by Miller in one of his latest publications [55], the cost of Nafion®-based polymers calls for alternative membrane materials along with higher operating temperatures closer to 150 °C. These membranes are required to improve kinetics and obtain higher conversion efficiencies in solid polymer electrolyte (SPE) electrolyzers and new solid electrolytes are needed.

Effective and affordable membranes are very important for the commercialisation of PEM water electrolyzers as it is both easier to manufacture and safer as neither acid nor electrolyte are in liquid phase.

3. Construction materials for high temperature PEM water electrolysis (bipolar plates and current collectors)

Elevated working temperatures involve increased demands for corrosion resistance of catalysts and current collectors, while the contact resistance in the GDL should remain reasonable.

High temperature PEM cell cannot be build from the same materials as a cell working below 100 °C. Among new materials to be developed are polymer membranes, as commercial Nafion® membranes lose their conductivity at temperature above 100 °C due to membrane dehydration [5]. This means that different membranes should be used for this temperature range. It will be further discussed in Section 2. Elevated temperatures as well create more severe corrosion media for other components in the cell.

The anodic compartment of electrolyzer is expected to have stronger corrosive conditions than cathodic due to high positive polarization in combination with presence of evolving oxygen. This will be even more severe when the temperature is elevated. It is therefore an important task to choose materials which possess sufficient corrosion resistance. This demands further development of all materials from which electrolyzer cells are built.

One of the important components in PEM stack is a bipolar plate. Bipolar plate is a multifunctional and expensive part in a electrolysis stack as it collects and conduct current from cell to cell, permits an adequate gas flow, and the flow channels in the plate carry off produced gases, as well as providing most of the mechanical strength of the stack. In a typical

PEM electrolysis stack, bipolar plates comprise most of the mass, and almost all the volume. Usually they also facilitate heat management in the system. These complex requirements make a task of finding proper materials difficult [12]. The highly oxidising acidic conditions in the oxygen electrode compartment pose a serious challenge to the materials used in these systems [19].

The most crucial demands for bipolar plate materials are resistance to spalling, dimensional stability and resistance to corrosion in electrolyte media under anodic/cathodic polarization. Numerous research projects have been devoted to bipolar plate materials in fuel cells [6, 25, 32, 37, 50, 71, 72, 76]. However, the number of suitable materials for PEM electrolyzers is still limited because of high requirements for corrosion resistance at the oxygen electrode, where high overpotentials are combined with low pH media of electrolyte.

In Nafion®-based systems, titanium is the most widely used bipolar plate material, which is ideal in terms of corrosion resistance and conductivity [17, 24, 41, 67]. Porous sintered titanium powder commonly serves as a GDL material [22, 23].

The conductivity of Nafion® membranes decreases dramatically at temperatures above 100 °C (Section 2). Thus, PBI membranes doped with phosphoric acid are typically used in PEM fuel cells at elevated temperatures [24]. However, materials like steels corrode easily in phosphoric acid solutions and therefore it is important to study other alloys and materials for current collectors [9, 62]. Tantalum and nickel alloys show better corrosion resistance than stainless steels partly due to higher corrosion potentials and partly due to the formation of passive oxide layers on the metal surface [57, 65]. Titanium generally has rather limited resistance to phosphoric acid [33]. Previous studies showed that titanium current collectors would considerably suffer from corrosion at temperatures above 80 °C in concentrated phosphoric acid environments [34, 59].

Different types of stainless steels can be used as bipolar plates, and they have advantages of being good heat and electricity conductors, can be machined easily (e.g. by stamping), are non-porous, and consequently very thin pieces are able to keep the reactant gases apart.

A possible alternative to stainless steel bipolar plates can be the use of nickel-based alloys [65]. Ni-based alloys are widely applied in process industry and energy production in nuclear power plants. When compared to conventional stainless steels, generally a higher degree of resistance against corrosion is observed for these materials. This can be explained partly by more noble corrosion potential of Ni and by different properties of the oxide films formed on Ni-based alloys [65]. Also, it has been proposed recently that nickel and stainless steel alloys can be used as a construction material in PEM water electolysers, but at temperatures no higher than 100 °C [34].

In order to simulate corrosion conditions at the anodic compartment of a PEM water electrolysis cell during half-cell experiments, it is necessary to choose a proper electrolyte. Investigating systems including membranes based on perfluorinated sulfonic acid, e.g. Nafion®, 0.5M sulphuric acid is commonly used as an electrolyte, simulating the electrolyzer cell conditions [52, 68]. Similarly, H_3PO_4 can be used to model systems based on membranes doped with H_3PO_4. 85% solution of H_3PO_4 can be chosen to study the limiting case of corrosion, considering that in working electrolyzer systems the actual concentration of active acid at the electrode-electrolyte-water three-phase boundary would by much less than in this limiting case.

In a highly oxidizing media such as the anodic compartment of high temperature steam electrolysis stack, it is essential to characterize the effect of different parameters on the behaviour of the protective oxide films. To date, no works have been addressed to the study of Ni-based alloys for use as bipolar plates in high temperature PEM steam electrolyzers.

In this work, metal alloys, namely austenitic stainless steels AISI 316L, AISI 321, AISI 347 and Ni-based alloys Hastelloy®C-276, Inconel®625, Incoloy®825, as well as titanium and tantalum were tested in terms of their corrosion resistance in the conditions, simulating those in the PEM electrolyzer systems, operating at temperatures above 100 °C. Platinum and gold were also investigated for studying the potential window of concentrated H_3PO_4. All samples were subjected to anodic polarisation in 85% phosphoric acid electrolyte solution at 120 °C. The corrosion speed of metal alloys was investigated additionally for 30 °C and 80 °C to show the influence of temperature on corrosion resistance.

3.0.1. Metal coatings and a CVD technique

As the requirements to construction materials are quite severe and the price for materials which fulfil these requirements tends to be rather high, one of the approaches can be use of a coating on the less expensive and available material. If the technology is robust and affordable, the price of the materials can be significantly reduced, as expensive material use is restricted to the surface. Tantalum was shown to have superior resistance towards acidic solutions [39]. This is attributed to the formation of a thin Ta_2O_5 passivating film. As the cost of this material is rather high, its use is often limited to the coatings.

Cardarelli et. al showed that IrO_2 electrodes, prepared on copper base material, which is coated with tantalum by molten salt electroplating present much better corrosion stability than coatings, made on pure titanium [11].

Chemical vapour deposition (CVD) process can be used for tantalum coatings preparation, where product is deposited on the surface of a substrate inside the reaction chamber. The most common process of chemical vapour deposition (CVD) coating is hydrogen reduction of a metal chloride [53].

In this work a commercial CVD "Tantaline" coating on stainless steel AISI 316L, provided by Tantaline A/S (Denmark) was tested for corrosion in high temperature PEM electrolyzer cell (HTPEMEC) environment [30].

3.1. Assessment of materials for their corrosion stability

3.1.1. Steady state polarisation and corrosion studies

Traditionally, the weight loss technique has been used to determine the corrosion rates of different materials [9, 48, 49, 57, 64]. It involves the periodic weight loss measurements after the defined time intervals having a sample immersed in an electrolyte. This technique is straightforward and does not require any knowledge of corrosion reactions that are occurring, however, prolonged test periods are needed (over 200h) for reasonable accuracy in this technique [18].

The electrochemical techniques potentially offer a faster way of determining corrosion rates, as nature of corrosion in electrolyte solutions is electrochemical. Therefore, generally considering corroding species of valence n, will give the following equation:

$$M \rightarrow M^{n+} + ne^{-} \tag{6}$$

The Tafel equation 7 describes the current density as a function of the electrode potential and can be used for the study of corrosion speed and mechanisms. The rate of the uniform corrosion can be calculated through the exchange current density value directly to the mass loss rates or penetration rates (corrosion rate). It is made by means of the Tafel extrapolations [3].

$$\begin{aligned} \frac{i}{i_0} &= -e^{-\alpha f \eta} \Leftrightarrow -ln\left(\frac{i}{i_0}\right) = -\alpha f \eta \Leftrightarrow \\ \eta &= \frac{1}{\alpha f} ln\left(\frac{i}{i_0}\right) \Leftrightarrow \eta = \frac{1}{\alpha f}(lni - lni_0) \Leftrightarrow \\ \eta &= \frac{RT}{nF\alpha}(lni - lni_0) = \alpha - (1-x)lni \end{aligned} \tag{7}$$

The overpotential η is plotted against the $logi$ value through the Tafel plot (Figure 3). The intersection of two branches of the plot defines the corrosion current (corrosion current density i_{cor}), which is attributed to the main corrosion reaction taking place (equation 6). i_{cor} equals to i_0 at E_{cor} (equations 8-13). The values of the measured exchange current will show the maximum possible rate of corrosion in these conditions, as the effect of passivation is not taken into account.

$$for\ E >> E_{cor},\ i \simeq i_a = i_0 \cdot \exp\left\{\frac{(1-\alpha) \cdot n \cdot F}{RT} \cdot (E - E_{cor})\right\} \tag{8}$$

$$for\ E << E_{cor},\ i \simeq i_c = -i_0 \cdot \exp\left\{\frac{-\alpha \cdot n \cdot F}{RT} \cdot (E - E_{cor})\right\} \tag{9}$$

$$\ln i_a = \ln i_0 + \frac{(1-\alpha) \cdot n \cdot F}{RT} \cdot (E - E_{cor}) \tag{10}$$

$$\ln -i_c = \ln i_0 + \frac{-\alpha \cdot n \cdot F}{RT} \cdot (E - E_{cor}) \tag{11}$$

$$\ln i_a = \ln -i_c \rightarrow E = E_{cor} \tag{12}$$

$$i_a(E_{cor}) = i_0 \tag{13}$$

The corrosion potential (E_{cor}) (Figure 3) is another key parameter which gives an indication of how noble an electrode material is, or what is the minimum potential at which it starts to undergo corrosion.

Another alternative is to use the so-called "Cyclic Tafel voltammetry" technique [59, 74]. In this case after the forward polarisation, the scanning direction is reversed and the corrosion potential and current are measured for the "passivated" material.

The corrosion current i_{cor} is found from the slope of the anodic polarisation curve, presented in the coordinates "electrode potential" vs. "log of the current density", as shown in Figure 3. The detailed ASTM technique description can be found elsewhere [3].

The assumption needs to be taken that the current distributes uniformly across the area of the electrode while using this technique. In this case, the current density equals:

$$i_{cor} = \frac{I_{cor}}{A} \tag{14}$$

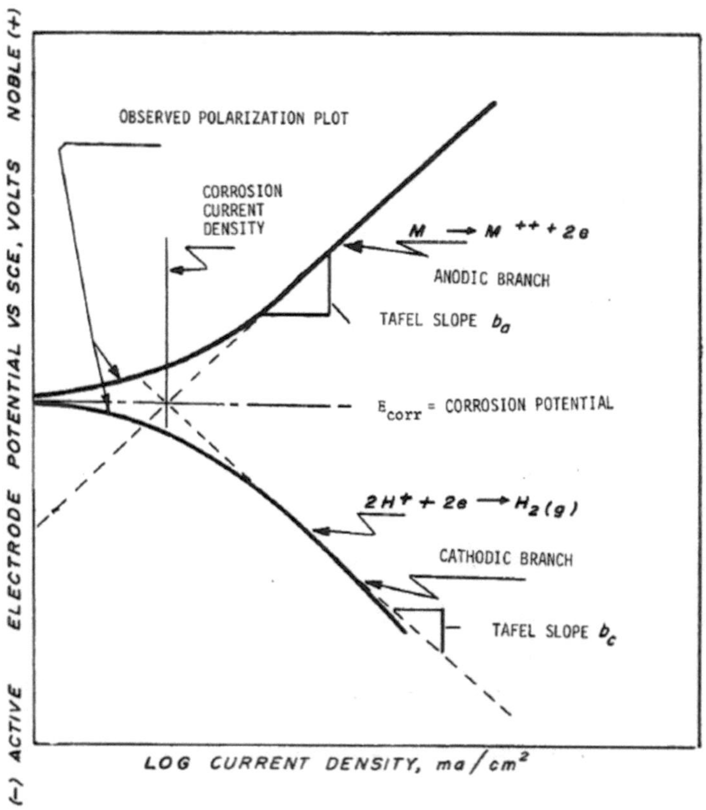

Figure 3. Hypothetical cathodic and anodic polarization diagram [3].

where A is the exposed specimen area, cm^2

Further, the Faraday's Law is used for the calculation of the corrosion rate. For the penetration rate (corrosion rate (CR)) the derived from Faraday's Law equation is:

$$CR = K_1 \frac{i_{cor}}{\rho} EW \tag{15}$$

- CR is given is mm/year
- i_{cor} in $\mu A/cm^2$
- $K_1 = 3.27 \cdot 10^{-3}$, $mm \cdot g/\mu A \cdot cm \cdot year$
- ρ is the density of material
- EW is considered dimensionless in these calculations ans stands for the Equivalent Weight. For the pure elements the euivalent weight is given by:

$$EW = W/n \tag{16}$$

where

- W = the atomic weight of the element, and
- n = the number of electrons required to oxidize an atom of the element in the corrosion process, that is, the valence of the elevent. Details can be found in [2]

3.1.2. Materials and reactants for the experiment

For preparation of the samples and the electrochemical experiments, the following substances were used:

- Demineralised water
- H_3PO_4 85%, Sigma Aldrich, puriss. p.a. (analytical purity)
- Ta plate electrode, Good Fellow Cambridge Ltd.
- Austenitic stainless steel plate (AISI 316L, AISI 321, AISI 347, annealed type of temper), by Good Fellow Cambridge Limited, England
- Ti foil, by Good Fellow Cambridge Limited (England)
- Hastelloy®C-276, Inconel®625 and Incoloy®825 plates, by T.GRAAE SpecialMetaller Aps (Denmark)
- CVD tantalum coated stainless steel AISI 316L was provided by Tantaline A/S, Denmark
- SiC abrasive paper, by Struers A/S (Denmark)
- Diamond powder polishing suspension, particle size less than 0.25 μm, by Struers A/S (Denmark)
- PolyFast phenolic hot mounting resin with carbon filler, provided by Struers A/S (Denmark)

3.1.3. Materials and sample preparation

Typical chemical compositions of stainless steels and nickel-based alloys investigated in this work are given in Table 1.

	Chemical composition of alloys (elements, weight%)											
Alloy type	Ni	Co	Cr	Mo	W	Fe	Si	Mn	C	Al	Ti	Other	Nb+Ta
AISI 347	9.0-13.0	-	17-19	-	-	Bal.	1.0	2.0	0.08	-	-	-	0.8
AISI 321	9.0-12.0	-	17-19	-	-	Bal.	1.0	2.0	0.08	-	0.4-0.7	-	-
AISI 316L	10.0-13.0	-	16.5-18.5	2.0-2.5	-	Bal.	1.0	2.0	0.03	-	-	N less 0.11	-
Hastelloy®C-276	57	2.5	15.5	16.0	3.75	5.5	0.08	1.0	0.02	-	- -	V 0.35	-
Inconel®625	62	1.0	21.5	9.0	-	5.0	0.5	0.5	0.1	0.4	0.4	-	3.5
Incoloy®825	44	-	21.5	3.0	-	27	0.3	1.0	0.05	0.1	1.0	Cu 2.0	-

Table 1. Alloy chemical composition.

All specimens were cut into round plates of 15 mm in diameter. Afterwards the surfaces of all samples, apart from CVD tantalum coated SS316L, were manually ground prior to testing to eliminate any mill finish effects. Abrasive paper was used, followed by polishing with diamond powder. Finally, surfaces were degreased with acetone.

3.1.4. Characterisation

A high temperature electrochemical cell (Figure 4) was specially designed for corrosion studies at elevated temperatures. The working electrode was designed to hold a disk sample, with a geometric area of opening ca. 0.2 cm^2. A coil of platinum wire was used as a counter electrode to ensure a good polarization distribution. A calomel electrode was used as a reference electrode, connected to the system through a Luggin capillary. 85% phosphoric acid (analytical purity) was used as an electrolyte. Tests were performed at 30, 80 and 120 °C at air atmosphere. In this work the electrochemical cyclic Tafel voltammetry technique is employed [3, 4, 74]. The experimental apparatus used for electrochemical studies was potentiostat model VersaSTAT 3 and VersaStudio software by Princeton Applied Research. After open-circuit potential was established, scanning was initiated with a scan rate of 1 mV/s. The potential window was 1.5 V, starting at a potential of 400 mV less than the reference electrode potential and going up to 1.1 V more than the reference electrode potential. Reversed polarization was performed afterwards.

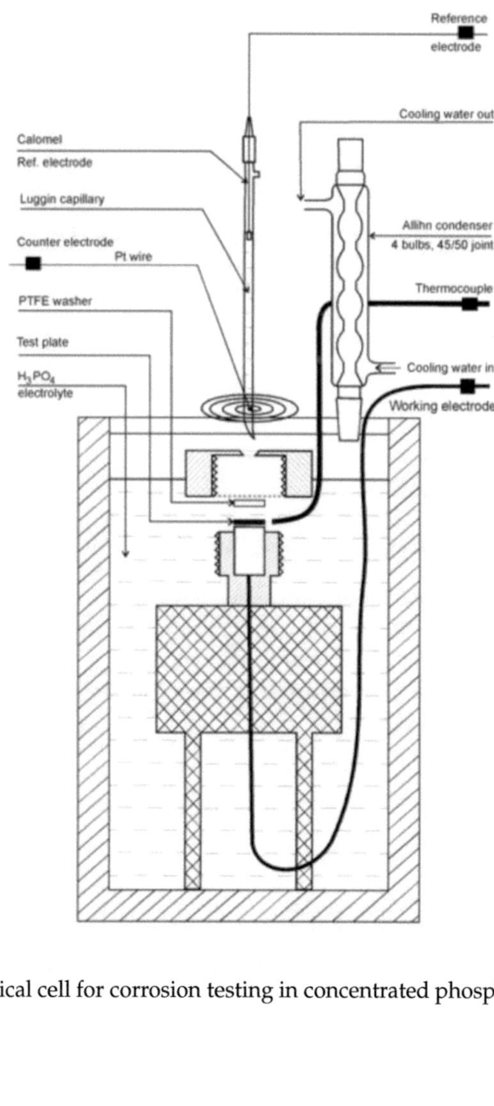

Figure 4. The electrochemical cell for corrosion testing in concentrated phosphoric acid.

Cross-sections of the samples before and after voltametric measurements were studied using scanning electron microscopy (SEM) and energy dispersive X-ray spectroscopy (EDX). The cut was made for all the samples before and after immersion in 85% H_3PO_4 at 120 °C for the time of electrochemical experiment. Duplicate plates were prepared for the cross-section investigation before the exposition. All samples were mounted in the hot mounting resin. SEM measurements were made with an JEOL JSM 5910 scanning electron microscope. The EDX-system used was INCA from Oxford Instrument (accelerating voltage 20.00 kV, working distance 10 mm).

3.1.5. Results and discussion

Figure 5 represents the polarisation curves for platinum and gold foil, which show the electrochemical stability window for these materials in concentrated phosphoric acid at 120 °C. It can be seen that platinum is a better catalyst for O^{2-} oxidation then gold (figure 5).

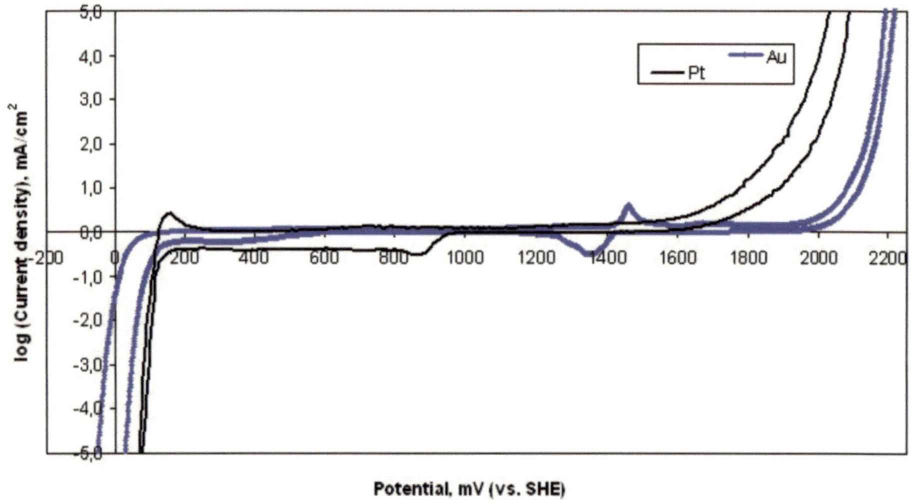

Figure 5. Potential window for Pt and Au in 85% H_3PO_4, 120 °C, 1 mV/s (vs. SHE).

The method of corrossion rate calculation, described in Section 3.1.1 was applied to all experiments. Figures 8-11 present Tafel plots for the materials tested, obtained at 80 °C and 120 °C. Anodic exchange current density values were obtained from cyclic Tafel plots [3]. Corresponding corrosion currents and approximate corrosion rates were calculated as described in Section 3.1.1 and are presented in Table 2. Approximate CRs were calculated in terms of penetration rate, using the Faraday's Law [2].

For all studied materials there is a dramatic influence of temperature on corrosion rate, which grows with increasing temperature.

It can be seen from cyclic Tafel behaviour, that for all of the studied alloys corrosion is of a local type, i.e. pitting or intergranular Figures 8-11.

The analysis of the shape of cyclic Tafel voltammograms can give useful information about possible corrosion mechanisms [74]. Particularly, data related to pitting behaviour can be

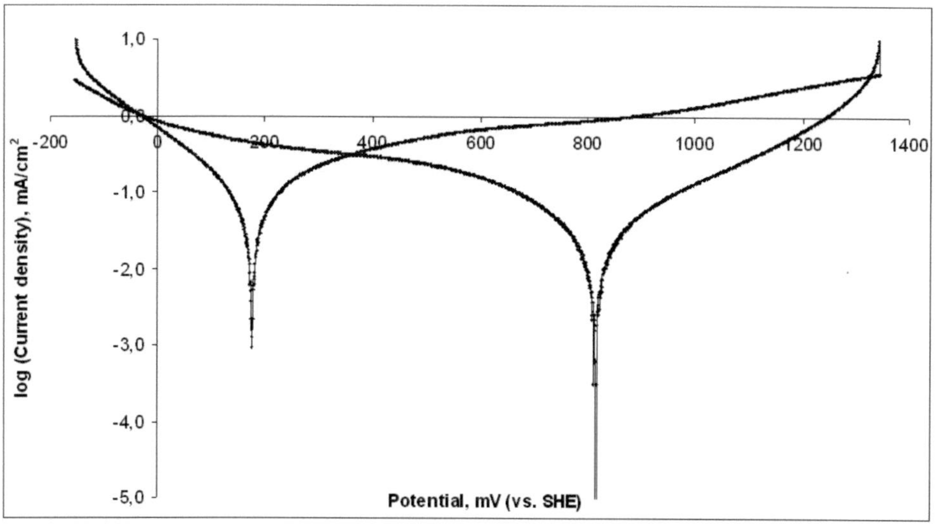

Figure 6. Tafel plot for AISI 321 in 85% H_3PO_4, 80 °C, 1 mV/s (vs. SHE).

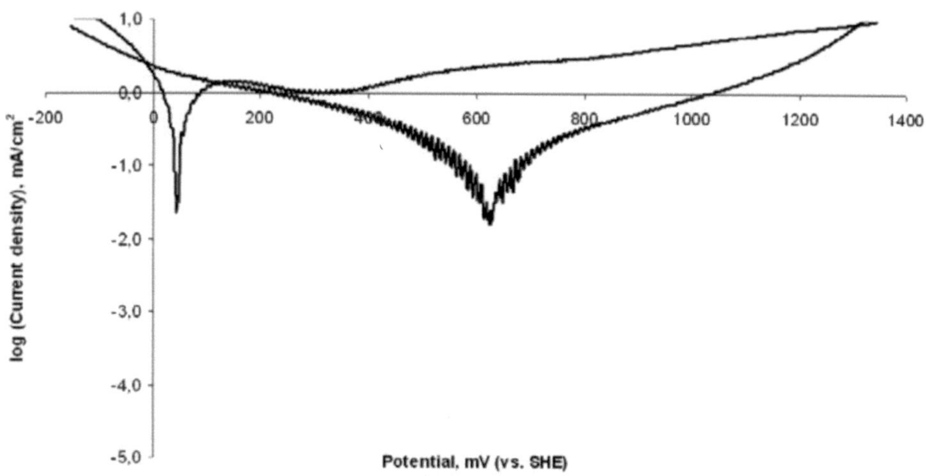

Figure 7. Tafel plot for AISI 321 in 85% H_3PO_4, 120 °C, 1 mV/s (vs. SHE).

obtained using a method, proposed by Pourbaix [63]. In this case, the anodic polarisation scan is not terminated at high anodic potential, but is reduced at the same scan rate until reverse E_{cor} is reached. Usually, this kind of graph is called "The pitting scan". Using this technique, it can be assumed that if any pits arise during forward anodic polarisation, any further initiation or propagation then ceases and the surface is covered with an oxide film.

For all materials investigated, besides titanium, the repassivation occurs easily. After changing the direction of polarisation in the highly anodic region, the reverse scan shows more positive corrosion potentials, and lower currents are recorded for the same values of potential. After the reverse voltametric curve crosses the forward one (closing the hysteresis loop), current continues to drop. In most cases, the loop is very small or does not exist, which

Sample	i_{corr}, mA (CR, mm/year) 30 °C	80 °C	120 °C
Stainless steel AISI 316L	$3.16 \times 10^{-3}(0,037)$	$6.3 \times 10^{-2}(0.73)$	$1.3 \times 10^{-1}(1.46)$
Stainless steel AISI 321	$1.26 \times 10^{-4}(< 0.01)$	$1.0 \times 10^{-2}(0.12)$	$4.0 \times 10^{-2}(0.46)$
Stainless steel AISI 347	$3.02 \times 10^{-4}(< 0.01)$	$2.5 \times 10^{-2}(0.29)$	$7.9 \times 10^{-2}(0.92)$
Inconel®625	$1.58 \times 10^{-4}(< 0.01)$	$5.3 \times 10^{-4}(< 0.01)$	$2.0 \times 10^{-2}(0.23)$
Incoloy®825	$1.58 \times 10^{-4}(< 0.01)$	$2.0 \times 10^{-2}(0.23)$	$3.2 \times 10^{-2}(0.37)$
Hastelloy®C-276	$1.95 \times 10^{-4}(< 0.01)$	$4.0 \times 10^{-3}(0.05)$	$2.4 \times 10^{-2}(0.28)$
Tantalum			$6.3 \times 10^{-5}(< 0,001)$
Titanium			$6.3(73,3)$

Table 2. The comparison of corrosion currents (approximate CRs) of different materials at T=30, 80 and 120 °C.

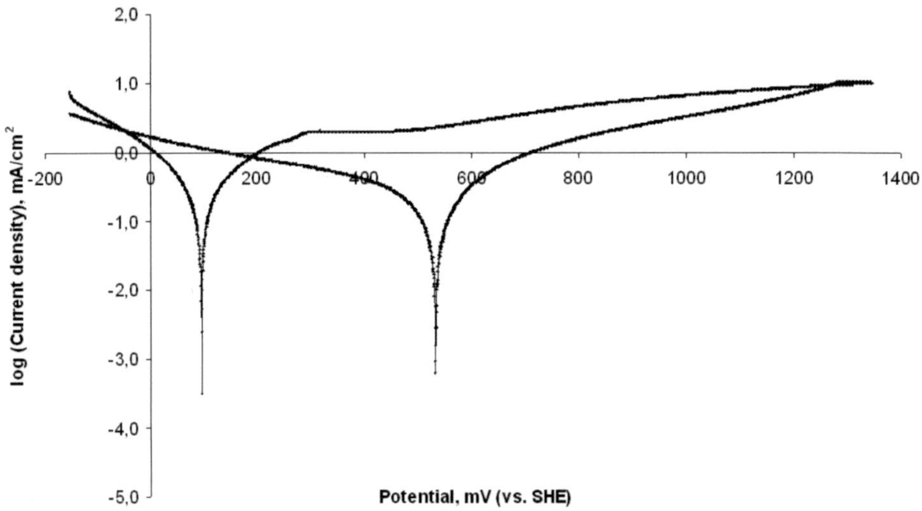

Figure 8. Tafel plot for AISI 316L in 85% H_3PO_4, 80 °C, 1 mV/s (vs. SHE).

usually indicates high resistance to pitting type of corrosion. In other words, if any break in the passive layer occurs, it easily "heals" itself, preventing any further development of pits. Thus, it is expected that the pitting resistance is excellent for all tested alloys, because surface protection eliminates local active sites. For titanium the hysteresis loop is very wide, and lasts for almost the whole anodic part of the polarisation curve. Reverse scanning repeats forward with higher values of currents, indicating the absence of "healing" passivation.

In Table 3 the comparison of corrosion potentials for forward and back scans is given. In most cases, there is an obvious dependence between corrosion rate and E_{cor}. For example, more positive value of E_{corr} for AISI 321 stainless steel during reverse scan corresponds to the lowest corrosion speed of AISI 321 among other tested stainless steels.

Alloy AISI 321 exhibited the largest difference between forward (starting) and reverse corrosion potentials, as well as the most positive repassivation potential among the tested alloys at 120 °C. This corresponds to the lowest corrosion rate of AISI 321 among the stainless steels.

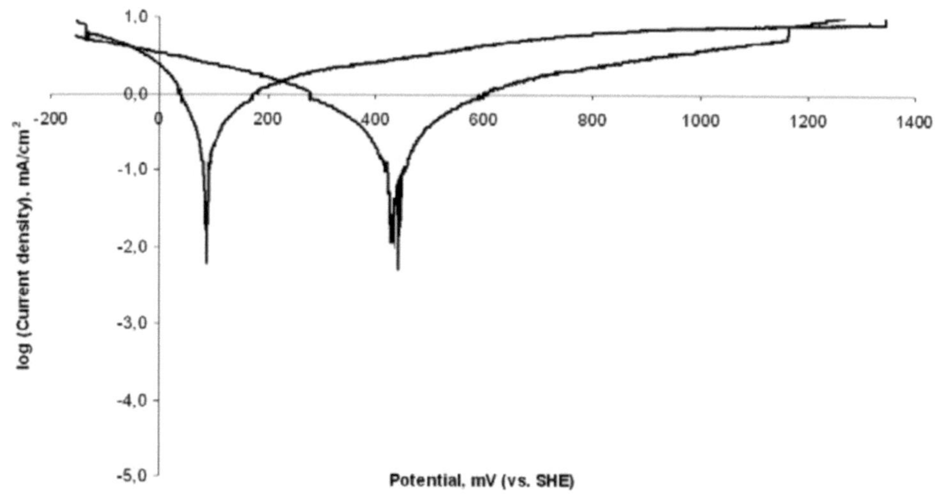

Figure 9. Tafel plot for AISI 316L in 85% H_3PO_4, 120 °C, 1 mV/s (vs. SHE).

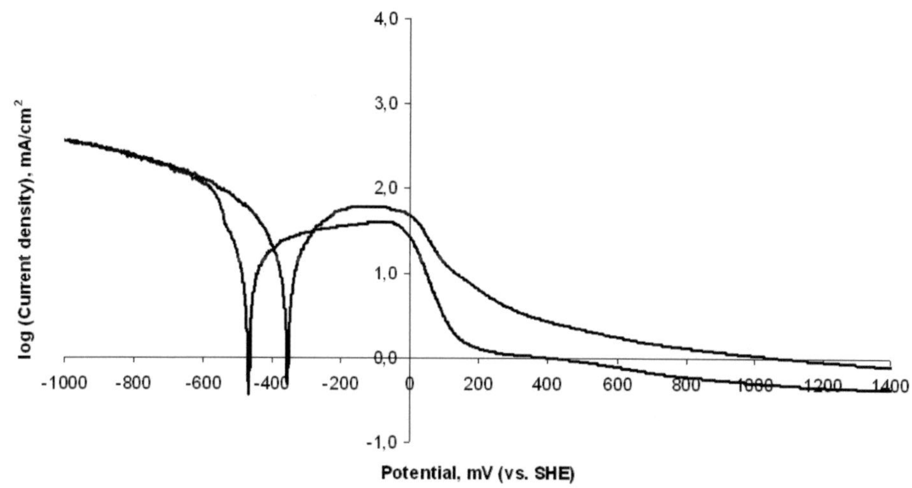

Figure 10. Tafel plot for titanium in 85% H_3PO_4, 120 °C, 1 mV/s (vs. SHE).

Titanium showed the poorest corrosion resistance. At 120 °C and open corrosion potential, the dissolution of titanium was observed visually, followed by intensive evolution of hydrogen gas. Under positive polarisation, it was partly passivated, but still the rates of dissolution were much higher than for austenitic stainless steels.

CVD tantalum coating on stainless steel showed an outstanding corrosion resistance, with the CRs being similar to earlier published data on this material [40]. The SEM image of the CVD-tantalum coated sample and the corresponding EDX spectra are shown in Figure 12 and Table 4 correspondingly. The coating appears to be homogeneous for the both sides of the plate, being around 5 and 50 μm on the contrary sides of the sample.

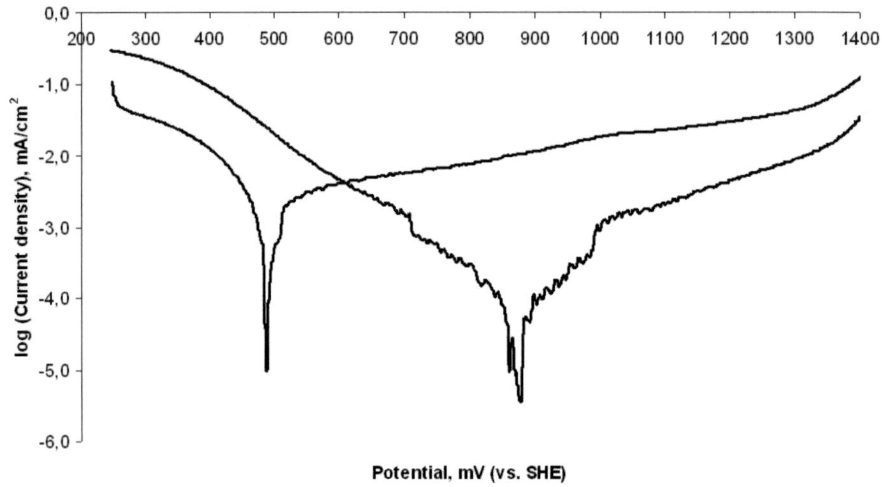

Figure 11. Tafel plot for tantalum in 85% H_3PO_4, 120 °C, 1 mV/s (vs. SHE).

Material	E_{cor}, mV (vs. SHE)			
	80 °C		120 °C	
	forward	reverse	forward	reverse
Stainless steel AISI 316L	100	530	80	430
Stainless steel AISI 321	175	820	40	640
Stainless steel AISI 347	320	770	320	500
Inconel®625	125	635	90	490
Incoloy®825	105	540	60	595
Hastelloy®C-276	440	620	120	580
Tantalum			490	875
Titanium			−465	−357

Table 3. Measured corrosion potentials for forward and back polarisation.

Spectrum	Composition, wt.%						
	O	Cr	Mn	Fe	Ni	Ta	Total
1		19.2	1.0	70.7	9.1		100
2	4.8					95.2	100
3				1.3		98.7	100

Table 4. EDX data for the CVD-tantalum coated stainless steel sample, in wt.%.

The corrosion resistance at 120 °C increases in the following sequence in our series:

Titanium < AISI 316L < AISI 347 < AISI 321 < Incoloy®825 < Hastelloy®C276 < Inconel® 625 < Tantalum

It can be clearly noticed, that for alloys the corrosion stability grows with the increasing content of nickel in this media, as shown in Table 5.

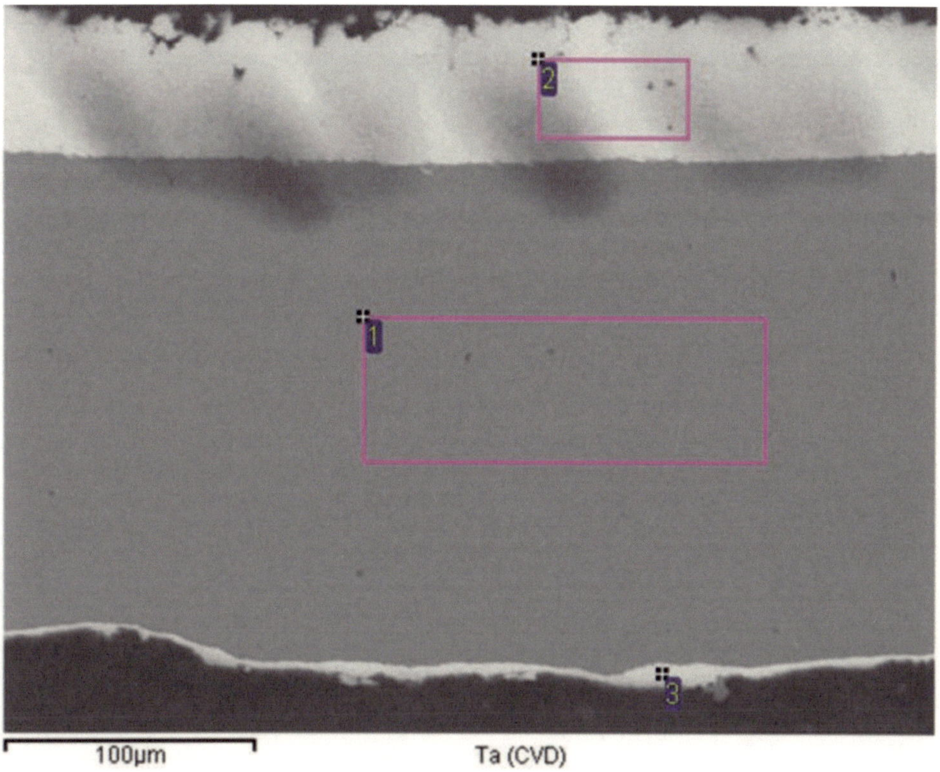

Figure 12. SEM image of the CVD-tantalum coated stainless steel sample. Numbers refer to EDX points and areas measured.

Generally, nickel based alloys show better corrosion stability than austenitic stainless steels in highly acidic media and elevated temperatures [77]. This tendency is also observed in our series.

Nickel's high degree of corrosion resistance is partly explained by the higher positive standard potential among the studied alloy compounds. Comparing with less resistant iron, nickel has 250 mV more positive standard corrosion potential. But compared to pure nickel metal, nickel-chromium-iron-molybdenum alloys show considerably better resistance to corrosion in all inorganic acid solutions [65].

It can also be seen from Table 5, that titanium has a positive effect on the corrosion resistance of the alloys tested, even though its own resistance is much lower. This effect can be explained by the EDX data, obtained from AISI 321 and Inconel®625 before and after the electrochemical tests.

Sample	AISI 347	AISI 316L	AISI 321	Inconel®625	Incoloy®825	Hastelloy®C-276
Ni content, wt.%	9-13	10-13	9-12	62	44	57
Ti content, wt.%	-	-	0.4-0.7	0.4	1.0	-

Table 5. The content of Ni and Ti, in the tested alloys.

It is visible from Figures 13(a) (spectrum 3,4) and 13(b) (spectrum 2,4)and Tables 5(a) and 5(b) that before the corrosion test, Ti is not spread evenly on the surface of AISI 321, it is localized at definite points, unlike the other elements, distributed more homogeneously. It is safe to assume that points of Ti location are situated on intergranular boundaries. It follows from the data that the content of titanium in the intergranular region dropped after the electrochemical experiment, indicating that corrosion in AISI 321 develops along the intergranular boundaries in this media. Titanium tends to be distributed along these boundaries during the severe anodic attack, thus preventing the formation of chromium carbides in these areas, which could promote chromium concentration drop resulting in a loss of passivity in these regions.

(a)

	Composition, wt.%, before the electrochemical tests							
	Si	Ti	Cr	Mn	Fe	Co	Ni	Total
Spectrum								
1	0.6	0.3	17.7	1.8	69.8	0.2	9.7	100
2	0.5	0.5	18.0	1.4	69.5	0.8	9.4	100
3	0.4	27.3	14.5	1.4	49.5	0.5	6.4	100
4	0.3	31.2	14.8	1.4	46.6		6.1	100

(b)

	Composition, wt.%, after the electrochemical tests							
	Si	Ti	Cr	Mn	Fe	Co	Ni	Total
Spectrum								
1	0.3	0.5	17.7	1.6	70.4	1.0	8.5	100
2	0.1	29.2	15.3	1.5	48.5	0.3	5.2	100
3	0.5	0.1	18.2	1.2	69.8	1.0	9.2	100
4	1.2	9.0	17.5	0.9	62.4	0.6	8.4	100

Table 6. EDX analysis data of AISI 321 before(a) and after(b) the electrochemical tests.

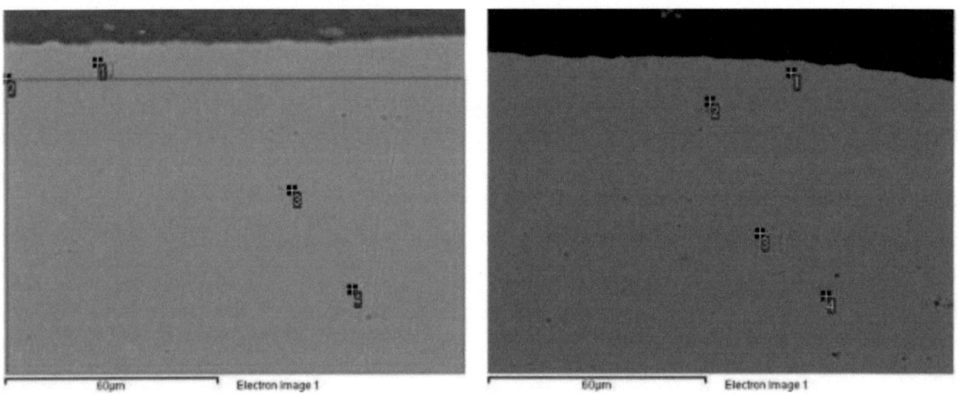

(a) (b)

Figure 13. SEM of AISI 321 before(a) and after(b) the electrochemical tests. Numbers refer to EDX points and areas measured.

The same behaviour is observed for another alloy, containing titanium as an addition, protecting the material from intergranular corrosion. Figures 14(a) (spectrum 3,4), 14 (spectrum 1,3,5) and Tables 6(a) and 6(b) show SEM and EDX data for Inconel®625. The same tendency is even more significant for this alloy. The titanium is distributed irregularly and its content decreases after the corrosion experiment.

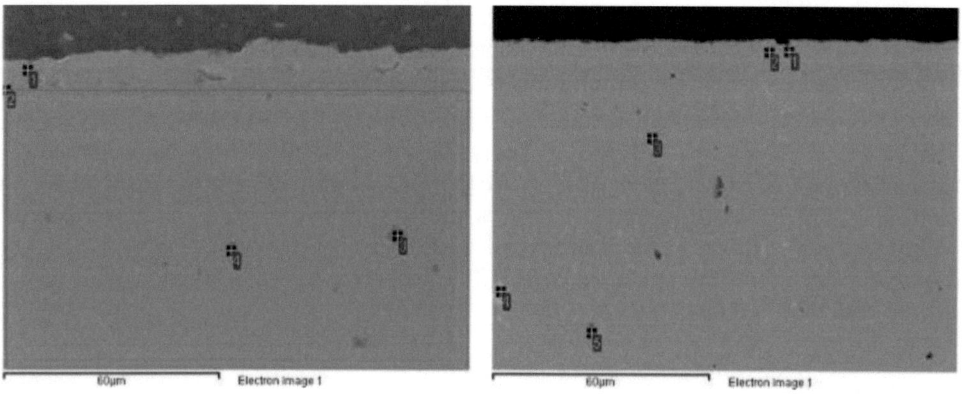

(a) (b)

Figure 14. SEM of Inconel®625 before(a) and after(b) the electrochemical tests. Numbers refer to EDX points and areas measured.

(a)

Spectrum	Composition, wt.%, before the electrochemical tests										
	Al	Si	Ti	Cr	Mn	Fe	Co	Ni	Nb	Mo	Total
1	0.3	0.3		21.7		0.9		63.1	4.0	10.1	100
2	0.3	0.4	0.3	22.1		0.9	0.2	63.4	3.4	9.4	100
3		0.1	63.6	5.8		0.1		7.0	23.1	0.6	100
4	0.6	0.3	54.9	10.6	0.1	0.3	0.1	18.9	12.2	2.0	100

(b)

Spectrum	Composition, wt.%, after the electrochemical tests										
	Al	Si	Ti	Cr	Mn	Fe	Co	Ni	Nb	Mo	Total
1	0.5	0.8	20.5	13.7	0.3	0.4		34.9	21.4	7.9	100
2		2.2	0.1	15.8	0.5	0.5		40.1	14.0	27.0	100
3	0.2	0.1	48.6	12.3	0.0	0.8		24.4	10.0	3.9	100
4	0.3	2.1	0.1	13.7	0.7	0.1	0.1	29.4	26.1	27.5	100
5		0.1	43.4	11.1		0.4	0.3	21.2	19.8	4.0	100

Table 7. EDX analysis data of Inconel®625 before(a) and after(b) the electrochemical tests.

The discussion above proves the extremely important role of doping the investigated alloys with titanium in this media, thus protecting them from the most apparent intergranular type of corrosion.

Molybdenum is more soluble in nickel than in austenitic stainless steels, and higher levels of alloying are possible with a higher content of nickel. Therefore, the molybdenum content limit grows with nickel content and high contents of molybdenum are only possible in high nickel alloys [10].

Generally, the addition of molybdenum to stainless steels and alloys is used for enhanced corrosion resistance. For instance, the addition of even one or two percent of molybdenum to ferritic stainless steels significantly increases the corrosion resistance of these material.

Pure nickel-molybdenum alloys, namely alloy B-2, contain approximately 28% molybdenum and and about 1,7% iron. The very high molybdenum content gives excellent resistance to reducing acids, i.e. hydrochloric and sulphuric [20]. For sulphuric acid, this alloy shows good resistance, even at concentrations close to 90% and temperatures up to 120 °C. Non-oxidant conditions, however, must certainly exist in this case. Either the presence of oxygen or aeration will significantly accelerate corrosion rate [77]. However, the role of molybdenum is not clearly noticed in this series.

4. Conclusions

The corrosion stability of the chosen stainless steels and nickel-based alloys is insufficient for these materials to be used in HTPEMECs. However, CVD-tantalum coating showed outstanding stability in the selected media. Therefore, such coatings on the bipolar plates and gas diffusion layers are recommended for long term tests of working HTPEMECs.

Tantalum coated AISI 316L stainless steel and Inconel® 625 are the most suitable materials for bipolar plate in high temperature steam electrolyzers with H_3PO_4 doped membranes. It has also been found that small addition of titanium to the alloys increases the corrosion stability in this media. Among austenitic stainless steels, AISI 321 has the lowest corrosion rate.

Anodic passivation with decreased rate of dissolution was observed from the Tafel plots for all alloys and metals studied indicating the formation of a protective oxide layer. The best corrosion resistance was found for tantalum. The titanium content was found to be an important parameter in the performance of the steels. The accumulation of titanium on the intergranular boundaries was assumed to inhibit the growth of chromium carbides on these regions, preventing intergranular corrosion of the samples. However, pure titanium showed the poorest resistance to corrosion, accompanied by the lowest corrosion potentials in the series and highest corrosion currents. Therefore, these facts exclude it as a possible material for use in bipolar plates in high temperature PEM steam electrolyzers, which operate on membranes doped with phosphoric acid.

Acknowledgements

Authors would like to express their gratitude to the Center for renewable hydrogen cycling (HyCycle), Denmark, contract No. 2104-07-0041 and the WELTEMP project under EU Seventh Framework Programme (FP7), grant agreement No. 212903.

Author details

Aleksey Nikiforov, Erik Christensen, Irina Petrushina, Jens Oluf Jensen and Niels J. Bjerrum
Proton Conductors Group, Department of Energy Conversion and Storage, Technical University of Denmark

List of Acronyms

AISI American Iron and Steel Institute

ASTM american society for testing and materials

CR corrosion rate

CVD chemical vapour deposition

EDX energy dispersive X-ray spectroscopy

GDL gas diffusion layer

HTPEMEC high temperature PEM electrolyzer cell

HTPEMFC high temperature PEM fuel cell

MEA membrane electrode assembly

OEE oxygen evolution electrode

PBI Poly[2,2'-(m-phenylene)-5,5'-bibenzimidazole

PEM polymer electrolyte membrane

PTFE polytetrafluoroethylene

SEM scanning electron microscopy

SHE standard hydrogen electrode

SOEC solid oxide electrolyzer cell

SPE solid polymer electrolyte

URFC unitized regenerative fuel cells

5. References

[1] Andolfatto, F., Durand, R., Michas, A., Millet, P. & Stevens, P. [1994]. Solid polymer electrolyte water electrolysis: electrocatalysis and long-term stability, *International Journal of Hydrogen Energy* 19(5): 421 – 427.

[2] *Annual book of ASTM standarts 10.05, G 102-89 (Reapproved 1999), p. 446-452* [n.d.].

[3] *Annual book of ASTM standarts 10.05, G3-89, p. 42-47* [n.d.].

[4] *Annual book of ASTM standarts 10.05, G5-94, p. 60-70* [n.d.].

[5] Antonucci, V., Di Blasi, A., Baglio, V., Ornelas, R., Matteucci, F., Ledesma-Garcia, J., Arriaga, L. G. & Arico, A. S. [2008]. High temperature operation of a composite membrane-based solid polymer electrolyte water electrolyser, *Electrochimica Acta* 53(24): 7350–7356.

[6] Antunes, R. A., Oliveira, M. C. L., Ett, G. & Ett, V. [2010]. Corrosion of metal bipolar plates for PEM fuel cells: A review, *International Journal of Hydrogen Energy* 35(8): 3632–3647.

[7] Appleby, A. & Vandrunen, C. [1975]. The oxygen evolution reaction on rhodium and iridium electrodes in 85% orthophosphoric acid, *Journal of Electroanalytical Chemistry* 60(1): 101–108.

[8] Barbir, F. [2005]. PEM electrolysis for production of hydrogen from renewable energy sources, *Solar Energy* 78(5): 661–669.

[9] Benabdellah, M. & Hammouti, B. [2005]. Corrosion behaviour of steel in concentrated phosphoric acid solutions, *Applied Surface Science* 252(5): 1657–1661.

[10] Bil'chugov, Y. I., Makarova, N. L. & Nazarov, A. A. [2001]. On limit of molybdenum content of pitting-corrosion-resistant austenitic steels, *Protection of Metals* 37(6): 597–601.

[11] Cardarelli, F., Taxil, P., Savall, A., Comninellis, C., Manoli, G. & Leclerc, O. [1998]. Preparation of oxygen evolving electrodes with long service life under extreme conditions, *Journal of Applied Electrochemistry* 28(3): 245–250.

[12] Carl W. Hamann, Andrew Hamnelt, W. V. [1997]. *Electrochemistry*, second edn, Wiley-VHC.

[13] Chen, G., Bare, S. R. & Mallouk, T. E. [2002]. Development of supported bifunctional electrocatalysts for unitized regenerative fuel cells, *Journal of The Electrochemical Society* 149(8): A1092–A1099.

[14] Chin, D. & Chang, H. H. [1989]. On the conductivity of phosphoric acid electrolyte, *Journal of Applied Electrochemistry* 19(1): 95–99.

[15] Connolly, D., Longwood & Gresham, W. [1966]. Fluorocarbon vinyl ether polymers.

[16] Daghetti, A., Lodi, G. & Trasatti, S. [1983]. Interfacial properties of oxides used as anodes in the electrochemical technology, *Materials Chemistry and Physics* 8: 1–90.

[17] Di Blasi, A., D'Urso, C., Baglio, V., Antonucci, V., Arico', A. S., Ornelas, R., Matteucci, F., Orozco, G., Beltran, D., Meas, Y. & Arriaga, L. G. [2009]. Preparation and evaluation of RuO_2 -IrO_2, IrO_2-Pt and IrO_2-Ta_2O_5 catalysts for the oxygen evolution reaction in an SPE electrolyzer, *Journal of Applied Electrochemistry* 39(2): 191–196.

[18] Divakar, R., Seshadri, S. G. & Srinivasan, M. [1989]. Electrochemical techniques for corrosion rate determination in ceramics, *Journal of the American Ceramic Society* 72(5): 780–784.

[19] Dutta, S. [1990]. Technology assessment of advanced electrolytic hydrogen production, *International Journal of Hydrogen Energy* 15(6): 379–386.

[20] et. al., A. [1989]. United states patent 4846885.

[21] Ghany, N. A. A., Kumagai, N., Meguro, S., Asami, K. & Hashimoto, K. [2002]. Oxygen evolution anodes composed of anodically deposited mn-mo-fe oxides for seawater electrolysis, *Electrochimica Acta* 48(1): 21 – 28.

[22] Grigoriev, S., Millet, P., Korobtsev, S., Porembskiy, V., Pepic, M., Etievant, C., Puyenchet, C. & Fateev, V. [2009]. Hydrogen safety aspects related to high-pressure polymer electrolyte membrane water electrolysis, *International Journal of Hydrogen Energy* 34(14): 5986–5991.

[23] Grigoriev, S., Millet, P., Volobuev, S. & Fateev, V. [2009]. Optimization of porous current collectors for PEM water electrolysers, *International Journal of Hydrogen Energy* 34(11): 4968–4973.

[24] He, R., Li, Q., Xiao, G. & Bjerrum, N. J. [2003]. Proton conductivity of phosphoric acid doped polybenzimidazole and its composites with inorganic proton conductors, *Journal of Membrane Science* 226(1-2): 169–184.

[25] Hermann, A., Chaudhuri, T. & Spagnol, P. [2005]. Bipolar plates for PEM fuel cells: A review, *International Journal of Hydrogen Energy* 30(12): 1297–1302.

[26] http://hycycle.dk/ [2009]. Hycycle. center for renewable hydrogen cycling.
URL: *http://hycycle.dk/*

[27] *http://www.ginerinc.com/* [n.d.].
URL: *http://www.ginerinc.com/*

[28] *http://www.hydrogenics.com/* [n.d.].
URL: *http://www.hydrogenics.com/*

[29] *http://www.protonenergy.com* [n.d.].
URL: *http://www.protonenergy.com*

[30] *http://www.tantaline.com/* [n.d.].
URL: *http://www.tantaline.com/*

[31] Hu, W., Cao, X., Wang, F. & Zhang, Y. [1997]. A novel cathode for alkaline water electrolysis, *International Journal of Hydrogen Energy* 22(6): 621 – 623.

[32] Hung, Y., EL-Khatib, K. M. & Tawfik, H. [2005]. Corrosion-resistant lightweight metallic bipolar plates for PEM fuel cells, *Journal of Applied Electrochemistry* 35(5): 445–447.

[33] I.Kreysa, G. & Eckermann, R. [1993]. *DECHEMA corrosion handbook: corrosive agents and their interaction with materials*, Vol. 12. Chlorinated hydrocarbons-chloroethanes, phosphoric acid, VCH Verlagsgesellschaft, Weinheim (Germany) and VCH Publishers, New York, NY (USA).

[34] *International patent application, 03.01.2008, WO 2008/002150 A1, PCT/NO2007/000235* [03.01.2008].

[35] Jensen, J., Bandur, V., Bjerrum, N., Højgaard, S., Ebbesen, S. & Mogensen, M. [2008]. Pre-investigation of water electrolysis, *Technical report*.
URL: *http://130.226.56.153/rispubl/NEI/NEI-DK-5057.pdf*

[36] Jensen, J. O., Li, Q. F., Pan, C., Vestbo, A. P., Mortensen, K., Petersen, H. N., Sorensen, C. L., Clausen, T. N., Schramm, J. & Bjerrum, N. J. [2007]. High temperature PEMFC and the possible utilization of the excess heat for fuel processing, *International Journal of Hydrogen Energy* 32(10-11): 1567–1571.

[37] Joseph, S., McClure, J. C., Chianelli, R., Pich, P. & Sebastian, P. J. [2005]. Conducting polymer-coated stainless steel bipolar plates for proton exchange membrane fuel cells (PEMFC), *International Journal of Hydrogen Energy* 30(12): 1339–1344.

[38] Jung, H. Y., Park, S., Ganesan, P. & Popov, B. N. [2008]. Electrochemical studies of unsupported PtIr electrocatalyst as bifunctional oxygen electrode in unitized regenerative fuel cell (urfc), *Proton Exchange Membrane Fuel Cells 8, Pts 1 and 2* 16(2): 1117–1121.

[39] Keijzer, M., Hemmes, K., VanDerPut, P. J. J. M., DeWit, J. H. W. & Schoonman, J. [1997]. A search for suitable coating materials on separator plates for molten carbonate fuel cells, *Corrosion Science* 39(3): 483–494.

[40] Kouřil, M., Christensen, E., Eriksen, S. & Gillesberg, B. [2011]. Corrosion rate of construction materials in hot phosphoric acid with the contribution of anodic polarization, *Materials and Corrosion* .

[41] Labou, D., Slavcheva, E., Schnakenberg, U. & Neophytides, S. [2008]. Performance of laboratory polymer electrolyte membrane hydrogen generator with sputtered iridium oxide anode, *Journal of Power Sources* 185(2): 1073–1078.

[42] Lage, L. G., Delgado, P. G. & Kawano, Y. [2004]. Thermal stability and decomposition of Nafion® membranes with different cations using high-resolution thermogravimetry, *Journal of Thermal Analysis and Calorimetry* 75(2): 521–530.

[43] Lessing, P. A. [2007]. Materials for hydrogen generation via water electrolysis, *Journal of Materials Science* 42(10): 3477–3487.

[44] Li, Q. F., He, R. H., Jensen, J. O. & Bjerrum, N. J. [2003]. Approaches and recent development of polymer electrolyte membranes for fuel cells operating above 100 degrees C, *Chemistry of Materials* 15(26): 4896–4915.

[45] Li, Q., Jensen, J. O., Savinell, R. F. & Bjerrum, N. J. [2009]. High temperature proton exchange membranes based on polybenzimidazoles for fuel cells, *Progress in Polymer Science* 34(5): 449–477.

[46] Linkous, C. [1993]. Development of solid polymer electrolytes for water electrolysis at intermediate temperatures, *International Journal of Hydrogen Energy* 18(8): 641–646.

[47] Linkous, C. A., Anderson, H. R., Kopitzke, R. W. & Nelson, G. L. [1998]. Development of new proton exchange membrane electrolytes for water electrolysis at higher temperatures, *International Journal of Hydrogen Energy* 23(7): 525–529.

[48] Lu, J.-s. [2009]. Corrosion of titanium in phosphoric acid at 250 °C, *Transactions of Nonferrous Metals Society of China* 19: 552–556.

[49] Lukashenko, T. A. & Tikhonov, K. I. [1998]. Corrosion resistance of a series of group IV-VI transition metal carbides and nitrides in concentrated solutions of sulfuric and phosphoric acids., *Zhurnal Prikladnoi Khimii (Sankt-Peterburg)* 71(12): 2017–2020.

[50] Makkus, R. C., Janssen, A. H., de Bruijn, F. A. & Mallant, R. K. [2000]. Use of stainless steel for cost competitive bipolar plates in the SPFC, *Journal of Power Sources* 86(1-2): 274–282.

[51] Marshall, A., Børresen, B., Hagen, G., Tsypkin, M. & Tunold, R. [2006]. Electrochemical characterisation of $Ir_xSn_{1-x}O_2$ powders as oxygen evolution electrocatalysts, *Electrochimica Acta* 51: 3161–3167.

[52] Marshall, A., Børresen, B., Hagen, G., Tsypkin, M. & Tunold, R. [2007]. Hydrogen production by advanced proton exchange membrane (PEM) water electrolysers-reduced energy consumption by improved electrocatalysis, *Energy* 32: 431–436.

[53] M.Bengisu [1963]. *Engineering Ceramics*, Springer-Verlag Berlin Heidelberg 2001.

[54] McElroy, J. F. [1994]. Recent advances in SPE® water electrolyzer, *Journal of Power Sources* 47(3): 369–375. Proceedings of the Fourth Space Electrochemical Research and Technology Conference.

[55] Millet, P., Mbemba, N., Grigoriev, S., Fateev, V., Aukauloo, A. & Etiévant, C. [2011]. Electrochemical performances of PEM water electrolysis cells and perspectives, *International Journal of Hydrogen Energy* 36: 4134–4142.

[56] Millet, P., Ngameni, R. & Grigoriev, S. [2009]. PEM water electrolyszers: From electrocatalysis to stack development, *International Journal of Hydrogen Energy* 35: 5043 – 5052.

[57] Mitsuhashi, A., Asami, K., Kawashima, A. & Hashimoto, K. [1987]. The corrosion behavior of amorphous nickel base alloys in a hot concentrated phosphoric acid, *Corrosion Science* 27(9): 957–970.

[58] Ni, M., Leung, M. & Leung, D. [2007]. Energy and exergy analysis of hydrogen production by solid oxide steam electrolyzer plant, *International Journal of Hydrogen Energy* 32(18): 4648–4660.

[59] Nikiforov, A., Petrushina, I., Christensen, E., Tomás-García, A. L. & Bjerrum, N. [2011]. Corrosion behaviour of construction materials for high temperature steam electrolysers, *International Journal of Hydrogen Energy* 36(1): 111–119.

[60] Oberlin R, F. M. [1986]. Status of the membral process for water electrolysis, *Hydrogen energy progress VI, proceedings of the sixth world hydrogen energy conference, Oxford: Pergamon Press* pp. 333–40.

[61] Oi, T. & Sakaki, Y. [2004]. Optimum hydrogen generation capacity and current density of the pem-type water electrolyzer operated only during the off-peak period of electricity demand, *Journal of Power Sources* 129(2): 229 – 237.

[62] Onoro, J. [2009]. Corrosion fatigue behaviour of 317LN austenitic stainless steel in phosphoric acid, *International Journal of Pressure Vessels and Piping* 86(10): 656–660.
[63] Pourbaix, M., Klimzack-Mathieiu, L., Mertens, C., Meunier, J., Vanleugenhaghe, C., de Munck, L., Laureys, J., Neelemans, L. & Warzee, M. [1963]. Potentiokinetic and corrosimetric investigations of the corrosion behaviour of alloy steels, *Corrosion Science* 3(4): 239–259.
[64] Robin, A. & Rosa, J. L. [2000]. Corrosion behavior of niobium, tantalum and their alloys in hot hydrochloric and phosphoric acid solutions, *International Journal of Refractory Metals and Hard Materials* 18: 13–21.
[65] Rockel, M. [1998]. *Corrosion behaviour of nickel alloys and high-alloy stainless steel*, Nickel Alloys, Marcel Dekker Inc, New York.
[66] Shin, Y., Park, W., Chang, J. & Park, J. [2007]. Evaluation of the high temperature electrolysis of steam to produce hydrogen, *International Journal of Hydrogen Energy* 32(10-11): 1486–1491.
[67] Song, S. D., Zhang, H. M., Ma, X. P., Shao, Z. G., Zhang, Y. N. & Yi, B. L. [2006]. Bifunctional oxygen electrode with corrosion-resistive gas diffusion layer for unitized regenerative fuel cell, *Electrochemistry Communications* 8(3): 399–405. Electrochemistry Communications.
[68] Song, S., Zhang, H., Ma, X., Shao, Z., Baker, R. T. & Yi, B. [2008]. Electrochemical investigation of electrocatalysts for the oxygen evolution reaction in PEM water electrolyzers, *International Journal of Hydrogen Energy* 33: 4955–4961.
[69] Suffredini, H. B., Cerne, J. L., Crnkovic, F. C., Machado, S. A. S. & Avaca, L. A. [2000]. Recent developments in electrode materials for water electrolysis, *International Journal of Hydrogen Energy* 25(5): 415–423.
[70] Swette, L. L., LaConti, A. B. & McCatty, S. A. [1994]. Proton-exchange membrane regenerative fuel cells, *Journal of Power Sources* 47(3): 343–351.
[71] Tawfik, H., Hung, Y. & Mahajan, D. [2007]. Metal bipolar plates for PEM fuel cell-A review, *Journal of Power Sources* 163(2): 755–767.
[72] Tian, R. J., Sun, J. C. & Wang, L. [2006]. Plasma-nitrided austenitic stainless steel 316L as bipolar plate for PEMFC, *International Journal of Hydrogen Energy* 31(13): 1874–1878.
[73] Trasatti, S. [1980]. *Electrodes of conductive metallic oxides*, Studies in physical and theoretical chemistry, Elsevier, 335 Jan van Galenstraat P.O. Box 211, 1000 AE Amsterdam, The Netherlands.
[74] Trethewey, K. R. & Chamberlain, J. [1988]. *Corrosion*, Longman Group UK Limited.
[75] Turner, J., Sverdrup, G., Mann, M. K., Maness, P., Kroposki, B., Ghirardi, M., Evans, R. J. & Blake, D. [2008]. Renewable hydrogen production, *International Journal of Energy Research* 32(5): 379–407.
[76] Wang, Y. & Northwood, D. O. [2007]. An investigation of the electrochemical properties of PVD TiN-coated SS410 in simulated PEM fuel cell environments, *International Journal of Hydrogen Energy* 32(7): 895–902.
[77] W.Z.Friend [1980]. *Corrosion of Nickel and Nickel-Base Alloys*, J.Wiley&Sons, New York-Chichester-Brisbane-Toronto.
[78] Yamaguchi, M., Okisawa, K. & Nakanori, T. [1997]. Development of high performance solid polymer electrolyte water electrolyzer in WE-NET, *Proceedings of the Intersociety Energy Conversion Engineering Conference* 3-4: 1958–1961.

[79] Zhang, Y., Zhang, H., Ma, Y., Cheng, J., Zhong, H., Song, S. & Ma, H. [2009]. A novel bifunctional electrocatalyst for unitized regenerative fuel cell, *Journal of Power Sources* 195: 142–145.

Chapter 5

Voltammetric Characterization Methods for the PEM Evaluation of Catalysts

Shawn Gouws

Additional information is available at the end of the chapter

http://dx.doi.org/10.5772/47248

1. Introduction

1.1. Electrolytic units

Electrolysis is a chemical reaction that produces electrical energy via an electrochemical process. This electrical energy is the driving force needed for the chemical reactions. This phenomenon was first observed by Nicolson and Carlisle in 1789. Hydrogen is considered to be one of the most promising energy carriers for providing clean, reliable and sustainable energy systems (Balaji et al., 2011). It could be beneficial in meeting the global threat of climate change; and it could eliminate those issues associated with the use of fossil fuels. Although the production of hydrogen is more expensive than fossil fuel, it is an inexhaustible resource that could meet most of our future energy needs (Van Ruijven et al., 2007)

During the process of water electrolysis, the water molecule is decomposed into hydrogen and oxygen when an electric current is passed through the system. Electrical current causes positively charged ions to migrate to the negatively charged cathode, where the reduction takes place in order to produce hydrogen gas (Zoulias et al. 2004,).

At the other electrode – the anode – oxygen is produced and escapes as a gas. The stoichiometry of the chemical reaction is that two molecules of hydrogen are produced for every one molecule of oxygen formed.

$$\text{Anode reaction: } 2H_2O \rightarrow 4H^+ + 4e^- + O_2$$

$$\text{Cathode reaction: } 4H^+ + 4e^- \rightarrow 2H_2$$

$$\text{Overall reaction: } 2H_2O \rightarrow 2H_2 + O_2$$

Electrolysis is considered to be one of the cleanest methods of producing hydrogen (Qi, 2008). The following section introduces briefly different kinds of electrolytic units that are currently being used to generate hydrogen for use in PEM fuel cells. There are two possible methods of interest for producing pure hydrogen on a large scale. The first method of interest is the alkaline electrolytic units that are currently being used in industry to generate hydrogen. The reason for the use of alkaline electrolysis is that it is easier to control the corrosion effects compared with acid-based electrolysis technologies, such as the phosphoric acid electrolyser. Another method, and one of more interest, is the proton exchange membrane (PEM) electrolyser that has been developed (Atlam & Kolhe 2011) over recent years.

These PEM electrolytic units are less corrosive than the alkaline electrolytic medium that uses large quantities of caustic soda. Another advantage is that due to the membrane used in PEM electrolysers, the explosion rate between hydrogen and oxygen is at a minimum compared with that in alkaline electrolysers. Another advantage to PEM electrolysers over traditional technologies are there higher production rates and more compact design (PSO-F&U, 2008, p66).

Because of oxygen and hydrogen generated in the electrolytic unit, and the subsequent high pressures generated, safety procedures are required for the storing and disposal of the large amounts of potassium hydroxide. Currently, the fields of application of PEM electrolysers at different capacities have been widening over the spectrum of applications. PEM electrolyser are utilized not only for the production of hydrogen from PEM fuel cells, but they can also be used for online hydrogen production to analytical laboratories (equipment such as gas chromatography, hydrogen supply for laboratory usage), hydrogen welding, metallurgy of specially pure metals and alloys, as well as the production of pure substances for the electronic industry.

In terms of economic limitations, PEM water electrolysis remains an expensive technology; and further research and development need to be done to reduce the noble metal content in a PEM water electrolyser (Millet et al., 2011). This could open a commercial route for the domestic usage of PEM electrolyser units. PGM metals – in particular Pt and Ir metal oxides – are still being widely used in this highly acidic environment found in perfluorinated proton-conducting materials used in the manufacture of the membrane electrode assembly (MEA).

One possibility for reducing PGM costs in a PEM electrolyser unit, is to reduce the catalytic content on the MEA to as low as 0.3 – 0.5 mg cm^{-2}. This can be achieved by depositing Pt nano-particles on the surface of carbon carriers of large surface areas (Millet et al., 2011). It must be noted that the cost is not only due the noble metal, but also in their precursor salt costs that are used in the initial plating process.

The focus of this chapter will be on the characterization and feasibility of PEM electrolyser catalysts used in the anode and cathode compartment. In order to achieve this characterization, it will be necessary to discuss cyclic voltammetry, kinetic aspects such as the comparison of Tafel slopes determined from the electrolyser I-V curves, in addition to thermodynamic studies to determine the hydrogen efficiencies.

1.1.1. Alkaline electrolysers

The theory of alkaline water electrolysis is illustrated in Figure 1 in the example of a monopolar arrangement electrolyser. Two molecules of water are reduced to one molecule of hydrogen and two hydroxyl ions at the cathode. The hydrogen ions escape from the cathode and recombine to produce gaseous hydrogen. The hydroxyl ions migrate between the anode and cathode compartments through a porous diaphragm (Zoulias et al., 2004).

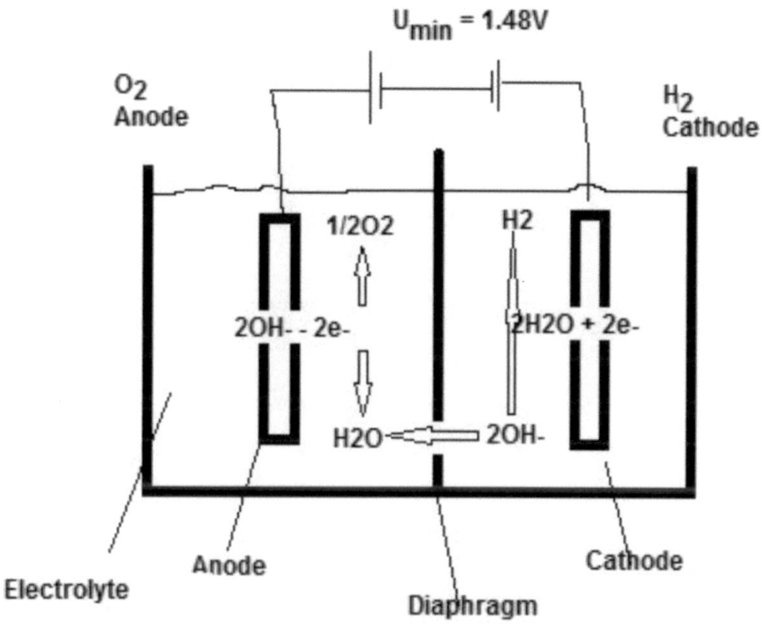

Figure 1. Alkaline electrolysis (Kreuter and Hofmann, 1998)

Half of a molecule of oxygen escapes at the anode surface; then it recombines and leaves the system as a gas. Some of the advantages of alkaline electrolysers are that the construction materials and electrolytes are relatively cheap in comparison with those required in acid electrolytic technologies. The current densities to operate these systems are relatively low; and consequently, they are economically feasible. However, these types of water electrolysers do have several disadvantages

These alkaline electrolysers are limited in their ability to be constructed into multiple cell configurations that have typically low current densities. These types of electrolysers need to be purged with an inert gas, such as nitrogen or argon, to prevent the system reaching explosive limits between the oxygen and the hydrogen

Alkaline electrolysers also utilize large concentrations of potassium hydroxide, which is a very corrosive material – especially at elevated temperatures.

1.1.2. Proton-exchange membrane electrolyser

The proton-exchange membrane electrolyser is based on the use of a polymeric proton exchange membrane as the solid electrolyte that can produce carbon-free hydrogen from water electrolysis (Marshall et al., 2007). The first PEM was introduced by General Electric for fuel cell application – and later for electrolyser application (Millet et al., 1996). PEM electrolyser systems offer numerous advantages over other types of electrolyser technologies. These advantages include greater energy efficiency, higher production rates, and more compact design (Oberlin & Fisher, 1996). A schematic representation of a PEM electrolysis cell is shown in Figure 2.

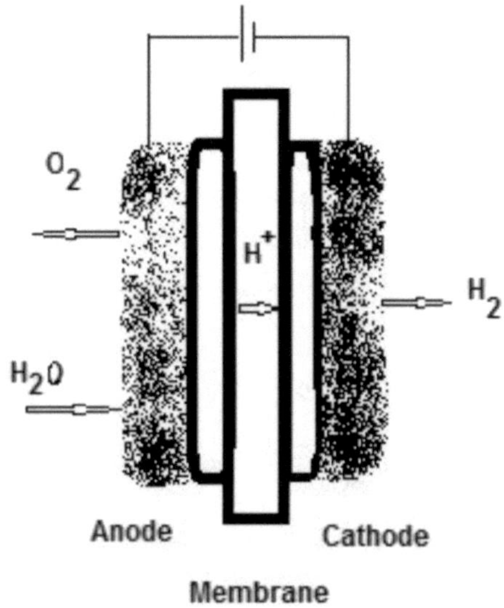

Figure 2. PEM water electrolyser

The PEM electrolyser consists of a membrane electrode assembly (MEA) to which an anode and a cathode are bonded. Typical catalyst particles for the anode utilization could be Pt, IrO_2, or mixtures, such as $Ir_xSn_{1-x}O_2$ (Marshall, 2004, 2005). The normal catalyst particles utilized for the cathode could be, for example, Pt. The electrical contact and mechanical support are established by means of porous metallic meshes. Hydrogen is produced in a PEM electrolyser by supplying water to the anode, where it splits into oxygen, hydrogen ions and electrons. The hydrogen ions migrate through the MEA to the cathode, where the protons and electrons re-combine to produce hydrogen gas.

$$\text{Anode: } H_2O \rightarrow 4H^+ + O_2 + 4e^-$$

$$\text{Cathode: } 2H_2O + 4e^- \rightarrow 2H_2$$

$$\text{Overall cell: } 2H_2O \rightarrow 2H_2 + O_2$$

The first PEM water electrolyser was installed in 1987 at Stellram, SA, a metallurgical specialty company, in Nyon, Switzerland. This unit was designed to produce up to 20Nm^3 h^{-1} of hydrogen at a pressure of 1-2 bars. The second designed PEM water electrolyser was demonstrated at Solar Wasserstoff Bayern (SWB); and it was slightly different from the first design used at the Stellram plant. The main difference was that the thickness of the bipolar plates was reduced. The stacks consisted of 3 modules of 40 cells each (Zoulias et al., 2004). Presently, the use of PEM electrolysers produced by Proton, are only sized for the home or small electrical grid systems.

The high capital costs of PEM electrolysis units has resulted in their limited viability in the large hydrogen production market, when compared to alkaline electrolysis systems, with their lower capital costs, producing a range of hydrogen capabilities (Ivy, 2004 and PSO-F&U 2008).

However, the development of PEM electrolyser systems has been slow, due to the cost of components, such as the precious metals used as electrocatalysts. Currently, there is still no non-noble metal that could be used as an electrocatalyst with satisfactory results. It is of great importance to reduce the noble metal content in the development of catalysts for PEM water electrolysers. Most of the overpotential losses that occur happen at the anode electrode, where oxidation takes place. Because of the acidic environment and high anode potential during water electrolysis, non-metal catalysts, lke Ni and Co, cannot be used because of corrosion (Millet, 1994).

High overpotentials for the oxygen evolution reaction (OER) were experienced for Pt electrocatalyst, making a Pt catalyst a poorly conducting oxide. Other possible noble metal catalysts, such as Ir or Ru, make very good conductors. However, Ru is unstable and needs to be used with other catalysts, such as IrO_2 for example.

Tseung and Jasem (Tseung and Jasem, 1977) and Trasatti (Trasatti, 1991) give good guidelines for the choice of oxide electrocatalysts needed for OER. The OER can only proceed when the electrode potential is higher than the potential of the metal/metal oxide couple, or the lower metal oxide/higher metal oxide couple, which indicates that the OER is governed by the surface electrochemistry (Song et al., 2008). The membrane material for the PEM electrolyser unit is Nafion™ 117, and this is manufactured by DuPont.

Individual cells can be stacked in bipolar modules with graphite separator plates that would provide the manifold for the water feed and the gas evacuation.

2. Cyclic voltammetry

All electrochemical techniques involve the use of electricity as either an input or an output signal. The function of an electrochemical instrument is to generate an input electrical signal as a function of time, and to measure the corresponding output signal as a function of either the voltage, current or charge (Qi, 2008). Charge is the integration of current with time. The input electrical signal is generated by an electrochemical reaction at the working electrode and the counter electrodes. This can be illustrated in Figure 3 for a typical CV waveform produced. When the voltage is swept past a potential corresponding to an active electrochemical reaction, the initial forward scan for the current will give a spiked response, which will consume the electrons involved in that reaction. On the reverse voltage scan, the reverse electrochemical reaction will be observed. The shape and size of the peaks give information about the relative kinetic rates and diffusion coefficiencies of the electrochemical system.

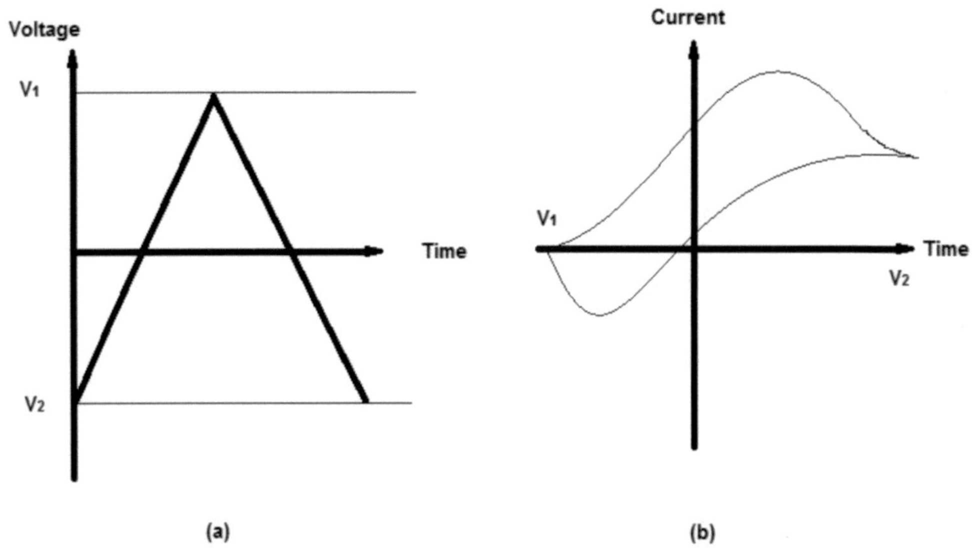

Figure 3. Schematic representation of a CV waveform and typical resulting current response (a). In a CV experiment, the voltage is swept linearly back and forth between two voltage limits [V_1 and V_2] (b). The resulting current is plotted as a function of voltage.

2.1. Volumetric studies

During the electrolysis of water, hydrogen and oxygen are produced at the surface of the polymer electrode membrane. These catalyst layers on the membrane need voltammetric characterization. The most important techniques used in electrochemical characterization include potential cycling, potential sweep, and rotating disk electrodes.

An electrochemical reaction step consists of at least one of the following steps: the transport of the reactants to the surface of the electrode, and adsorption of the reactants onto the surface of the electrodes. Charge transfer occurs through either oxidation or reduction on the surface of the electrode, and the transporting of product(s) from the surface of the electrode follows. The purpose would be to characterize each of these steps that are taking place inside a PEM electrolysis cell (Qi, 2008).

The cyclic voltammetry technique is one of the most commonly used techniques in electrochemical analysis for the study of electro-active catalytical species and electrode surfaces (Kumpulainen et al., 2002).

The characterization of the electrochemically active species would be carried out in a typical three-electrochemical cell. These are typically: a conventional 3-electrode cell, a half cell and single cell. In these types of cells, the catalysts and the electrodes will be characterized in the form of the working electrode, the potential of which, or the current, will be monitored or controlled in a specific environment.

What happens on the surface of the working electrode will comprise the heart of the investigation. The second electrode is called the counter electrode in the circuit; and the current flowing through the circuit would cause a reaction on the counter electrode as well. However, the investigation has no interest in what happens on the counter electrode surfaces, so long as it does not interfere with the working of the electrode (Qi, 2008)

In order to minimize the impact of the electrolyte resistance to the potential of the working electrode, a third electrode is introduced – called the reference electrode. This is often used to form a second circuit with the working electrode. Ideally, this electrode should be non-polarizable and should maintain a stable potential. There is high impedance in the voltage measurement equipment, which makes the current in the circuit very small (Qi, 2008).

A typical 3-electrode cell is shown in Figure 4, where the working electrode is the exposed area of a glassy carbon electrode impregnated with the catalytic material being tested. A piece of Pt wire – in the form of a coil or sheet – could be used as the counter electrode. The reason for a large surface counter electrode is to ensure that the electrochemical reaction occurring is large enough so that there is no interference with the performance of the working electrode. The reference electrodes used are typically electrodes, such as $Pt/H_2/H^+$ (standard or dynamic hydrogen electrode), $Ag/AgCl/Cl^-$ (silver/silver chloride) and $Hg/Hg_2Cl/Cl^-$ (the calomel electrode).

When the counter electrode might possibly interfere, it is often placed in a separate compartment – away from the working electrode – to ensure that there is no interference. This is done by placing the counter electrode in a porous frit, such as a sintered porous glass or an electrolyte bridge.

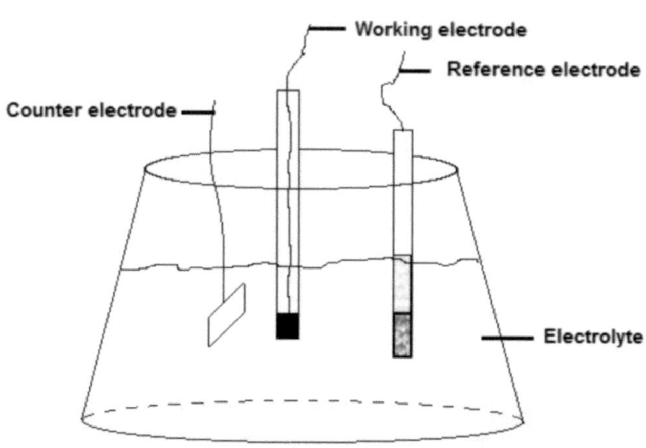

Figure 4. This demonstrates the basic setup of a conventional 3-electrode cell

This conventional 3-electrode cell was used to characterize the half-cell reactions of the catalyst that would occur on either side of the PEM electrolyser cell.

2.2. The basic principles of CV studies

The initial working potential is usually set at a potential that does not cause any electrochemical reaction. After the CV scan is started, the potential is either increased to a maximum, or decreased to a minimum, indicating either an oxidation or reduction reaction of an electrochemically active species; and then an anodic (or cathodic) current appears. The current response, as a result of this polarization, is then plotted as a function of the applied potential. The current voltage curve is referred to as the cyclic voltammogram; and it gives information about the electrochemical reactions taking place on the working electrode surface (Kumpulainen et al., 2002).

Other information regarding the adsorption and desorption of hydrogen could also be obtained.

As the kinetics of the reaction go faster, there will be an increase or decrease in the anodic (or cathodic) currents. The maximum is reached when all the electrochemically active species oxidized or reduced have been consumed in the electrochemical reaction. The highest current in Figure 5(b) is achieved at the moment that the mass transfer rate is at its maximum. This is driven between the gradient of the bulk concentration of the electrochemically active species and that of the surface concentration on the working electrode area.

When the potential increases beyond this point, the current starts to decrease due to the double layer thickness increases, resulting in a less-steep concentration gradient of the active species. When the potential reaches the set high or low potential, the scan is reversed, forming thereby a cyclic voltammogram.

The concentration gradients are linear, and the ratio of oxidized to reduce species for a reversible reaction, is given by the Nerst equation (Eqn. 1)

$$E_e = E_e^o + \frac{RT}{nF} ln \frac{C_O}{C_R} \tag{1}$$

Where E_e is the equilibrium potential (V) and E_e^o is the standard potential (V). R,T, n and F are constants and C_O and C_R the surface concentrations of the oxidized and reduced species. When the scan rates are increased, this shortens the timescale of the CV redox cycle and the concentration gradient has time to relax. This relationship between the peak current density and the scan rate is mathematically shown by the Randles-Sevcik equation (Eqn.2) (Heineman and Kissinger, 1996):

$$I_p = (2.69 \times 10^5) n^{3/2} A D^{1/2} C^o \upsilon^{1/2a} \tag{2}$$

At 25°C where i_p is the peak current (A); n is the number of electrons transferred; A is the electrode area (cm^2); D is the diffusion coefficient of the species being oxidized/or reduced (cm^2/s); C^o is the concentration of this same species in the bulk solution (mol cm^{-3}); and υ is the scan rate (V/s). There should be a linear relationship between the peak current density and the square root of the scan rate. Any deviation from this relationship can be related to a quasi-reversible or completely irreversible system.

In search of redox couples, the electrode potential sweeps rapidly (meaning moving back and forth). The characteristics of the reversible electrochemical reaction on CV record are as follows (Eqn. 3 – 4):

$$\Delta E = E_p{}^a - E_p{}^c = 59/n \text{ mV} \tag{3}$$

$$I\, i_p{}^a / i_p{}^c\, I = 1$$

$$I_p \,\alpha\, V \tag{4}$$

According to Eqn. 2, Ip is proportional to $v^{1/2}$. Although the peak currents increase with scan rate, the potential at which the peak occurs is invariant with scan rate. This is shown in Eqn. 4.

3. Preparation of the working electrode surface

Noble metal catalysts were prepared by reducing the chloride precursor by means of a fusion process, called the Adam's method (Adam & Shriner, 1923) with a solution of sodium nitrate, to produce $M(NO_3)$. This is then annealed at 340°C – to produce the metal oxide (Cheng, 2009 & Marshall, 2007). This method has been used successfully for the preparation of several anode catalysts in PEM water electrolytic systems (Hutchings et al., 1984); and it was found to be convenient and fast for the screening of different catalysts.

For example a reaction can be written as follows:

$$6\, NaNO_3 + H_2IrCl_6 \rightarrow 6NaCl + Ir(NO_3)_4\, 2HNO_3$$

$$Ir(NO_3)_4 \rightarrow IrO_2 + 4NO_2 + O_2$$

A typical method would be to charge a reactor with the metal precursor; and then to mix the precursor thoroughly with sodium nitrate mixed previously in water. After mixing, the water is evaporated; and the remaining slurry is heated to 340°C to produce the crude Iridium Oxide. This product is washed to remove the salts produced in the reaction, then dried and stored to be used in characterization and evaluation (Rasten et al., 2003).

For PEM electrolyser cell applications, the catalysts need to be annealed to the surface of a glass carbon electrode, in order to be used to screen various catalyst mixtures, or to study the reaction mechanisms and kinetics by coating the electrode with a layer of catalysts. A catalyst is first thoroughly mixed with other components, such as short-chain alcohols, for example water, ethanol, or isopropanol – through agitation and/or sonication. Other solvents such as 1, 2 propanediol may be added, in order to create a formulation that can assist the catalyst to disperse to the glass carbon surface.

Sonication and agitation can increase the temperature of the catalyst mixture, in order to make the mixture more viscose, and thereby to increase the dispersion onto the working electrode. The catalyst formula is then applied to the working electrode surface and dried. In order to ensure that the catalysts particles adhere to the surface of the working electrode, it is important to add convenient additives, such as a perfluorinated isomer like DuPont's or Dow's Nafion. A second ultrathin layer could be applied on the top of the catalyst layer to enhance adhesion. It is important to dry the catalyst layer above the glass transition temperature stage of the Nafion ionomer, so that it can adhere effectively. The surface of the electrode disk itself should be inactive to the electrochemical reaction.

For example, if the catalyst particles are platinum, the surface cannot be platinum as well. This is done by ensuring that the metal used on the surface is not the same as the metal under investigation.

4. The selection of the electrolyte

A dilute aqueous acid solution, for example a 0.1 – 1M solution, is typically used as the electrolyte. The reason for this is that there is an exchange of protons between the various reactions taking place inside the electrolytic cell. Sulphuric acid has been use extensively for these types of studies, and it resembles a typical solid electrolyte used in the membrane electrode assembly (MEA) of the PEM electrolyser. However, sulphate anions can adsorb on the surface of Pt catalysts, which alters the reaction kinetics of the cell. Instead, perchloric acid can be used as an electrolyte medium, which does adsorb onto the surface of the Pt catalyst.

5. What do we need to characterize for a PEM electrolyser cell to be most effective?

In this section, the various PEM electrolyser properties are listed that may need characterization:

- Overall performance of electrolyser (J-V curves)
- Kinetic properties (electrochemically active surface)
- Ohmic properties (electrolyte conductivity, contact residence, electrode resistance)
- Mass transport properties (pressure losses, reactant/ product homogeneity)
- Catalytic structure (catalyst loading, particle size, electrochemically active surface area)

This list gives an idea of the different kinds of properties that can be used to characterize PEM electrolyser catalysts that could contribute to the overall performance and behaviour of electrolysis. How do we know on what property to focus, or which one is important to characterize? In this chapter, the focus will be on just a few of the most widely used characterization techniques, such as for example, cyclic voltammetry, linear potentiometer and rotating disc electrode techniques.

Let us start with basic electrochemical techniques to characterize the catalysts for PEM electrolysers. These could provide quantitative information about the performance of PEM electrolysers.

There are two types of electrolyser characterization:

a. *Electrochemical characterization techniques* (in situ). These techniques use variables that measure various electrochemical properties, such as potential, current and time required to characterize the catalytic behaviour.
b. *Ex situ characterization techniques*. These techniques characterize the detail of the structural properties of the catalyst.

With respect to the first type of the PEM electrolyser characterization, I will discuss three major techniques:

1. *Cyclic voltammetry* (CV). This is a sophisticated technique that provides valuable information on the kinetics of the PEM electrolyser reactions. CV in general can be time consuming, and some of the results might be difficult to interpret. This can be overcome by specialised modified methods under argon or nitrogen.
2. *Current – Voltage (J-V) measurement.* The most needed PEM electrolysis technique is a J-V measurement that provides an overall quantitative evaluation of a PEM electrolyser's performance. The J-V curves shows that, depending on different types of electrolyser, where the current starts to flow would indicate where hydrogen gas starts to be released from the electrolyser unit. This galvanic cell has a certain polarization voltage which will set off the current. A further increase in the external voltage indicates the ongoing development of the gas. The minimum voltage at which the splitting of water begins is called the decomposition voltage. For PEM electrolysers, this voltage is 1.23V. The main difference between the theoretical and the experimental voltage could be the overpotential. The overpotential is a function of the electrode material, the texture of the electrode surface, the type and concentration of the electrolyte, the current density and the temperature. In practical applications, it is the aim to minimize the overpotential. This is important, in order to achieve a very good and active electrode surface.
3. *Demonstrating hydrogen efficiency* – To ensure that the PEM electrolyser performs correctly, it would be advantageous to measure the hydrogen efficiency in the electrolyser cell.

In the area of ex situ characterization, one can discuss the following methods:

1. **Porosity determination** is an effective characterization method, due to the high porosity that the catalyst structure must have, in order to be a good and effective catalyst for PEM electrolysers.
2. **Surface area measurements** could be achieved by either the Brunauer-Emmett-Teller (BET) analysis, or by means of XRD to determine the catalysis particle size; this, in turn, can be related to the electrochemically active surface area measurements.
3. **Gas permeability:** Even if the electrodes are highly porous, it does not mean that the gas will necessarily permeate through the membrane. Therefore, understanding mass transport in PEM electrolyser electrodes requires permeability measurements, in addition to porosity measurements. Gas permeability testing is an important factor in the development of ultra-thin membranes.

4. **Chemical determination** can assist in the physical characterization of the materials used in the PEM electrolyser cell.

6. Electrochemically active surface area

An electrochemically active surface area (ECSA) (Eqn 5) can be determined from the scanning of a cyclic voltammetry. This is needed in catalyst characterization, in order to obtain critical data regarding the surface activity of the catalysts. This technique for the determination of ECSA for PEM electrolyser cells has been utilized for several decades. The ECSA of the electro-active species is calculated from the current density Q (C/cm^2 electrode) obtained from the CV experiments, the charge required to reduce a monolayer of protons on Pt Gamma = 210 $\mu C/cm^2$, and the Pt content or loading in the electrode. L is recorded in g/cm^2 electrode (Kinoshita K. & Stonehart P., 1977 & Gloagen et al., 1997).

$$ECSA = \frac{Q_{pt}}{210\,,L} \qquad (5)$$

ECSA is only an indication of the catalyst particles that participate in the reaction medium. The fraction of the catalyst that participates in the electrode reaction is given by the ratio of the electrochemical surface area to the specific area of the catalyst. This is determined by an ex situ technique, such as XRD analysis or Brunauer-Emmett-Teller (BET) analysis. This ratio is referred to as the catalyst utilization; and the higher this value, the better is the catalyst. Figure 5 shows a CV scan for Pt.

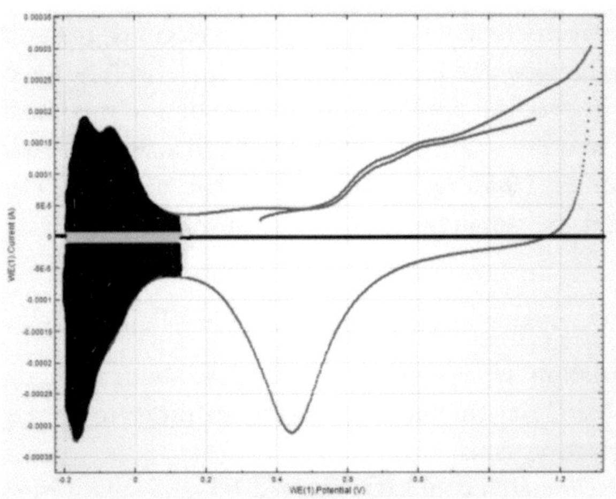

Figure 5. CV curve for a Pt catalyst. The grey areas represent the Q_{ads} and Q_{des} peak areas respectively.

This current contribution response is a non-linear response that corresponds to a hydrogen absorption reaction occurring on the electrochemically active catalyst surface. The grey areas in Figure 5 represent the Q adsorption (Q_{ads}) and Q desorption (Q_{des}) peak areas of the Pt electrolysis catalyst surface, respectively. The electrochemical active surface area can be calculated from the area under the Q_{ads} and Q_{des}.

It could be noted that a highly porous, well-made MEA electrode may have an active surface area that is several orders of magnitude larger than its geometric area (O'Hayre et al., 2002). This could be expressed as the ECSA.

Half-cell experiments are convenient and relatively fast methods of analysis for the screening of various catalysts; but they are not suitable for the evaluation of PEM electrodes in-situ, since these are the conditions for an electrolytic cell setup.

7. Kinetic measurements

Kinetic parameters can be obtained from the steady state polarization curve also called a current-voltage graph (J-V). The J-V curve shows the voltage output as a function of current density loading in a PEM electrolyser. The performance of the characteristic J-V curve can be used to calculate the Tafel slope in the low current density (high load resistance) region, where the mass and ohmic transport effects do not interfere with the data. A typical polarization curve is shown in Figure 6.

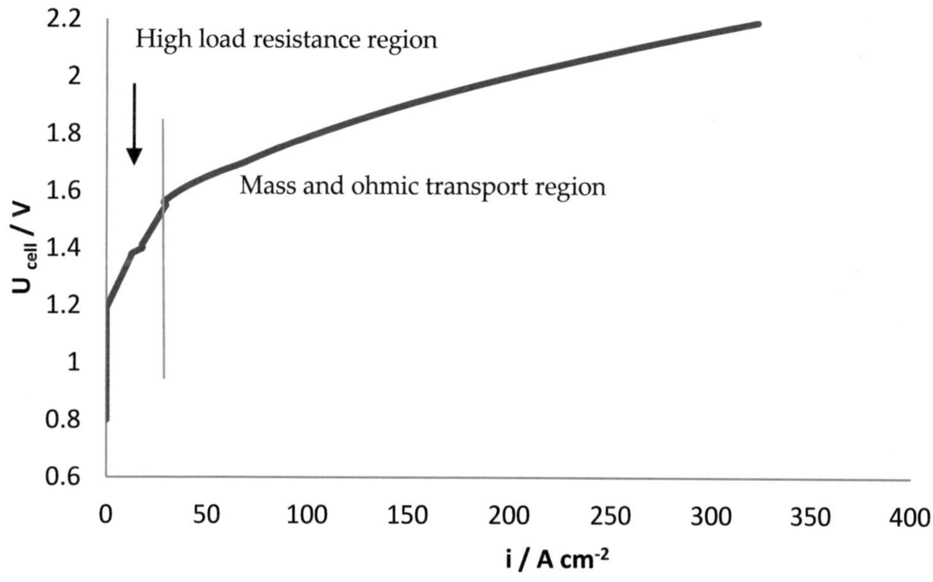

Figure 6. Typical PEM electrolyser J-V curve

By controlling the current or the potential across the electrode surface, the steady state polarization behaviour of the electrochemical reaction can be measured. The steps are performed in small increments; and the response is measured typically after 10 minutes, when the equilibrium conditions of the electrode reaction can be assumed. The J-V steady-state polarization curve includes effects of thermodynamic potentials, the overpotential due to surface reactions, ohmic losses and diffusion terms. The required potential for the electrolyzer can be expressed in the following equation (Eqn. 6).

$$V = E + n_{ohmic} + n_{act,a} + n_{act,c} \tag{6}$$

Where E is the equilibrium voltage; n_{ohmic} *is the ohmic potential across the PEM,* $n_{act,ai}$ and $n_{act,c}$ are the activation overpotentials at the anode and cathode, respectively. The overpotential concentrations are significantly small due to gas transport limitations in the thin electrode; and these can therefore be neglected. Kinetic rate can be determined directly from the electrolysis cell by measuring the current. The overall electrochemical reaction at the anode could be expressed by the Butler-Volmer equation (Eqn. 7), if it may be assumed that there are no transport limitations.

$$i = i_{A0}\,[\exp\left(\frac{\propto_A V_e - F n_a}{RT}\right) - \exp(-\frac{(1-\propto_A) v_e - F n_A}{RT})] \tag{7}$$

Where i_{A0} is the anode exchange current density [A cm^{-2}]; V_{e-} is the stoichiometric coefficient of electrons in the anode reaction; $\propto A$ is the transfer coefficient; n_A is the anode overpotential (Bockris et al., 2000). The current inputs are directly proportional to the kinetic rate of the electrochemical reaction. To determine the kinetics and the thermodynamics of electrochemical reactions in a PEM electrolysis cell can be done by means of a Tafel analysis, which is a simplified Butler-Volmer equation. The Tafel slope equation (Eqn. 8) describes the J-V steady-state polarization curve in the kinetic controlled area.

$$n^{act} = \beta \log[i] - A \tag{8}$$

n^{Act} = the voltage loss (mV) due to the slow kinetics, I = current density (mA cm^{-2}), and A and β are the kinetic constant parameters, while β, in particular, is the Tafel slope (Eqn. 9) (Wang, 2001).

$$\beta = 2.303\ RT/\alpha F \tag{9}$$

α is the transfer coefficient, ranging between a value of 0 and 1, but it is assumed that the cathode side would show a 0.5 degree of coefficiency. The Tafel slope is a graph plotted by n^{act} (E-V) as a function of the log current density; and the slope of the graph can be measured in the kinetically controlled region of the J-V polarization curve. If a high Tafel slope is obtained, this could be due to mass transport or ohmic losses in the system. Typical Tafel slopes for Ir, ranging between 30-40 mV dec^{-1} at low current densities against 120 mV dec^{-1} at high current densities (Andolfatto et al., (1993).

8. Hydrogen efficiencies

In this section, the objective of the PEM electrolyser catalyst characterization is to improve catalyst hydrogen efficiencies and to bring down the high material costs involved in these electrolytic systems. The main focus area required to bring down the material costs of PEM electrolysers could be to further develop and improve the noble metal catalysts. The following could be done to reduce the amounts of noble metals used in the PEM water electrolyser:

- Smaller amounts of PGM metals,
- Maintaining higher energy outputs with lower catalyst loadings,

- Higher hydrogen production capacity,
- Effectively increasing the lifetime of the PEM electrolyser.

In order to facilitate these improvements in today's technologies, the system must be characterized and evaluated, in order to understand and reveal the most important limitations; and further optimization of the electrode catalysts must be carried out.

Thermodynamics could be used to calculate the hydrogen efficiencies from the overall chemical reaction, by using Faraday's law (Voigt et al., 2005). At stoichiometry, water decomposes into one part oxygen for every two parts of hydrogen. This amount of electricity (n.F.E), where n= moles of the water molecule, F= Faraday's constant, and E is the thermodynamic potential associated with the decomposition of a water molecule. This equitation is referred to as Gibbs free energy ΔG_d, as shown in Eqn. 10 of the water dissociation reaction:

$$\Delta G_d - nFE = O \text{ and } \Delta G_d > 0 \tag{10}$$

Where n =2 the number of electrons exchanged during the electrochemical decomposition of water; Gibbs free energy is a function of both the operating temperature and the pressure of the electrolysis cell, and thus Eqn. 11 is:

$$\Delta G_d\,(T;P) = \Delta H_d\,(T;P) - T\Delta S_d\,(T;P) > 0 \tag{11}$$

ΔH_d (T;P) and $T\Delta S_d$ (T;P) are respectively the enthalpy change (J mol^{-1}) and entropy change (J $mol^{-1}K^{-1}$) associated with the decomposition of water. To be able to decompose one molecule of water, heat is required. The thermodynamic potential is defined according to Eqn. 12:

$$E(T;P) = \frac{\Delta G_d(T:P)}{nF} \tag{12}$$

The thermo-voltage V is defined, as described in the following equation, Eqn. 13:

$$V(T;P) = \frac{\Delta H\ (T:P)}{nF} \tag{13}$$

The hydrogen efficiency could be derived from Eqn. 14 where $n_{efficiency}$ is as follows:

$$n_{efficience} = \frac{E_{waterstof}}{E_{theoreticl}} = \frac{V_{H2} * H}{U.I.t} \tag{14}$$

H_o= Energy for hydrogen fuel = 12.745 x 10^6 J m^{-3}
V_{H2}= Volume of hydrogen produced in m^3
U= Potential in V
I= Current loading in A
T= time in seconds

According to literature studies, the PEM electrolyser cell potential is usually reported to be in the region of 2V, and the commercial type PEM electrolysers usually have an efficiency ranging from 65% to 80% (Barbir, 2005). This efficiency could increase to almost 95% at higher operating temperatures; this condition requires a lower cell voltage, which also lowers the current that passes through the electrolyte, while the hydrogen production rate increases. The problem could be overcome by utilizing a stack in the electrolyser setup to increase the efficiency.

9. Concluding remarks

This chapter discusses the difference between two electrolytic units (e.g. alkaline versus polymer electrode membrane (PEM) electrolyser units). There is a third type still under research investigation namely, the solid oxide electrolyser unit. Currently there are six major suppliers identified to produce hydrogen by an alkaline electrolytic unit and they are:

Norsk Hydro in Norway, Hydrogenics in Belguim, Iht in Switzerland, AccaGen in Switzerland, Erre Due in Italy represented by H2Indistrial in Denmark, and Uralkhimmash in Russia. In contrast to these bigger units, PEM units supplied by Proton are sized for home or small neighborhood grid systems.

This trend comes into effect because of the high capital cost associated with PEM electrolyser units, which is being pushed by expensive materials that are used to manufacture the membrane, and the costs associated with the PGM metals used as catalysts in PEM electrolyser units. This compared to the lower capital cost for the manufacturing of alkaline units. Other factors that give PEM's an advantage are that these units are safer to use when compared to alkaline units and PEM units do not utilize caustic liquids such as potassium hydroxide when compared to the solid electrolyte used in PEM units.

PEM electrolyser catalysts and MEA components can be characterized by a variety of electrochemical methods. Quick-screen cyclic voltammetry, rotating disc and linear potentiometry can be used to characterize and study the electrocatalytic behaviour. These studies can provide valuable information regarding the kinetics, the controlled potentials, and mechanisms at play during half-cell measurements of the catalysts. An electrochemically active surface area, specific activity, and electronic resistance are all needed to optimize PEM electrolysis, in order for it to operate and give optimal performance.

J-V steady-state polarization curves can provide information regarding the open circuit voltage, thermodynamic (hydrogen efficiencies) and kinetics (Tafel slopes). Typical Tafel slopes found for Ir catalyst were between 30-40 mV dec^{-1} at low current densities against 120 mV dec^{-1} at high current densities.

Hydrogen efficiencies are calculated based on Faraday's equations and give a good overall performance of the PEM electrolyser unit. Typical units reported give a potential output of about 2V with hydrogen efficiencies ranging between 65-80%. With the correct catalyst materials and electrolyser design, this efficiency can be increased to 95% at higher operating temperatures.

Author details

Shawn Gouws
Nelson Mandela Metropolitan University, South Africa

Acknowledgement

The author wishes to thank the Nelson Mandela Metropolitan University for facilitating that this research could be conducted at an Institute; for financial funding from NRF (National Research Foundation) and HYSA (Hydrogen South Africa) who both supported this programme for a number of years.

10. References

Adams R. & Shriner R.L.: (1923), Platinum oxide as a catalyst in the reduction of organic compounds. iii. Preparation and properties of the oxide of platinum obtained by the fusion of chloroplatinic acid, *Journal of American Chemical Society,* Vol. 45, Issue 9, ISSN 1520-5126 pp2171-2179

Andolfatto F., Durand R., Michas A., Millet P. and Stevens P., (1994), Solid Polymer Electrolyte Water Electrolysis: Electrocatalysis and Long-term Stability, *International Journal of Hydrogen Energy,* Vol. 19, Issue 5, pp421-427

Atlam O. & Kolhe M. (2011), Equivalent electrical model for a proton exchange membrane (PEM) electrolyser, *Energy Conservation and Management,* Vol. 52, pp2952-2957

Balaji R., Senthil N., Vasudevan S., Ravichandran S., Mohan S., Sozhan G., Madhu S., Kennedy J. & Pushpavanam S., (2011), Development and performance evaluation of Proton Exchange Membrane (PEM) based hydrogen generator for portable applications, *International Journal of Hydrogen Energy,* Vol. 36, pp1399-1403

Barbir F., (2005), PEM electrolysis for the production of hydrogen from renewable energy sources, *Solar Energy,* Vol. 78, pp661-669.

Bockris J. O'M., & Reddy A.K.N., (2000), *Modern Electrochemistry,* Kluwer Academic / Plenum Publishers, New York.

Cheng J., Zhang H., Ma H., Zhong H. and Zou Y., (2009), Preparation of Ir0.4Ru0.6MoxOy for oxygen evolution by modified Adam's fusion method *International Journal of Hydrogen Energy,* Vol.34, pp6609-6613.

Gloaguen F., Leger J-M. and Lamy C., (1997), Electrocatalytic Oxidation of Methanol on Platinum Nanoparticles Electrodeposited onto Porous Carbon Substrates, *Journal of Applied Electrochemistry,* DOI:10.1023/A:1018434609543, Vol.27, pp1052 - 1060.

Hutchings R., Muller K., Kotz R. & Stucki S., (1984), A structural investigation of stabilized oxygen evolution catalysts, *Journal of material sciences,* DOI:10.1007/BF00980762, Vol. 19, p3987-3994

Ivy J,; (2004), Summary of electrolytic hydrogen production, *NREL/MP-560-36734, p4*

Kinoshita K. & Stonehart P., (1977), Preparation and Characterization of Highly Dispersed Electrocatalytic Materials, In. *Modern Aspects of Electrochemistry,* Bockris J.O.M. and Conway B.E., Eds., and Plenum Press: New York, Vol. 12, and Chapter 4, pp183-266

Kissinger P.T. & Heineman W.R., Large-Amplitude Controlled-Potential Techniques, In. *Laboratory Techniques in Electro-analytical Chemistry,* Kissinger P.T. & Heineman W.R., 2nd Ed, Marcel Dekker, Inc, New York, ISBN,0-8247-9445-1, p81.

Kreuter W, and Hofmann H, (1998), .Electrolysis: the important energy transformer in a world of sustainable energy, *International Journal Hydrogen Energy,* Vol. 23, Issue 8, pp 661-666.

Kumpulainen H., Peltonen T., Koponen U, Bergelin M., Valkiainen M. and Wasberg M., (2002) In-situ volumetric characterization of PEM Fuel Cell Catalyst layer, ESP00

Marshall A., Borresen B., Hagen G., Tsypkin M. and Tunold R., (2007), Hydrogen production by advanced proton exchange membrane (PEM) water electrolysers – Reduced energy consumption by improved electrocatalysis, *Energy*, Vol 32, pp431- 436.

Marshall A., Borresen B., Hagen G., Tsypkin M. and Tunold R., (2004), Nanocrystaline $Ir_xSn_{1-x}O_2$ electrocatalysts for oxygen evolution in water electrolysis with polymer electrolyte – effect of heat treatment, *Journal New Material electrochemistry Systems*, Vol 7, pp197-204.

Marshall A., Borresen B., Hagen G., Tsypkin M. and Tunold R., (2005), Preparation and characterization of nanocrystalline $Ir_xSn_{1-x}O_2$ electrocatalytic powders, *Material Chemical Physics*, Vol 51, pp226-232

Marshall A.T., Sunde S., Tsypkin M. & Tunold R., (2007), Performance of a PEM water electrolysis cell using IrxRuyTazO2 electrocatalysts for the oxygen evolution electrode *International Journal of Hydrogen Energy*, Vol. 32, pp2320-2324

Millet P., (1994), Water electrolysis using PEM technology: electrical potential distribution inside a Nafion membrane during electrolysis, *Electrochim Acta*, Vol. 39, Issue: 17, pp2501-2506.

Millet P., Andolfatto F. & Durand R., (1996), Design and performance of a solid polymer electrolyte water electrolyser, *International Journal of Hydrogen Energy, Vol.* 21, Issue 2, pp87-93

Millet P., Ngameni R., Grigoriev S.A. & Fateev V.N., (2011), Scientific Engineering issues related to PEM technology: Water Electrolysers, Fuel Cells and unitized Regenerative Systems, *International Journal of Hydrogen Energy*, Vol. 36, pp4156-4163

O'Hayre R., Lee S-J., Cha S-W. & Prinz F. B., (2002) A sharp peak in the performance of sputtered platinum fuel cells at ultra-low platinum loading, *Journal of Power Sources*, Vol. 109, pp483-493.

Oberlin R. and Fisher M., (1986), Status of the membrane process for water electrolysis, In. Hydrogen energy progress VI, proceedings of the sixth world hydrogen energy conference, Veriroglu T., Getoff N. and Weinzeirl P., Oxford, Pergamon Press, pp 333-340.

PSO-F&U,; (2008), Pre-Investigation of Water electrolysis, *NEI-DK-5057*

Qi Z., (2008) Electrochemical methods for catalyst activity evaluation, In., *PEM Fuel cell electrocatalysts and catalyst layers*, Zhang, J, Springer, ISBN 978-1-84800-935-6, Spain

Rasten E., Hagen G. and Tunold R., (2003), Electrocatalysis in water electrolysis with solid polymer electrolytes, *Electrochim Acta*, Vol. 48, pp3945-3952

Song S, Zhang H., Ma X., Shao Z., Baker R.T., Yi B., (2008), Electrochemical investigation of electrocatalysts for the oxygen evolution reaction in PEM water electrolysers, *International Journal of Hydrogen Energy*, Vol. 33, pp 4955-4961.

Trasatti S. (1991) Physical electrochemistry of ceramic oxides, *Electrochim Acta*, Vol.36, pp 225-241

Tseung A.C.C., & Jasem S., (1977) Oxygen evolution on semiconducting oxides, *Electrochim Acta*, Vol. 22, pp31-34

Van Ruijven B.J., *(2007)*, The potential role of hydrogen in energy systems with and without climate policy, *International Journal of Hydrogen Energy*, Vol. 32, Issue 12 pp 1655-1675

Wang X., Hsing I.M. and Yue P.L. (2001), Electrochemical characterization of binary carbon supported electrode in polymer electrolyte fuel cells, *Power Sources*, Vol.96, pp282-287

Zoulias E.,Varkaraki E., Lymberopoulos N., Chritodoulou C.N. & Karagiorgis G.N., (2004) A review on water electrolysis, *TCJST*, Vol. 4, Issue 2, pp 41-71

Section 2

Industrial Electrolysis

Chapter 6

Direct Electrolytic Al-Si Alloys (DEASA) – An Undercooled Alloy Self-Modified Structure and Mechanical Properties

Ruyao Wang and Weihua Lu

Additional information is available at the end of the chapter

http://dx.doi.org/10.5772/52962

1. Introduction

Aluminum became attractive only after the invention of Hall-Heroult electrolysis process in 1886. In the earlier part of last century, the usage of aluminum products was restricted in decorative parts. After World War II, a dramatic expansion of the aluminum casting industry occurred. Many new alloys were developed to comply with the engineering requirements. Among the commercial aluminum alloy castings, Al-Si alloy is the most commonly used and constitutes 85-90% of the total aluminum cast parts produced. Al-Si alloys containing silicon as the major alloying element offer excellent castability, good corrosion resistance and machinability. Small amounts of Cu, Mg, Mn, Zn and Ni are being added to achieve strengthening of Al-Si alloys.

Al-Si alloys have been made for a long time by simply adding crushed silicon metal or a high-silicon aluminum base master alloy to molten aluminum in reduction cell or smelting furnace. In those processes pure silicon and aluminum are needed, and both metals are reduced from oxides in electrolytic cell. The idea of direct electrolytic reduction of silica dissolved in the cryolite bath in electrolytic cell has been developed at the end of nineteenth century. The idea to produce alloys in electrolytic process is not new. For several years before Hall-process the Cowles process, by which Cu-Al alloys in range of 30-40% Al were directly reduced from a mixture of Al_2O_3 and CuO or Cu by electric arc at high temperature, was used [1].

1891 Menit firstly conducted the experiment to reduce the silica to silicon metal in Hall cell. In 1911 Frilley [2] achieved the production of Al-Si alloys containing less than 5% silicon by direct electrolytic reduction of alumina-silica and 5-96% silicon by aluminum-thermal reduction in laboratory. Frilley also obtained Mn-Si, Cr-Si, Fe-Si, Cu-Si and Si-Ni in electrolytic cells. Moreover, he found that the silicon appearance in Al-Si alloy with less than 10% Si was very fine and different from the existed alloy, but no attention had been paid on the change of structural characteristics of silicon due to limited usage of aluminum in industry at that time. Fridley's discovery revealed that electrolytic process is a powerful potential measure to improve the quality of alloy.

In the middle of last century a number of works had been reported to electrolyze Al-Si alloy in Hall cells, to which pure silica, quartzite containing more than 99% SiO_2 [3], sand stone with about 90% SiO_2 [4] glass scrap having 72% SiO_2 [5]. bauxite with 11%SiO_2 [6], sand and clay [7] were added. Recently the refractories from spent potlining were successfully introduced to alumina reduction cells to produce Al-Si alloys [8]. As well known, the purity of molten aluminum is of major concern in electrolytic reduction process. The impurity is considered as a negative factor, deteriorating operation conditions. Hence, the direct electrolytic reduction of silica in Hall cell is a difficult process. There are two severe problems related with silica added into molten cryolite, in which silica must be easily dissolved. One of them is how to compensate for alumina generated by the reaction of aluminum with the added silica for achieving a desired chemical composition of alloy. Other is that direct addition of silica or other silicates often results in the formation of the heavy ridges of silicate along the bottom of the cell, as a result the cell becomes inoperable, so limiting the size and placement of the ridge is a major concern in production. In 1970s C. J. McMinn and A.T. Tabereaux [9, 10] provided a procedure to strictly control the feed of alumina and silica into the cell, stabilizing the electrolytic process and successfully producing Al-Si alloys with up to 16%Si in Hall cell. However, they viewed this process to be economical when the price of silicon greatly increases. Production of Al-Si alloys in electrolytic reduction cell had not found industrial application.

Since 1970s many works have been carried out on direct electrolytic production of Al-Si alloys(DEASA) in China [11]. Most Chinese bauxites contain high content of silica, titania and small amount of rare earth oxides. It is very difficult to extract the pure alumina from bauxite by the Bayer process [12]. In electrolytic process the charge is composed of bauxite, from which the iron oxide is removed, and alumina, using which to regulate the proportion of bauxite added into salt bath in terms of the desired chemical composition of Al-Si alloy. Note that bauxite tested is easily to be dissolved into molten electrolyte compared to the commercial bauxites. It would be an important factor to successfully produce Al-Si alloys in alumina reduction cells. At the end of last century several thousand tons of DEASA ingots containing Si content from 6% to 12% have been used in foundries to produce car parts such as engineering block and head, wheel and piston [12- 14]. Table 1 lists the chemical compositions of some DEASA ingots, which contain higher level of impurities such as Na, Sr, Ti and rare earth elements compared to commercial alloys. Undoubtedly it is related with bauxite composition. .

Since 1980s author has focused attention on the microstructure of DEASA and its mechanical properties [15]. It has been found that the microstructure and fracture surface of

DEASA ingots are very fine and similar to impurity-modified Al-Si alloy. Hence this phenomenon is characterized as self-modification due to no impurity- modifier added. The further research indicated that self-modification is attributed to the eutectic undercooling during solidification of DEASA. To answer the question why self-modified microstructure occurs and how it links with the electrolytic process, we must discuss some events related with electrolysis process. This chapter restricts the consideration into the structural characteristics of alloys and its original, which is related with electrolytic process. The details of electrolysis process can be referred to References [11,12].

Alloy*		Si	Cu	Mg	Mn	Ni	Zn	Fe	Cr	Ti	Na	RE	Sr
EZL101	top	7.9	<0.01	—	0.01	<0.01	0.01	0.25.	<0.01	0.33	—	0.002	0.001
No1	bottom	8.2	<0.01	—	0.01	0.01	0.01	0.24	<0.01	0.35	—	0.002	0.001
EZL101 No2		7.3	<0.01	0.36	0.01	0.02	<0.01	0.11	—	0.11	<0.0001	—	0.001
ESi 9**	No1	9.5	<0.02	0.010	0.060	0.15	0.03	0.65	0.02	0.48	0.0045	0.038	0.0034
	No2	9.2	<0.02	0.02	0.005	0.12	0.02	0.44	0.18	0.66	0.014	0.037	0.0026
EZL102		12.2	<0.01	0.15	0.005	—	<0.01	0.50	—	0.12	—	—	—
ZL108		11.60	1.95	0.65	0.62	0.30	—	0.25	—	0.20	0.0020	—	0.000
EZL109		12.1	<0.01	0.91	0.01	0.81	<0.01	0.25	—	0.09	0.0023	—	0.000
ZL101 (A356)		6.7	—	0.39	0.01	0.005	—	0.06	0.016	0.12	—	0.0005	0.002

E is abbreviation for electrolysis. ZL represents "cast aluminum alloy" in Chinese:

* Alloy designations are to Chinese specification.

** Alloy mark representing an electrolytic Al-9% Si-0.5%Ti alloy.

Table 1. Chemical analysis of DEASAs ingot wt%

2. Behavior of alloy melt in electrolytic process

The electrolysis cell runs at around 950°Cwith a voltage drop of 4.5-5.5 V across each cell[11]. The bauxite, from which iron oxide is removed, contains SiO_2, TiO_2, Fe_2O_3, Na_2O, CaO and rare earth oxides (RExOy) ,besides the Al_2O_3. During electrolysis process those compounds are reduced to Al, Si, Ti, Fe, Na, Ca and RE, respectively, which in atomic form continuously remove from electrolyte to the carbon bottom of the pot, forming a homogeneous Al-Si alloy melt with several impurities, as shown in Tab.1. Then the melt is siphoned out of the reduction cell at 24h intervals and held in a 10 ton insulated metal-mixer for homogenizing the composition, then poured into ingot mould with dimension of 100x60 x600mm^3 and weight of 10kg, without any impurity-addition or treatment. Hence, there are four factors i.e. homogeneous melt, superheating, impurity and electric field (current density and anode potential), influencing the structure of DEASA melt and its crystallographic characteristics and properties in solid state.

In many years a lot of studies have been done on the structure of liquid metals, including Al-Si alloy. The liquid metals can be considered as a system composed of ions and electrons, which are moving through the disordered liquid [16-18]. Below we discuss how superheating and electric field change the structure of Al-Si melt and its crystallization

2.1. Effect of superheating on the crystallographic characteristics of Al-Si alloys

As well known melt superheating is a powerful factor influencing the microstructure and properties of commercial Al-Si alloys. The effect of superheating is associated with the temperature, holding time and cooling rate during solidification [19-24]. In 1990s many researchers [21,25] investigated the regularity of variety of viscosity and density of family of Al-Si alloy in liquid state with temperature, revealing that as temperature exceeding about 1000°C, these physical properties dramatically change. Therefore, they suggested that for the near eutectic Al-Si alloy containing 10-14% Si there is a critical temperature in range of 1050-1150°C, as shown in Fig.1, above which the silicon grains and other heterogeneous substances such as iron-rich particles are dissolved in melt, resulting in a homogeneous melt, which will change the crystallographic characteristics of alloy. This event has been proved by recent studies. At the beginning of this century X.F. Bian et al studied Al-13%Si alloy melt heated in the temperature range of 625-1250°C using high temperature X-ray diffractometer [22] and reported that when increasing the temperature to 875°C a sudden change of the atomic density and the coordination number of the Al-13%Si alloy melt occurs, demonstrating that the liquid structure has changed, which is caused by dissolving of Si-Si clusters into aluminum melt. In other study it has been found that at the temperature of about 1050°C the electrical resistivity of hypereutectic Al-16%Si alloy melt steeply changed and hereditary effect of different original structure can be eliminated after remelting, indicating that the change of liquid structure happened at temperature of 1050°C[26]. Hence Al-Si alloy melt at high temperature consists of two ion groups: Al-Si and Si-Si groups, which appear to consolidate the short-range order and the electrons are moving through the disordered melt [27-29]. Based on the experimental results P.J.Li [23] considered that in homogeneous Al-Si alloy melt the size of Si-Si and Al-Si micro-heterogeneous clusters range is from 10 to 100Å. M. Singh reported that in Al-Si alloys either hypoeutectic or hypereutectic silicon is present as silicon cluster essentially with the size of about 50-70 Å [27]. Moreover, as increasing the temperature, the size and number of ion groups simultaneously decrease.

P.C. Popel et al [21, 23] studied the influence of superheating on the crystallographic characteristics of alloys and revealed that superheating Al-Si alloy shifts its eutectic reaction toward higher level of silicon accompanying with the appearance of Al-dendrite. As temperature is higher than 780°C, eutectic silicon becomes finer with the fine α-Al dendrite. When heating temperature is in range of 900-1000°C, the size of silicon flake is less than 7μm. It is interesting that heating at temperature higher than 1000°C the modified silicon appears in eutectic alloy. The heating at 1000°C is capable of eliminating the occurrence of primary silicon and refining α-Al dendrite in Al-17%Si alloy. But when superheating hypereutectic Al-20%Si alloys at 950°C the primary silicon particles become finer [24, 30] . The higher the temperature, the finer the silicon grain. It would be expected that a higher superheating temperature is required for hypereutectic Al-Si alloys having higher silicon content to achieve a complete eutectic structure. It is worthy of note that if the holding time is insufficient to dissolve all the silicon particles present in original alloy, even the superheating at 1200°C does not significantly change the crystallographic characteristics of alloy, and the modified structure does not appear [31].

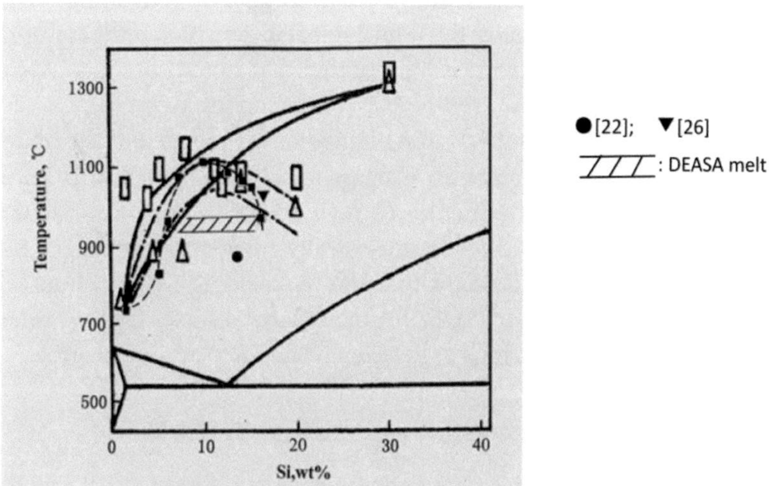

Figure 1. The dome of decay of metastable colloidal microheterogenity in Al-Si melts [21,23]

For hypoeutectic alloys as temperature rises to 950°C the dendrite arm spacing(DAS) steeply decreased and the dislocation density in α-Al dendrite increased. Moreover, the eutectic silicon tended to a fine fibrous structure [32].

Overheating significantly increases the content of silicon, magnesium and iron in α-Al-dendrite in hypoeutectic alloy [33]. As overheating Al-8%Si alloy at temperature of 950°C for 10min silicon and magnesium content solved in α-Al-dendrite increases to 1.9% and 0.3%, respectively, much higher than their solubility in Al-matrix at room temperature. Undoubtedly, overheating is one of the factors strengthening the mechanical properties of alloys.

Superheating also prompts the morphological variety in iron-bearing compound in alloys [34]. As heating Al-7%Si-Mg alloy at temperature higher than 800°C, AlSiFe compound appeared in Chinese script form instead of coarse needle-like shape, increasing the impact strength of alloy. It is apparent that superheating is a powerful mean greatly affecting the feature of microstructure in Al-Si alloys.

It is worth noting that the overheating effects on the change in structure significantly depends upon the cooling rate in freezing in alloy [23, 35]. For hypereutectic Al-17%Si alloy even heated at temperature in the range of 1000-1050°C the primary silicon grows faceted in sand castings, where the freezing rate is less than 10°C/sec. By contrast the formation of more equiaxed, nearly globular silicon crystal can be observed if the melt is quenched with the cooling rate of higher than 100°C/sec.

The reason why superheating leads to a change in crystallographic characteristics of alloys is associated with the undercooling generated by a variety of structure in molten alloy, where the size and number of Si-Si clusters acting as a nuclei of eutectic silicon in solidification of alloy greatly affect the crystallization of alloy[36]. Higher superheating decreases Si-Si cluster in size and amount, depressing the liquid-to-solid transition temperature, as a result a deep undercooling ocurs. A.Y.Gubinko[37] reported that superheating an Al-Si alloy melt to 100°C above its liquidus temperature offers an undercooling twice as great as for a melt superheated 35°C. The higher the superheating temperature, the greater the undercooling in freeze of alloy. Note that temperature in electrolysis cell is about 950°C lower than the critical temperature, above which structure of melt transits from microheterogeneous to homogeneous state (Fig.1) and DEASA is intrinsically homogeneous due to its reduction from oxides. Hence, it would be thought that the overheating in reduction cell does not affect the structure of melt, but holding DEASA melt in metal-mixer for long time causes the structural transition from homogeneous to heterogeneous state in some degree.

2.2. Role of electric field in the crystallization of Al-Si alloys

Over the past decades a lot of studies relating with the effect of electric field on the structure and properties of Al-alloys have been carried out [38-42]. Electric field either continuous current or pulse electric discharge deeply affects crystallographic character of alloy and its properties. In this chapter we will only focus our attention on the effect of direct current, which is related with electrolysis process.

By introducing the direct current into molten Al-10%Si alloy at 740°C for treating time of 50min, H.Li et al studied the effect of different current density on the structure and mechanical properties of alloy [43]. It was found that the electric field causes a morphological transition of eutectic silicon from flake to fibrous shape, accompanying with the reduction of second α-Al dendrite space. As increasing the electric current density to 30A/dm^2 silicon phase grows modified and finally primary Al-dendrite appears in near-nodular shape. As a result the elongation of alloy was raised by 100% and its tensile strength was improved by 15%. It is interesting that an increase in current density leads a rise in undercooling in freezing of alloy as shown in Fig.2. When increasing current density to 100A/dm^2 a deep undercooling of 15°C occurs, then undercooling grows slowly with current density. Undoubtedly, the deep undercooling is the reason of the change in morphology of silicon particles.

L. G. Huang et al introduced direct-current into melt poured into mould during solidification and investigated the effect of current density on the structural feature of Al-4%Si and Al-10%Si alloys, which were firstly heated at 700°C. It was found that silicon became finer with direct-current density and reached a limit as the density is increased to 283A/dm^2 and the size of α-Al dendrite arm space (DAS) also reduced with a minimum at density of 325A/dm². It is interesting that the effect of alternating current is the same to direct- current [44]. B.A. Timchenko et al [45, 46] studied the effect of high direct-current density (100-10 000A/dm²) on the quality of casting made of eutectic Al-12%Si alloy. When a large current is passed through alloy during its solidification, the solubility of silicon in α-Al matrix is raised to 20%, and its distribution becomes more homogeneous with a reduced size of silicon particles. In addition, the mold filling ability (fluidity) of casting alloys is greatly improved accompanying with a less tendency to gas porosity and shrinkage. As a result the tensile strength and hardness are increased by 10%. Recently A. Prodhan [47] reported that molten eutectic Al-12.16%Si alloy, which firstly was superheated to 750°C, can be degassed by direct- current treatment during solidification (semisolid state). The initial hydrogen level in alloy made from the ingot is about 2.5ppm, and under current treatment within 10min the hydrogen content is reduced to near 1.7ppm, which is necessary for producing a casting without porosity [47]. However, a large current density will cause an increase of hydrogen concentration. It is obvious that electric field, which is introduced into melt at more or less higher temperature or during solidification, improves the casting properties with an increase in mechanical index. This is attributed to the structural rearrangement of alloy melt generated by electric field. However, we are unable to clarify how the electric field affects the properties of DEASA melt due to the absence of experimental results at high temperature of above 900°C.

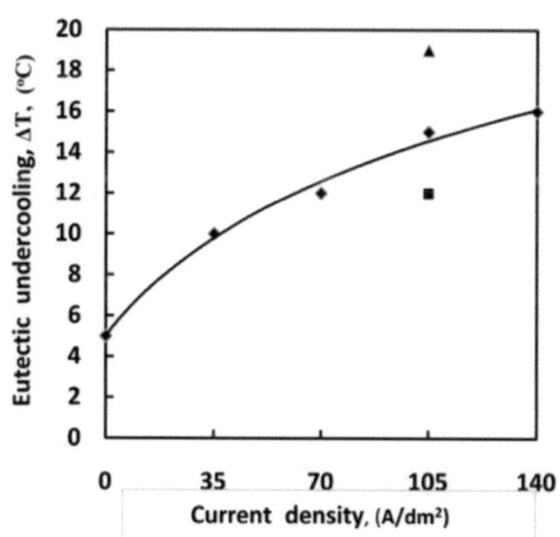

◆[22]; ■EZL101;▲EZL109

Figure 2. Eutectic undercooling in freezing against current density in Al-Si melt.

The major effects induced by electric field on the behavior of alloy melt include Joule's effect and electron-transport. Electric current causes the input of heat due to Joule's heating effect, which leads to an increase in solidification time, resulting in the improvement of fluidity of alloy, and hence the reduce of shrinkage and porosity[47]. Obviously, Joule's heating effect doesn't affect the properties of DEASAs, which solidify without electric field in present study.

In electrolysis process Al-Si alloy melt is ionized to macroscopic homogeneous Al^{3+} and Si^{4+} ions, and conduct electrons, which are moving through the melt. Under electric potential, positive ions migrate to the cathode and the electrons move toward the anode. The so-called electron- transport, which depends on ionization potential of constituent elements and its mobility in the applied field, is most important factor that reduces the solute distribution co-efficient and influences the rearrangement of elements on the solid-liquid alloy boundary during solidification, therefore reducing the constitutional undercooling and changing the crystallographic behavior of alloys. Under the current potential, the conduction electrons surrounding the aggregates of Si^{+}-rich or Al^{+}-rich groups are readily to be transferred to unlike atoms, making the groups unstable[17,28]. When the electron-drag applied to the ions, the unstable groups either Si^{+}-rich or Al^{+}-rich are capable of splitting into smaller one. The smaller Si^{+}-rich aggregators, which act as a nuclear center of silicon phase in Al-Si alloys as reported in reference [36] will promote a larger undercooling in eutectic reaction, which strongly change the crystallographic characteristic of silicon phase. Our data showed that compared to commercial unmodified Al-Si alloys eutectic arrest temperature in DEASA ingot drops to about 15-18°C[15], which is sufficient to modify the microstructure of Al-Si alloys either eutectic[48] or hypereutectic.

Summarizing the experimental results in literatures mentioned above, it is apparent that the effect of overheating on the microstructure and properties of Al-Si alloys is more or less same as electric current that leads the same variety in arrangement of melt in some degree, resulting in a large undercooling in solidification of alloy. It is worth noting that the structure of liquid DEASA is homogeneous in electrolysis process, and therefore, the effect of both of superheating and current field is weakened compared to the existing alloys. It is thought that in the electrolysis process the combination of both factors (superheating and electric field) provides Al-Si alloys a circumstance, where the ability of melt to stabilize the homogeneous structure is enhanced, hence the morphological transition of constituents of DEASA easily undergoes either under lower cooling rate during solidification or upon remelting compared to the common alloy. DEASA is an excellent undercooled alloy, of which the crystallographic behavior is same to alloy treated with electric field at high temperature and rapid cooling rate during solidification. This inference has been evidenced in present and previous studies [14,15, 49].

3. Crystallization feature of DEASA

3.1. Morphology of silicon phase and its inheritance upon remelting

As well known, silicon is the major alloying element in Al-Si alloys and its morphology is primary important factor affecting the mechanical properties, castability, machinability and other physical properties. In 1950s it has been found that for Al-Si alloy the growth of silicon crystal is temperature-dependent and dictated by the undercooling in freezing [50]. Since then a number of investigations have been done to clarify the relationship between its morphology and undercooling in solidification [48,51-53]. In general, an eutectic temperature undercooling of 6-8°C is necessary for appropriate modification for hypoeutectic or eutectic Al-Si alloys. If the combination of undercooling induced by cooling rate during solidification and modifier is below the critical value, an unmodified structure is obtained.

The relationship between temperature / undercooling in freezing and morphological transition including eutectic, primary silicon and aluminum phase in Al-Si alloys containing different silicon content can be described in quasi-equilibrium Al-Si diagram (Fig.3.)[54].

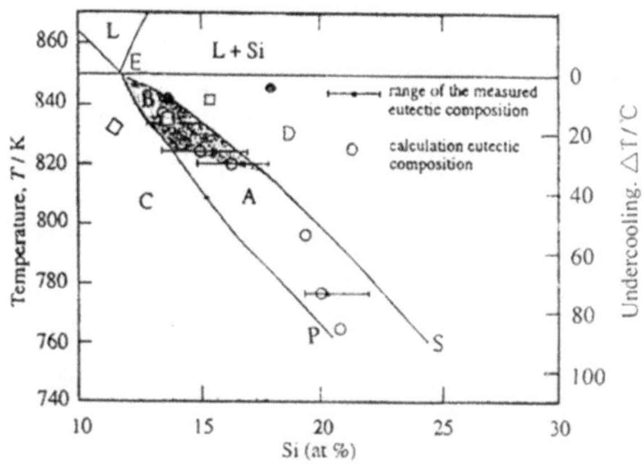

A,B: Quasi-eutectic zone; C:Al-dendrite + eutectic; D:Primary silicon + eutectic;

◇: Al-dendrtic + eutectic in present work;
□: Coupled eutectic in present work;
●: Primary Si + eutectic in present work.

Figure 3. Quasi-eutectic zone in the Al-Si system.[54]

Compared to equilibrium diagram, where eutectic reaction runs at a constant temperature and silicon content, the region of formation of quasieutectic structure exists, i.e.in a wide range of temperature/undercooling and silicon content the eutectic structure can be observed. For hypereutectic Al-Si alloys with an increase in silicon content the region shifts towards higher silicon concentration and depresses the eutectic temperature, implying that a higher undercooling is required to produce quasieutectic structure and, meanwhile, the silicon content in quasieutectic is much more than equilibrium. Whether hypereutectic alloy displays a quasieutectic structure or quasieutectic plus primary silicon grain depends upon undercooling. Obviously, the microstructure of eutectic alloy composes of eutectic plus primary α-Al dendrite in casting condition. On the other hand for hypoeutectic alloys due to eutectic shift toward higher silicon content the volume fraction of primary aluminum dendrite increases compared to the equilibrium Al-Si diagram with same silicon content, whereas with undercooling the volume fraction of Al-dendrite increases. In general, using the quasi-diagram the variety in crystallographic feature of Al-Si alloy with different silicon level and undercooling / temperature can be clearly explained.

In order to reveal this relationship between the crystallographic feature in DEASA and undercooling in freezing we observed the microstructure of DEASAs containing silicon content in the range 6- 18% and measured their cooling curves during solidification. Chemical analysis is listed in Table 1 and 2. The samples of eutectic (EZL102, EZL108 and EZL109) and hypoeutectic (EZL101 and ES9) alloys were cut from the center ingots. Hypereutectic alloys (EZL14, EZL16 and EZL18), of which the charge was composed of DEASA (EZL108)(Tab. 2.) ingot and Al-30%Si master alloy along with other master alloy additions., were melted in a 2 kg graphite crucible in an electric resistance furnace and heated to 850°C. After melting (Note: it is 1st remelting for EZL108) the molten alloy was held for 15 min to homogenizing the composition, then poured into a metallic mold, preheated to 250°C to form a casting 40x50x120mm^3 as shown in Fig.4. Pouring temperature is about 740°C for all alloys tested.

All tested alloys with different silicon content were repeatedly remelted to produce the unmodified structure with measured undercooling. This promotes to reveal the effect of undercooling on the structure in DEASA. Metallographic specimens were cut from the interiors of the casting near the site of a chromel-alumel thermocouple (Fig.4), by which the cooling curve was recorded. The cooling rate during solidification was about 1.0°C/sec.

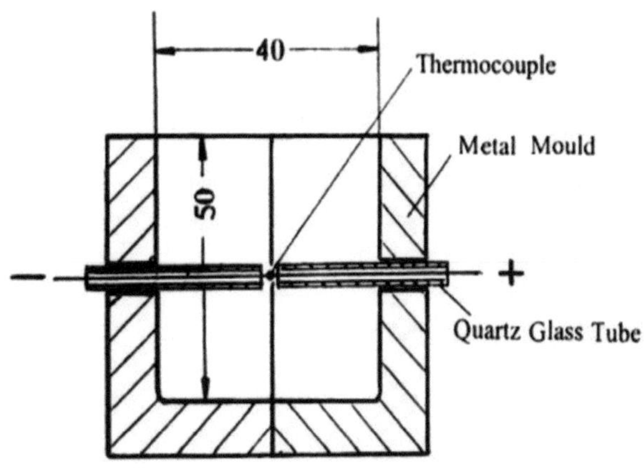

Figure 4. Mold and thermocouple

Alloy	Si	Mg	Cu	Mn	Ni	Fe	Ti	Sr	Ca	Zr	Remark
EZL108	11.60	0.65	1.00	0.60	0.25	0.25	0.10	<0.000	0.001	0.0070	DEASA
EZL14	13.70	0.55	0.80	0.31	<0.05	0.25	0.03	<0.0006	<0.001	0.0010	D*+AS30**
EZL16	15.70	0.59	1.00	0.34	<0.02	0.35	0.03	0.0006	0.001	0.0072	D+AS30
EZL18	17.60	0.39	0.75	0.25	<0.05	0.35	0.04	<0.0006	<0.001	0.0017	D+AS30

D: DEASA; AS50: Al-30%Si master alloy.

Table 2. Chemical Analysis of DEASA tested

As well known, the alloying elements such as Mg, Cu, Mn, Ni, Fe and Zn lower the eutectic arrest temperature, T_E, in Al-Si alloy [55-57]. In general the following equation (1) is used to estimate the change of T_E in commercial alloys where the total of %Al +%Si is high, near 99% [57, 58].

$$T_E = 577 - \left(12.5 \;/\; \%Si\right)\begin{bmatrix} 4.43\left(\%Mg\right) + 1.43\left(\%Fe\right) + 1.93\left(\%Cu\right) + 1.7\left(\%Zn\right) \\ + 3.0\left(\%Mn\right) + 4.0\left(\%Ni\right) \end{bmatrix} \tag{1}$$

In present work the estimated eutectic arrest temperatures, T_E,, range from 569°C to 573 °C depending upon the composition of alloys tested. Thus the undercooling, $\triangle T$, will be

$$\Delta T = T_E - T_E' \tag{2}$$

where T_E' is the measured eutectic temperature for given alloy.

Microstructure of eutectic DEASA (EZL102, EZL108 and EZL109) ingot is shown in Fig.5-8. A high volume proportion, 43-50%, of primary aluminum dendrite, which distributes evenly in modified eutectic matrix, can be found. and the eutectic undercooling is higher than 18°C that is significant different from the commercial eutectic alloy and similar to the impurity-modified alloys although the silicon content is just near to the eutectic composition.

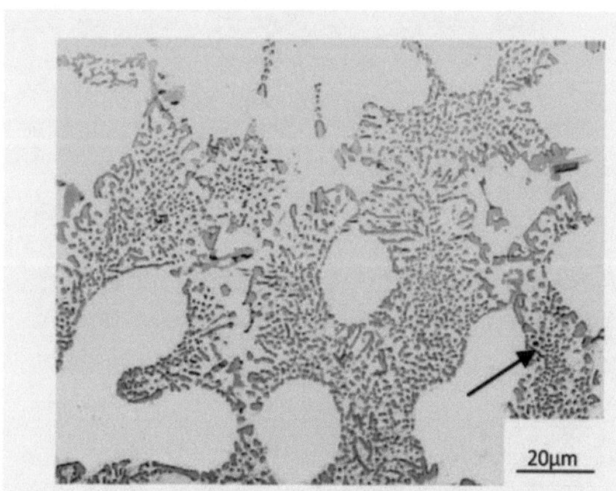

Figure 5. Optical micrograph of EZL102 ingot, showing as-cast self-modified structure. A few iron-rich crystal appears as a fine flake form as indicated by arrow.

The DEASA hypereutectic alloys, which contain 14% and 16%Si solidify with a completely modified eutectic microstructure (Fig.9 and 10) with undercooling of 12°C and 9°C, respectively. For the alloys with silicon content more than 17% (EZL18) the microstructure exhibits the coarse primary silicon crystals well distributed throughout the unmodified matrix as seen in Fig.11. In this case the eutectic temperature reached 568°C with undercooling of 5°C. In the hypoeutectic electrolytic Al-7%Si ingot the volume proportion of α-Al dendrite reach 72% accompanying with modified eutectic silicon phase and undercooling of 12°C similar to

Sr-or Na-modified Al-7%Si alloy (Fig.12). As increasing silicon level to 9% the fine silicon grows in modified mode with a high volume percentage of α-Al dendrite of 60% (Fig.13).

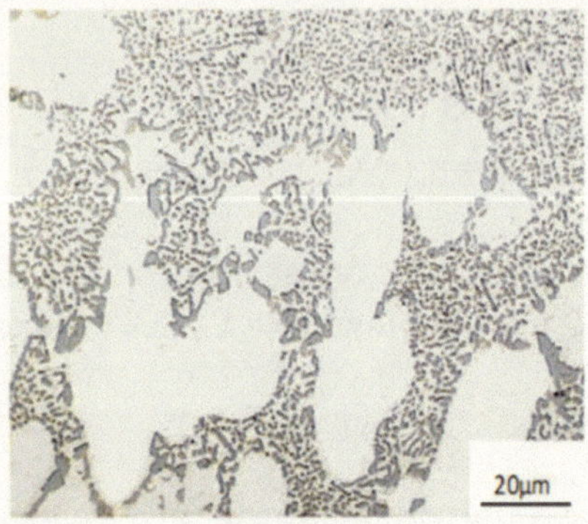

Figure 6. As-cast micrograph of EZL108 ingot, revealing self-modified structure. Optical

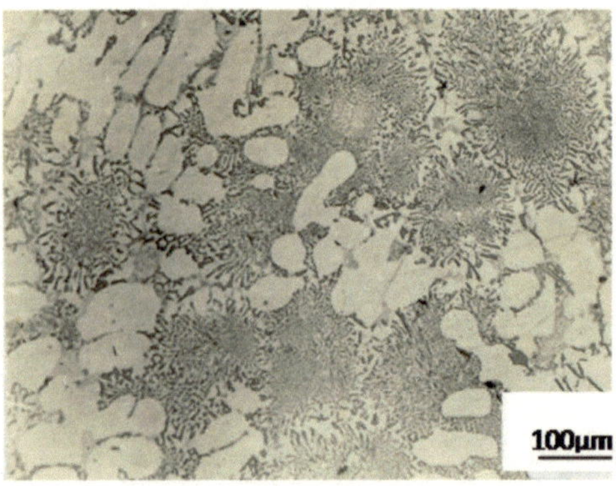

Figure 7. Microstructure in the top region of the electrolytic EZL109 ingot. The equiaxed coarse Al-Si eutectic cell appears, in which silicon grows in modification manner. Optical

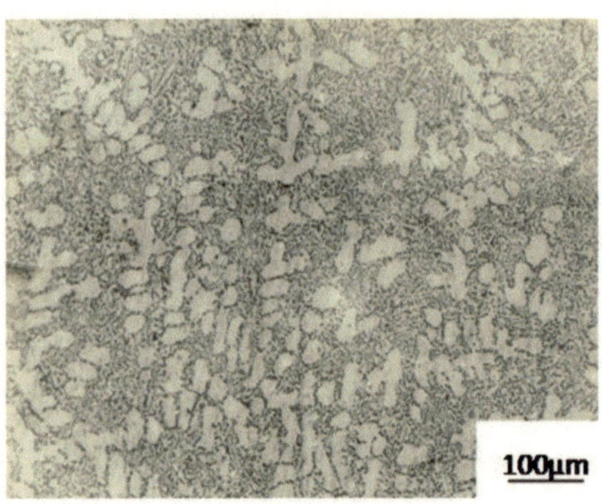

Figure 8. Optical micrograph in the bottom part of electrolytic EZL 109 ingot, indicating the fine self-modified structure.

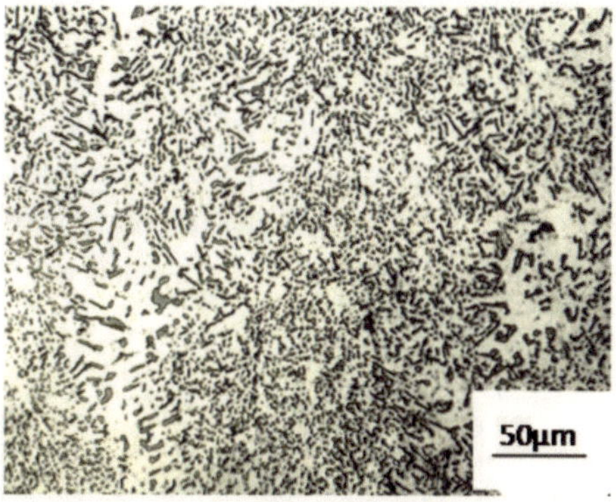

Figure 9. Self-modified microstructure of hypereutectic DEASA(EZL14),showing complete eutectic structure. On the boundary of eutectic cell some silicon flake can be observed. Optical.

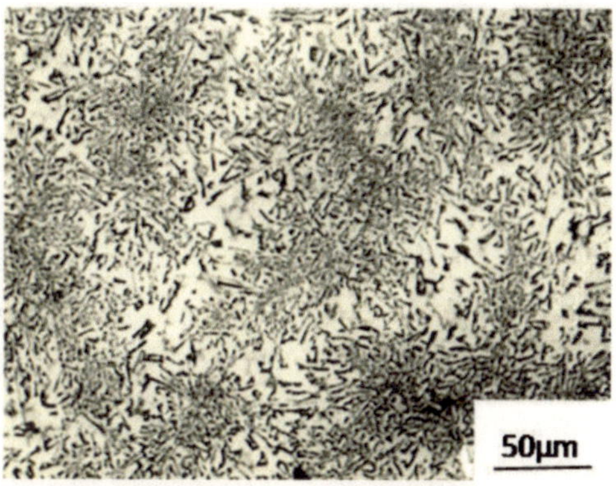

Figure 10. Optical self-modified structure of hypereutectic DEASA (EZL16). Complete eutectic structure appears accompanying some fine silicon flake on the boundary of eutectic cell.

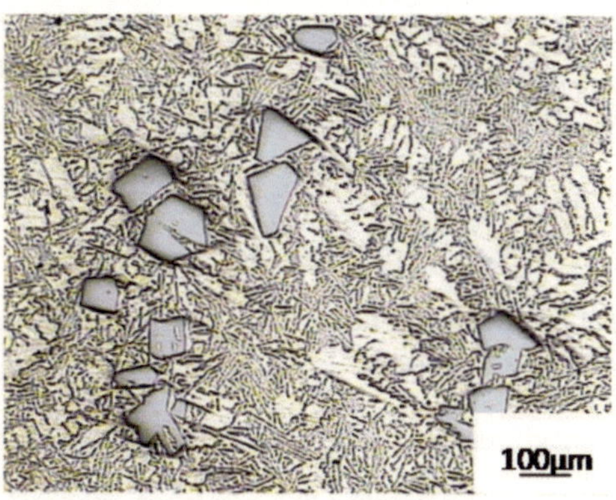

Figure 11. Microstructure of hypereutectic DEASA (EZL18). Coarse primary silicon distributes through the unmodified eutectic matrix. Some α-Al dendrites. occur. Optical

By combining with the regularity of morphological transition of silicon phase in Al-Si alloys treated by electric field at high temperature in literatures and our experimental results it would be expected that the variety in crystallographic feature is attributed to the undercooling in freezing. This inference has been strongly supported by the experimental results in remelting DEASAs. The very fine self-modified structure in DEASA such as EZL101 and EZL109 is fully inherited upon first remelting with a deep undercooling of 9°C and 13°C, respectively, as shown in Fig.14. As undercooling is higher than critical value of 6-8°C, the alloys solidify in modified manner. In contrast, an unmodified structure in Al-17%Si alloy appears due to lower undercooling of 5°C

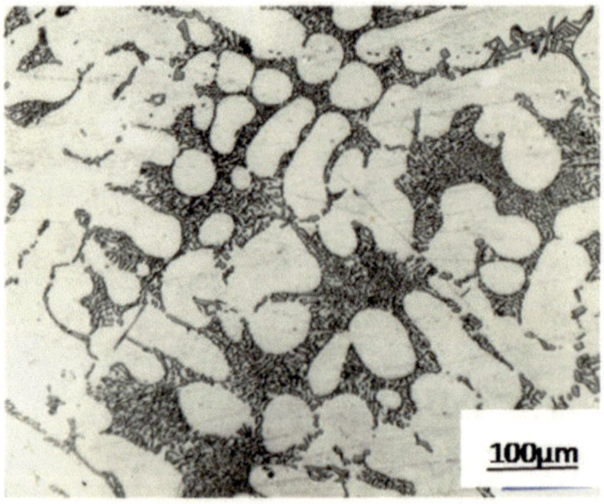

Figure 12. Microstructure of hypoeutectic DEASA(EZL101) ingot, demonstrating the self-modified structure with high volume fraction of α-Al dendrite. Optical.

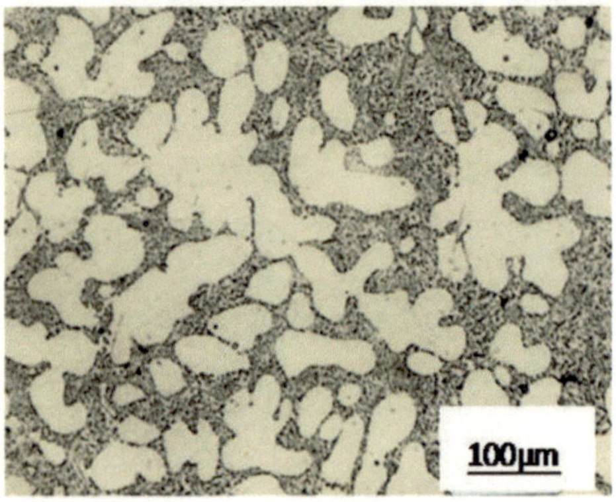

Figure 13. Optical micrograph of hypoeutectic DEASA(ES9) ingot, showing fine modified silicon phase with high volume percentage of α-Al dendrite.

It is interesting that the modified structure in EZL109 fades considerably slower and even upon 3-fold remelting the modified structure is inherited (Fig15) with an undercooling of 5°C. However, for EZL101with 7% Si or EZL102 having 12%Si after 3-fold remelting some silicon flake can be observed, displaying a decreased undercooling (Fig.16). It is thought that the alloying elements such as Cu, Mn, Ni and Mg prompt the occurrence of deep undercooling, strengthening the structural inheritance in EZL108,EZL109 and ES9 (Table I) In general, with 4-fold or more remelting the almost fully structural fading occurs and the undercooling disappears. In this case the microstructure in eutectic EZL108 and EZL10 is composed of eutectic with few, if any, α-Al dendrites. Note that upon first remelting the quasi-eutectic in DEASA hypereutectic alloys is subjected to fully fading, resulting in an appearance of coarse primary silicon grain distributed in unmodified eutectic matrix as shown in Fig.17, while the undercooling cannot be found. Fig.18 shows the variety of undercooling with remelting for DEASA (EZL101, EZL109). When undercooling is lower than 5°C, the inheritance of self-modified structure of EZL101 is subjected to significantly fading. In contrast, as undercooling decreases to 5°C the self-modified microstructure of EZL109 remains unchanged. Thus, it is reasonable to consider that for hypoeutectic and eutectic DEASAs the critical undercooling is 5°C, which is lower than critical value of 6-8°C for commercial Al-Si alloys. This phenomenon is thought to be associated with the homogeneous characteristics of DEASA melt, which cause silicon to solidify in modification mode at lower undercooling and cooling rate [23].

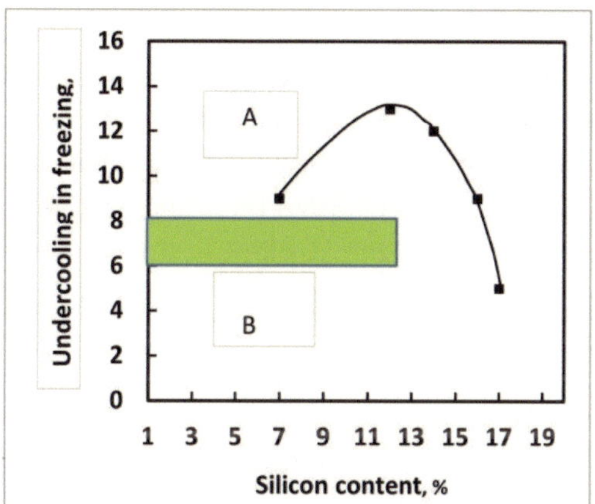

A: modified silicon B: unmodified silicon

Critical region

Figure 14. Relationship between eutectic undercooling and Si content in remelted DEASA.

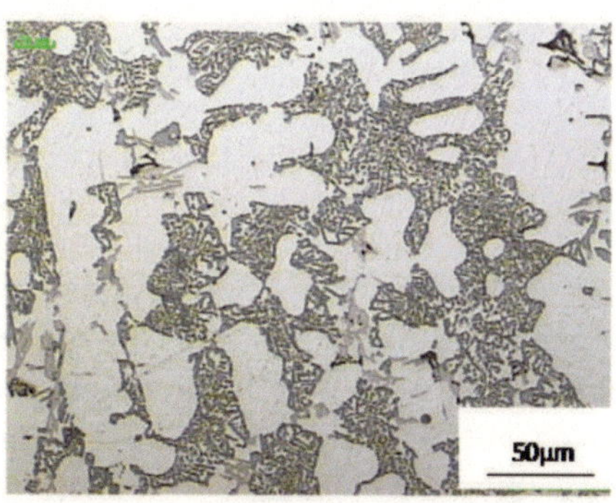

Figure 15. Optical structure of DEASA(EZL109) upon 3-fold remelting. Self-modified structure is fully inherited.

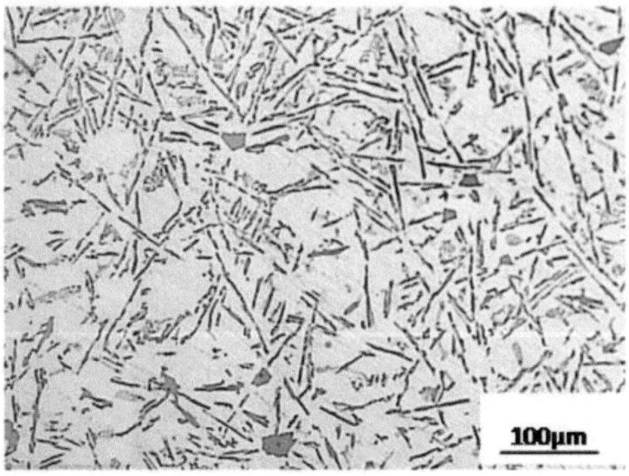

Figure 16. Microstructure of DEASA(EZL102) upon 3-fold remelting composed of unmodified structure. Some small faceted primary silicon appears. Optical.

It is worth noting that the self-modified structure is relatively insensitive to cooling rate as compared to commercial alloy. In general, the microstructure in top area of commercial ingot, where the cooling rate is very slow, displays coarse unmodified silicon flake but in the edge the fine silicon structure can be observed due to rapid cooling rate. Our observation reveals that there is no obvious difference in fineness of the eutectic between top and edge of eutectic DEASA ingot (EZL109) (Fig.7and 8). It is expected to be associated with the homogeneous melt, of which the stability is strengthened by the electric field in electrolytic process. That would be thought to be superiority over commercial alloy to produce complex castings.

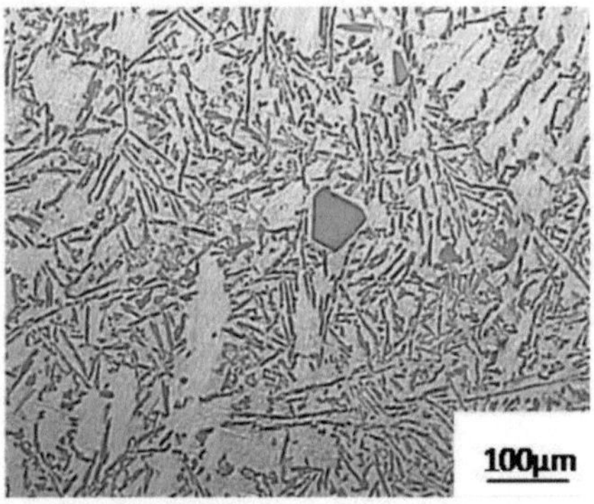

Figure 17. Optical micrograph of DEASA(EZL14) after first remelting. Self-modification is fully subjected to fading. Faceted primary silicon occurs

The origin of the variety of undercooling of DEASA is associated with the homogeneous character of its melt. The original DEASA melt either hypoeutectic or eutectic is intrinsic homogeneous, causing silicon to solidify at a large undercooling due to the lack of large silicon-rich clusters acting as nuclei in freezing. The repeated re-melting of DEASA, in which the large undissolved silicon particles exist, causes a lower undercooling, accompanying with the fading of modified microstructure The homogeneous character in hypereutectic DEASA melting can be partially survives or fully lost depending upon the silicon composition, because with increasing silicon content the undissolved silicon particles dramatically increases, nucleating silicon in solidification with a lower undercooling.

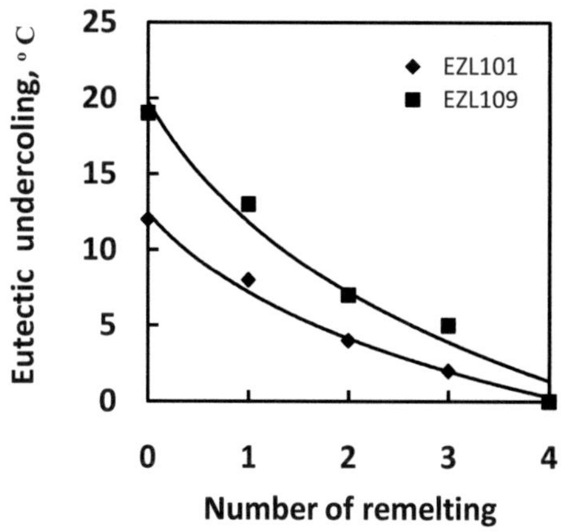

Figure 18. Relationship between undercooling and remelting undercooling and remelting

By combining the results in present and previous works we suggested the following growth mechanism of quasi-eutectic structure [49]. At initiation of the growth the silicon particle as nucleus would be assumed to be a nodule or irregular shape, with many different facets exposed in the melt [59-61]. Whether or not such a nucleus grows as polyhedron primary silicon crystal in freezing is determined by the degree of undercooling. As the nucleus grows, the boundary layer of eutectic composition starts to form around the growing nucleus and isolates it from the melting, thus preventing the further development of nucleus. With lower undercooling or higher silicon concentration the silicon atoms are capable of diffusing cross the layer to be trapped on the surface of the silicon nucleus, thus the silicon nucleus further grows, developing a primary silicon crystal and eutectic structure before the temperature of melt lowers down to the critical value shown by curve ES in Fig.3. Under high undercooling silicon atoms diffusion is limited, suppressing the primary silicon crystal to form. If the primary silicon cannot develop until the melt is cooled, reaching through the apparent eutectic temperature as curve ES shown in Fig.3, the quasi-eutectic structure occurs. In this case the silicon particles could act as the nucleus of eutectic, promoting the growth of eutectic structure.

As well known, whether the eutectic silicon grows in modified manner is attributed to the undercooling in solidification of alloy. This phenomenon is related with the entropy of melting and crystallographic structure as reported by A. Jackson in 1958 [62]. This relationship can be expressed as:

$$\alpha = \left(\Delta S / R\right) \left(N_S / N_V\right) \tag{3}$$

where α is Jackson criterion; ΔS is entropy of melting; R is gas constant; N_S and N_V are the number of an atom's nearest neighbors on the surface and within the body of a crystal. If α

is less than 2 cal/°C, crystal grows isotropically with an atomically rough interface. By contrast, if α is greater than 2, crystal is faceted with an atomically smooth surface. It is very interesting that for silicon the Jackson criterion for principal crystallographic planes varies in range from 0.89 for (110) plane to 1.87 for (100) plane, to 2.67 for (111) plane. Thus silicon crystal is a borderline material, of which the growth mode can easily change from faceted to non-faceted when the undercooling increases [63]. The variety in undercooling, which is induced by cooling rate during freezing or impurity element or others, will significantly cause the change in morphology of silicon either eutectic or primary. Generally speaking, for hypoeutectic or near eutectic DEASAs undercooling of 5°C is considered as a critical value to change the growth mode of silicon(Fig.18). Recently H.S.Kang et al [64] reported that the critical undercooling is a linear function of silicon content. For the higher silicon content an increased undercooling is required to change the morphology of eutectic silicon phase. They revealed that for Al-13%Si alloy at undercooling of 14°C the eutectic silicon morphology changes from flake to fibrous shape. However, for hypereutectic Al-20%Si alloy an increase in undercooling to 73°C is required. The different critical undercooling reported in literatures is thought to be associated with the different structure in liquid state. In current study DEASA melt is homogeneous, but the melt heating treated at 720°C in study by Kang is microheterogeneous, for which a deep undercooling /high cooling rate is needed to achieve the modified eutectic structure as evidenced in study[23].

Other important variety in structural feature of DEASAs is that iron-rich phase appears in fine flaky form instead of needle-like shape in the center of ingot containing iron concentration of 0.25% as shown in Fig.19. It is interesting that as Fe-level is more than 0.5% in EZL102 ingot the morphology of Fe-bearing precipitate also remain unchanged as shown in Fig.5. That is also attributed to effect of superheating the melt in electrolysis pot on the crystallization of iron-rich composition as reported in reference [33,34].

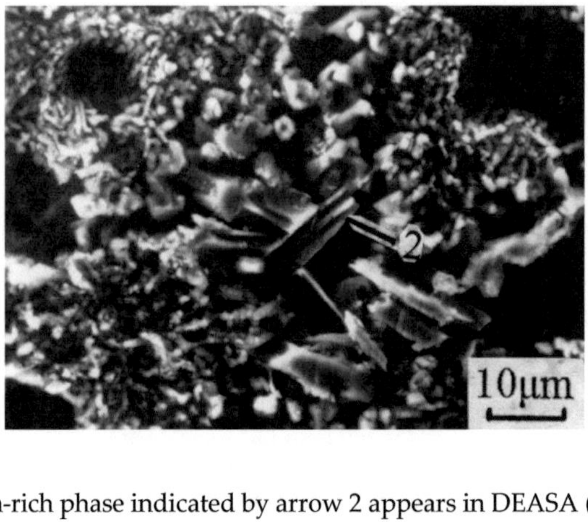

Figure 19. Fine flaky iron-rich phase indicated by arrow 2 appears in DEASA (EZL108) containing iron level of 0.25%.

3.2. Primary α-Al phase

Primary α-Al phase is an important phase constituent, of which the volume fraction, grain size and morphology, dendrite arm space (DAS) and alloying element greatly affect the mechanical and foundry properties of hypoeutectic Al-Si alloys[65,66]. In DEASAs either eutectic such as EZL102, EZL108 and EZL109 or hypoeutectic such as EZL101, the volume proportion of primary α-Al dendrite is higher than unmodified Al-7%Si alloy (ZL101) (Fig.20), and with increasing silicon content the volume percentage of Al-phase decreases. Undoubtedly, the increase of primary α-Al dendrite greatly affects the properties of eutectic DEASA castings.

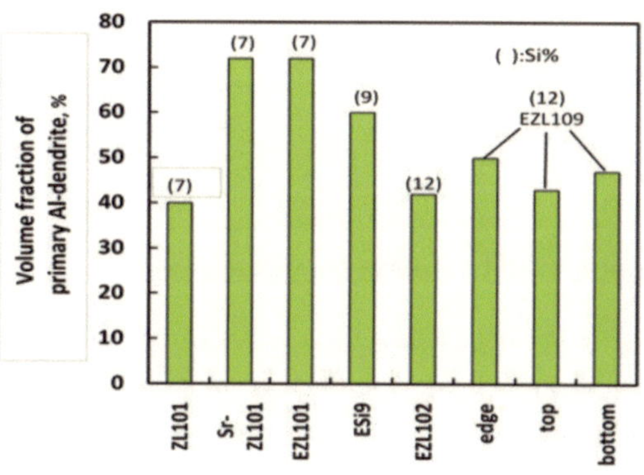

Figure 20. Volume fraction of α-Al dendrite in commercial and electrolytic Al-Si alloys.

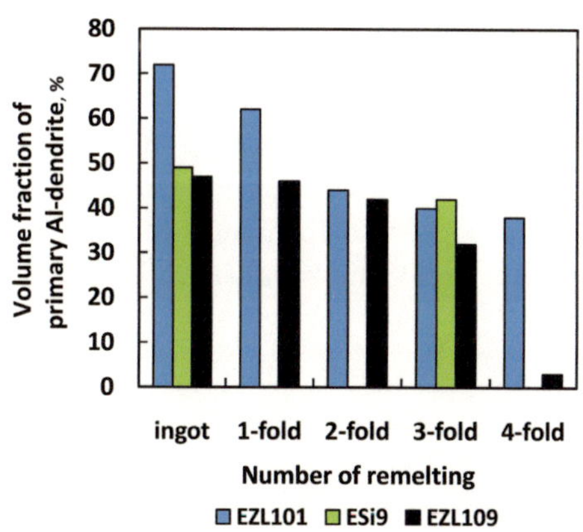

Figure 21. Volume percentage of α-Al dendrite in DEASA against the remelting.

The volume percentage of α-Al dendrite in the edge area of EZL109 ingot, where the cooling rate is much higher than bottom or top, is more or less larger than other areas (Fig.20). In addition, the volume proportion of aluminum dendrite decreases with remelting, which causes a decrease in undercooling in freezing (fig.18). After 3 or 4-fold remelting the volume percentage recovers to the value estimated from equilibrium Al-Si phase diagram (Fig.21) accompanying with an unmodified silicon structure. Apparently, the volume fraction of α-

Al dendrite is a function of undercooling, which can be clarified using Al-Si quasi-diagram (Fig.3).

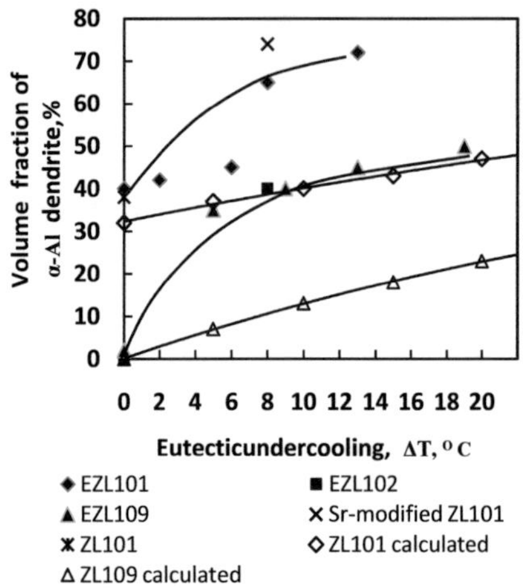

Figure 22. Volume percentage of α-Al dendrite in DEASA is a function of undercooling in freezing.

The fact that undercooling shifts the eutectic content toward to higher level during freezing and depresses the eutectic temperature, leading to an increase in temperature interval, in which the primary aluminum phase precipitates from melt, and, thus, the volume fraction of Al-phase is increased. Curve EP (Fig.3) represents the relation between undercooling and silicon content solved in Al-matrix. Therefore, we are able to estimate the volume fraction of α-Al dendrite in terms of undercooling in our tests (Fig.22). It is interesting that the volume fraction of α-Al dendrite measured in our study is much higher than the value calculated in terms of curve EP. By combining the successful achievement of quasi-eutectic structure in EZL16 with undercooling of 9°C, which is much smaller than the critical undercooling of 20° C shown on curve ES to obtain quasi-eutectic for commercial Al-16Si alloy (Fig.3), it is reasonably postulated that the region of formation of quasi-eutectic structure in DEASA moves toward higher silicon content and smaller undercooling due to the homogeneous DEASA melt.

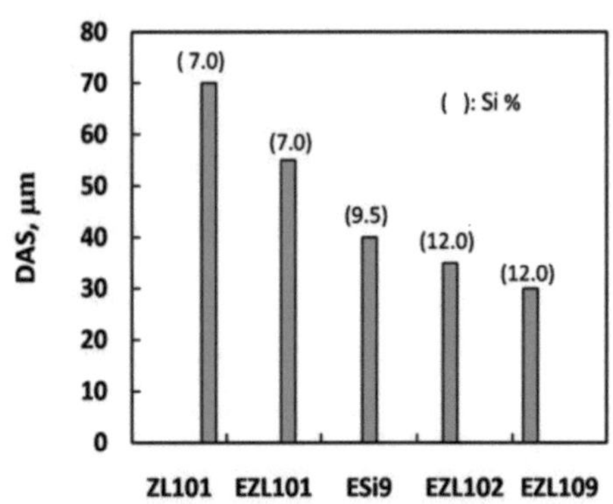

Figure 23. DAS of α-Al dendrite in DEASA and commercial ZL101(A356) ingot .

Dendrite arm space (DAS) is an important crystallographic feature in primary Al-phase, greatly affecting the mechanical properties of hypoeutetic Al-Si alloys. The DAS, which is not related in any way to the volume percentage of Al-dendrite, can be varied considerably by cooling rate. As far back as the 1960s it has been found that for commercial Al-Si alloy castings with cooling rate in range of 10^{-1}-10^{2} °C/sec such as cast in sand and in a metal mould, and continuous castings the DAS value is a function of cooling rate as follows [67]:

$$d = A \cdot V^{-n} \tag{4}$$

where d is DAS (μm), R is the cooling rate (°C/sec), A is related to the chemical composition and n=1/3-1/2.

In electrolytic EZL101ingot having 7% Si content the primary Al-dendrite displays the smaller DAS value than commercial Sr-modified ZL101 ingot as seen in Fig.23. Meanwhile the DAS decreases with silicon content. After 2 or 3-fold remelting DAS value doesn't change, if any. That is related with the same cooling rate in freezing of those samples [67].

Summarizing the structural characteristics of DEASAs we reach the conclusion that the electrolytic alloy castings either hypoeutectic or eutectic or hypereutectic exhibit very fine eutectic silicon grain with high volume fraction of α-Al dendrite, small DSA and small curved iron-bearing compound compared to commercial alloy. It is an advance superiority of DEASA over existing alloy for producing the high quality casting with excellent usage properties.

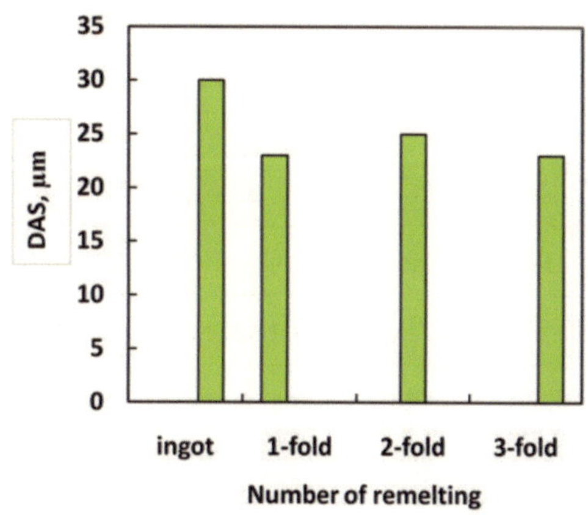

Figure 24. DAS in DEASA (EZL109) against remelting.

4. Technological parameters influencing self-modification in DEASA castings

In foundry several technological parameters including amount of DEASA ingot in metallic charge, furnace temperature and holding time, remelting and level of modifier added into melts, significantly affect the microstructure of castings and its mechanical properties. To optimize those parameters is an important event for producing high quality cast product with low cost.

In this chapter we have discussed the structural heredity of alloys upon remelting. It is concluded that remelting a fine metal easily produced a fine casting compared to a coarse metal at the same condition [68]. Experimental results demonstrated that at least 10% of a fine Al-Si ingot is required to achieve a casting with fine silicon grain [69]. Therefore, amount of DEASA ingot added in metal charge is an important factor for producing a casting with the fully modified microstructure.

In our test the metallic charge is composed of EZL101 ingot (Table 1.), pure aluminum ingot and Al-20%Si along with other master alloy additions. No modifier is added in melt. The percentage of DEASA ingot in metallic charge ranges from 10 to 50%. Table 3. lists the chemical analysis of alloys tested. Note that with either the holding time in furnace or remelting the strontium content is unchanged and much less than the critical level (0.004%) to create a modified eutectic in Al-Si alloys [70, 71]. Thus, the change in microstructure is associated with the amount of DEASA ingot used rather than strontium content in alloy.

Alloy	Amount of DEASA	Number of remelting	Holding time (min)	Chemical Analysis wt%								
	(%)			Si	Mg	Ti	Fe	Cu	Mn	Zn	Ni	Sr
EA50-0*	50	0	10	7.33	0.21	0.11	0.21	<0.01	0.01	0.003	0.016	0.0011
			10	7.08	0.41	0.19	0.27	<0.01	0.006	0.012	0.007	0.0012
EA30-0	30	0	120	6.90	0.34	0.17	0.30	<0.01	0.007	0.012	0.006	0.0013
			240	6.98	0.30	0.17	0.25	<0.01	0.007	0.012	0.006	0.0012
EA30-2		2	120	7.00	0.25	0.21	0.26	<0.01	0.006	0.012	0.009	0.0011
			10	6.71	0.39	0.053	0.25	<0.01	0.009	0.015	0.038	0.0016
EA10-0	10	0	120	6.84	0.28	0.050	0.25	<0.01	0.009	0.015	0.006	0.0016
			240	6.50	0.28	0.048	0.25	<0.01	0.009	0.015	0.006	0.0014
EA10-2		2	10	6.69	0.32	0.047	0.26	<0.01	0.006	0.013	-	0.0012

E: electrolytic; A: Al-Si alloy; 50: 50% of DEASA in charge; 0: no remelting.

Table 3. Technological parameters and chemical analysis of DEASA alloys wt%

When 10% of metallic charge is DEASA silicon crystals grow in modified manner in as-cast microstructure of EA10-0 alloy, but a few silicon flakes can be found (Fig.25). However, increasing the amount of DEASA to 30% (EA30-0) or more (EA50-0) results in a fully modified microstructure as seen in Fig.26. It would be expected that the self-modified silicon crystal in DEASA acts as a modifier for the commercial Al-Si alloy. Like Na, Sr and other modifiers there is a critical amount of DEASA, below which the eutectic is not modifiable. For ZL101(A356) alloy 30% DEASA ingot in metallic charge is needed to obtain a full modified microstructure. Obviously, the higher the amount of DEASA used in charge, the stronger the trend to modification in alloy. Note that there is no overmodification with increasing the amount of DEASA ingot. It is a superior characteristic of DEASA castings to existing Al-Si alloy.

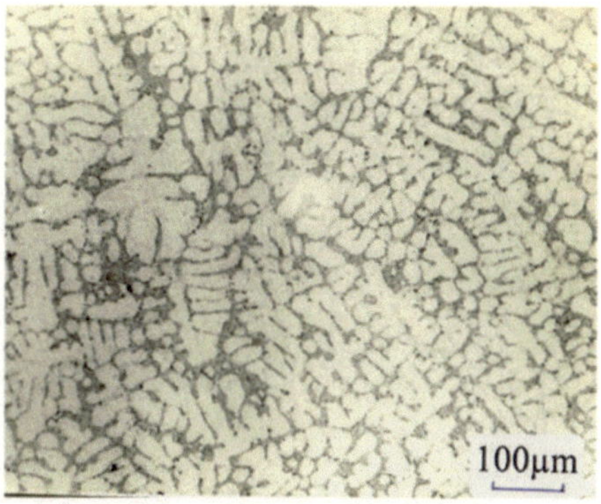

Figure 25. Modified microstructure of EA10-0 alloy with 10% of DEASA ingot in metallic charge and holding time of 10min.. Some silicon flake can be found. Optical

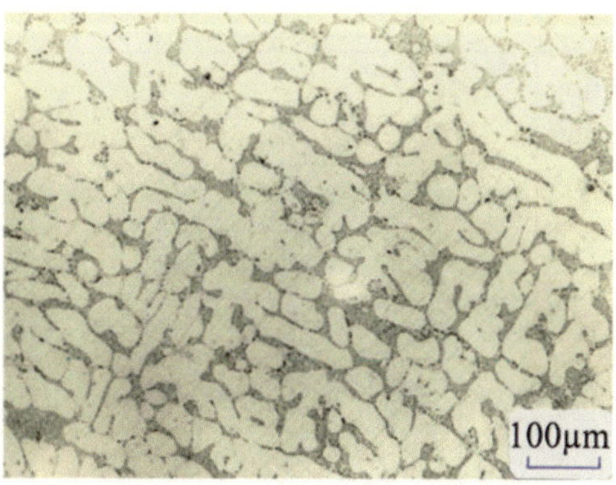

Figure 26. Optical microstructure of EA30-0 alloy with 30% of DEASA ingot in metallic charge and holding time of 10min, indicating a completely modified eutectic silicon..

Self-modified structure in EZL101(**A**356) alloys with either 10% or 30% of DEASA (EA10-0 and EA30-0) ingot strongly depends upon remelting and furnace holding time. When furnace holding time at 720°C is about 120min, both EA10-0 and EA30-0 alloys are subjected to partial structural fading as shown in Figs.27 and 28. Our test results reveal that for both alloys the inheritance of modified structure can be survived after first remelting. However, upon 2-fold remelting EA10-2 with 10% DEASA is subjected to partially fading in modified structure (Fig.29). As amount of DEASA increases to 30% (EA30-2), the modified structure can be maintained upon 2-fold remelting with holding time of 2 hrs, as shown in Fig.30, but when the holding time excesses 2 hours, some silicon flakes can be found.

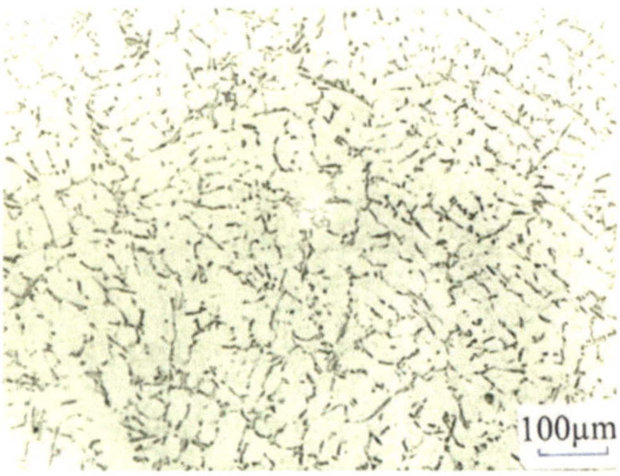

Figure 27. Optical micrograph of EA10-2 with 10% of DEASA ingot and holding time of 2hr at 720°C, indicating the unmodified structure.

In general, in foundries casting must be done within two hours after degassing aluminum melts in furnace. It is evident that strengthening the modification of DEASA alloys is needed for producing high quality castings. Below we will discuss the effect of small amount of strontium added into the molten alloy on the modified structure after different holding time. Table 4. lists the chemical analysis of Sr-modified DEASAs in tests.

Alloy	Amount of DEASA.%	Holding time. min	Chemical Analysis wt%									
			Si	Mg	Cu	Mn	Fe	Ti	Cr	Ni	Zn	Sr
SEA10*	10%	10	7.08	0.43	0.03	0.006	0.29	0.064	<0.00	—	0.014	0.003
		120	6.84	0.28	0.04	0.009	0.35	0.088	0.004	0.006	0.015	0.002
		240	6.95	0.30	0.04	0.015	0.33	0.070	0.004	0.008	0.017	0.002
SEA30	30%	10	6.81	0.39	0.032	0.007	0.36	0.16	0.002	0.004	0.012	0.003
		120	6.66	0.34	0.032	0.007	0.35	0.17	0.002	0.006	0.012	0.002
		240	6.50	0.32	0.033	0.007	0.35	0.17	0.002	0.003	0.012	0.002

*S: Sr-modification; E: electrolytic; A: Al-Si alloy; 10: 10% of DEASA in charge.

Table 4. Chemical analysis of EZL101 alloys tested

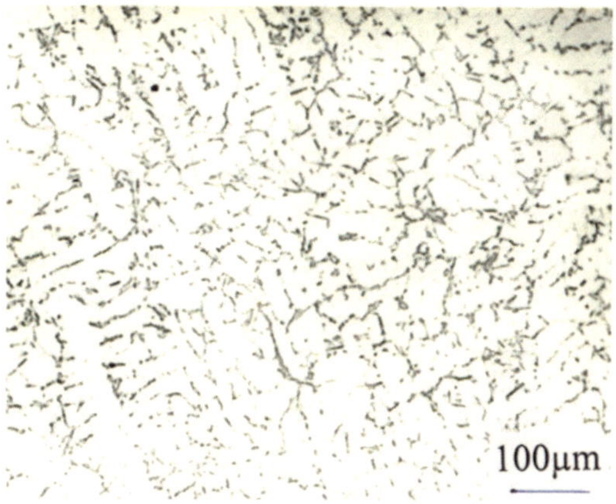

Figure 28. Micrograph of EA30-2 with 30% of DEASA ingot and holding time of 2hr at 720°C, indicating the unmodified structure. Optical

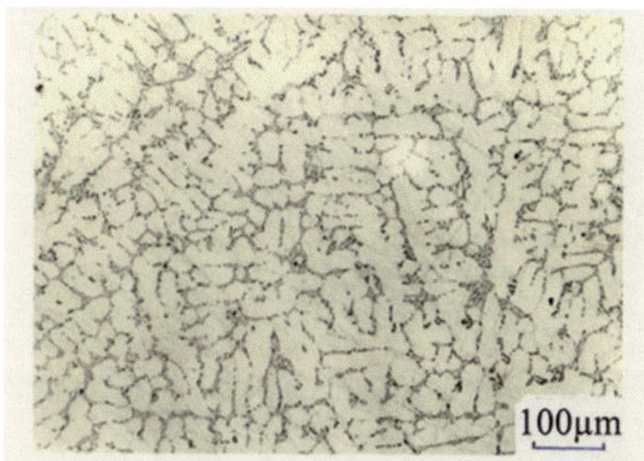

Figure 29. Optical micrograph of EA10-2 with 10% of DEASA ingot upon 2-fold remelting, indicating the partial fading of modified structure.

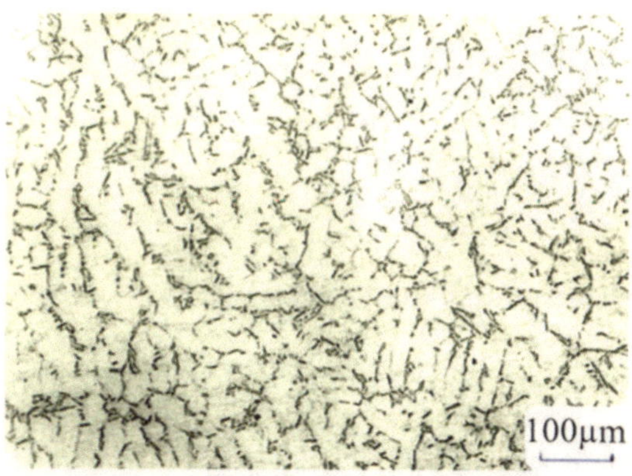

Figure 30. Micrograph of EA30-2 with 30% of DEASA ingot upon 2-fold remelting and holding time of 2 hrs, showing the modified structure. Optical.

Note that after adding Al-10%Sr master alloy into melts the level of Sr in alloys increases to about 0.003%, which is less than in commercial Sr-modified alloys, resulting in a fully modified structure in either SEA10 or SEA30 alloy (Fig.31). It is important that when holding time is four hours, in microstructure of SEA10 alloy some flaky silicon crystal can be found, but for SEA30 alloy the modification fading occurs in some degree as seen in Fig.32. Later is sufficient to meet the requirement in foundry. 30% of DEASA used in charge with 0.002-0.003% of Sr-level in alloy are necessary to produce high-quality Al-Si alloy. The process, in which the low level of strontium addition promotes a fully modified structure, is an advance advantage of DEASAs over Sr- or Na-modified Al-Si alloys.

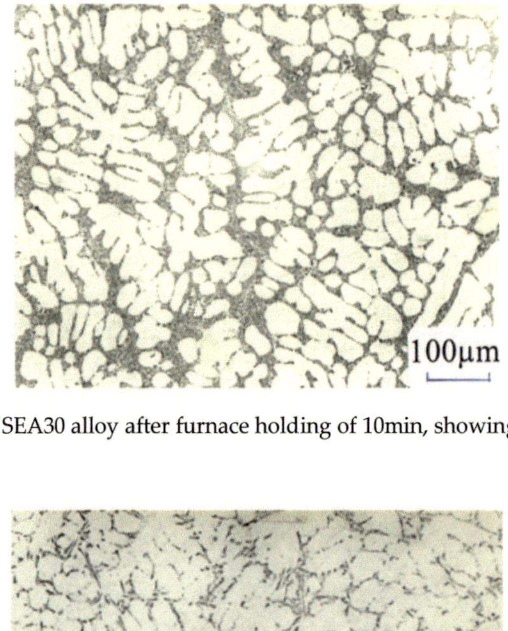

Figure 31. Micrograph of SEA30 alloy after furnace holding of 10min, showing the modified structure Optical.

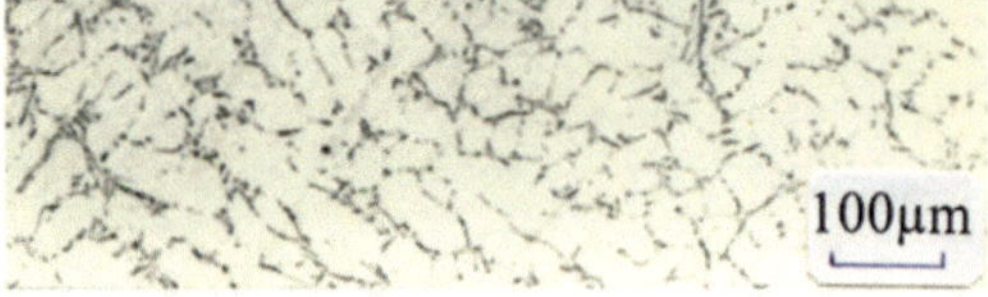

Figure 32. Partial modified microstructure of SEA30 alloy after furnace holding of 4hr at 720°C. Optical

5. High-quality automotive wheels made from DEASA

The wheel is an important part of a vehicle in terms of safety. The impact strength and fatigue life are on the top of the quality list of wheel characteristics. Al-7Si-0.3Mg alloy(A356), due to higher impact resistance and fatigue life, good castability and machinability, is a preferred choice to produce quality wheel. A lot of studies have demonstrated that the mechanical properties of A356 alloys are strongly affected by morphologies of eutectic silicon, iron content and porosity dispersed in castings, which is associated with the impurity modification and hydrogen level [71, 72]. For producing the quality wheel with higher impact resistance the maximum allowable iron level is limited to 0.20%. As mentioned above, DEASAs exhibit the excellent modified structure with a low level of modifier such as Sr, accompanying with the stringy Fe-rich precipitate of small size (Fig.19). Thus, the wheel made from DEASAs might be porous-free, resulting in higher impact resistance, ductility and tensile strength. At the end of last century we have examined the mechanical properties of wheel made from DEASA ingot (ES9, Tab.1) in the foundry [14].

A 600kg crucible was used to prepare the melts in an electric furnace. The metallic charge consisted of pure aluminum ingot, clean scrap of A356 alloy, other master alloy and DEASA (ES9) ingot pieces, of which the amount was a third of charge. Each melt was degassed with N_2 at temperature of 710°C. After degassing and holding for 15min, a small amount of Al-10%Sr master alloy was added into the melt to obtain Sr level in alloy below the critical value of 0.003%. Then the prepared melts were poured into a permanent mold to produce wheel casting and Y-shape plate castings with dimension of 22x150x220mm3. Finally, the castings were heat treated to a T_6 temper by solution at 535°C for 4 hrs, water quenching, and aging at 135°C for 6 hours. Table 5. lists the chemical analysis of DEASA (ES9) and EZL101 alloys, which have the different iron level, near or above the allowable value of 0.20% for wheel casting in order to clarify the effect of Fe-rich precipite on the mechanical properties of DEASA wheels.

Alloy	Si	Mg	Ti	Fe	Cu	Mn	Zn	Sr	RE
E S9	9.20-9.60	<0.01	0.40-0,60	0.44-0.65	<0.02	0.005	0.03	0.001-0.004	0.03
EZL101-17	6.63-6.80	0.28-0.30	0.12	0.16-0.17	0. 03	0. 1	0. 01	0.0016-0.0024	—
EZL101-21	6.50-7.00	0.24-0.27	0.10	0.19-0.22	0. 03	0. 01	0. 01	0.0022-0.0030	—
EZl01-27	6.90-7.30	0.28-0.30	0.10	0.26-0.27	0. 03	0. 008	0. 01	0.0031-0.0034	—
ZL101(A356)	6.90	0.30	0.12	0.10-0.13	0, 02	0.01	0. 01	0.0060-0.0080	0. 0005

Table 5. Chemical analysis of DEASAs for wheel tested wt%

Table 6. lists the mechanical properties of conventional (ZL101A) and electrolytic Al-7Si-Mg alloy (EZL101). As iron content is less than or near the maximum allowable limit of 0.20%, the superiority of DEASAs over existing alloy is very evident. DEASA alloys offer the mechanical properties higher than existing alloy (ZL101A) with lower iron content of 0.12%. As increasing Fe-level from 0.21 to 0.27% there is a slight tendency to decrease the mechanical indexes. But the tensile strength remains to be higher than conventional alloy (ZL101A), the elongation and impact strength are lower than existing alloy.

Alloy	Fe content wt%	Tensile strength MPa	Elongation %	Impact strength J/cm²	Hardness HB
ZL101A	0.12	(213-238)/225	(7-16)/12	(15-52)/32	(76-80)/78
EZL101	0.16	(217-260)/239	(7-18)/13	(31-52)/37	(74-80)/76
	0.21	(211-241)/231	(11-20)/14	(28-52)/38	(75-85)/79
	0.27	(218-240)/232	(7-10)/8	(16-52)/29	(70-85)/79

Note: ①Test samples are cut from Y-shape plate castings with dimension of 22x150x220mm3

②(Range of data)/Average data.

③Averaged data obtained from 4 Y-shape plate castings poured after holding time of 10 and 90min in furnace, respectively.

Table 6. Mechanical properties of conventional and electrolytic Al-Si alloys with different Fe-level

Wheel impact strength test is carried out at wheel shock testing apparatus (Fig.33). In general, the critical impact strength for automobile wheel is 230mm (height) x6000kN (weight) at 13degree of inclination. In production some 50% of the wheels made from ZL101 (A356) alloy exceed this minimum requirement by 10-20%. Addition of DEASA in charge exerts a significant improvement on the shock resistance. Testing results demonstrate that the wheels made of DEASA with different iron level offer the impact value exceeding 256mmx 6000kN at 13degree of inclination, and most of them are higher than 276mmx6000kN,which exceeds the critical requirement by 20%. Moreover, in the extreme test at 30 degree of declination two third of DEASA wheels tested exceed the shock resistance of 230mmx1010kN. However, none of wheel made from ZL101(A356) could pass this limit.

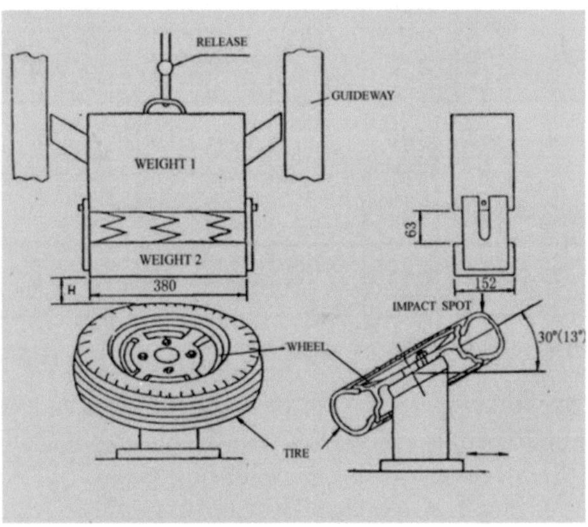

Figure 33. Schematic drawing of wheel shock testing apparatus

Fig.34 shows the wheel fatigue test apparatus, which uses torque of 3000N-M for loading with rotating speed of 1500rpm of shift. For wheel of 14 or 15 inches diameter the design fatigue lifetime, which is expressed in terms of the number of cycles-to–failure, is 10^5 cycles. Usually the lifetime for wheel made of ZL101(A356) ranges from 0.4 to 2.0×10^5 cycles. Some of them are not capable of exceeding the minimum lifetime. However, experimental data show that the dramatic improvement in impact resistance on DEASAs stated above is also evident in fatigue strength. DEASA wheels with iron content exceeding the allowable limit of 0.20% exhibit higher fatigue lifetime exceeding 2.0x105 cycles, except for E356-27 with higher iron content of 0.27% that has fatigue lifetime of 1.5×10^5 cycles.

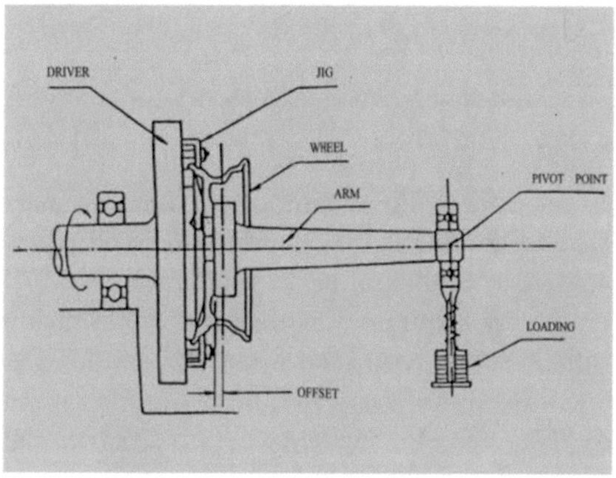

Figure 34. Schematic diagram of wheel fatigue testing apparatus

Summarizing the experimental results it is reasonable to conclude that as iron content exceeds the maximum allowable limit of iron level of 0.20% in some degree, for example, reaching 0.27%, the mechanical properties of DEASAs, especially impact strength and fatigue resistance, significantly are improved. Therefore, it is expected that the allowable iron content would be limited to more than 0.20%, which would save the cost of wheel.

The reason why DEASA wheel containing different iron level exhibits an excellent impact strength and fatigue resistance compared to conventional alloy, is believed to be attributed to porosity, if the difference between morphologies of silicon crystals and Al-dendrites in DEASA and existing A356 is difficult to find in micrography of the wheel as demonstrated in Fig.35 and 36. Porosity is an undesirable feature of the cast structure because pores, either surface or internal, acting as stress raiser during loading, seriously degrade the mechanical properties [73-75]. This inference is strongly supported by leak test for wheel, revealing that all the DEASA wheels were leakproof, while 10% of ZL101 (A356) wheels were not. Moreover, visual inspection showed that no pinholes and microporosities could be found on the surface of DEASA wheels compared to common ZL101, implying that DEASA wheels exhibit much less porosity than existing alloy. Undoubtedly, the sound alloy made from DEASA has higher mechanical properties.

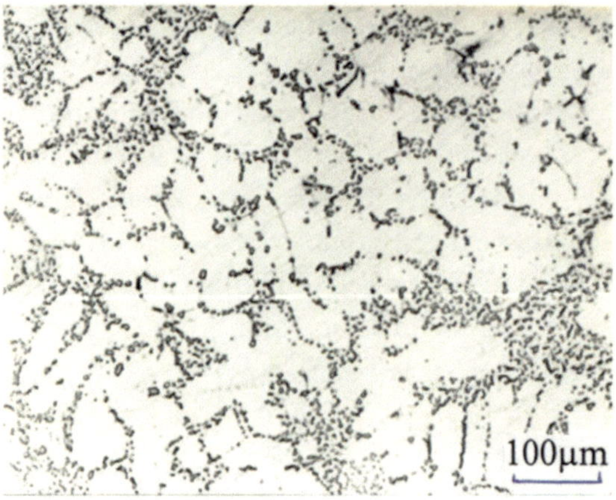

Figure 35. Opitical microstructure of hub in wheel made from DEASA EZL101-17(Table 6.) heat-treated by T_6

The origin of porosity is associated with two important, if not primary important, factors for Al-Si alloy in given casting condition, i.e. hydrogen dissolved in melt and amount of strontium or sodium added in molten alloy as modifier [71,76,77]. High hydrogen level causes an increase in porosity, resulting in decrease in mechanical properties [72-75]. Strontium or sodium increases the tendency to porosity of alloy [71]. In our study due to self-modification in DEASA ingot much less amount of modifier is required to be added into the DEASA molten alloy. Therefore, the tendency to porosity becomes weakened and sound castings are more easily obtained, resulting in higher impact resistance and fatigue strength in DEASA wheel.

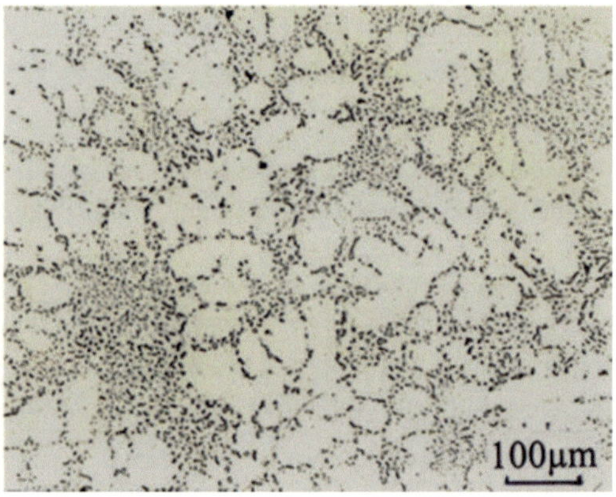

Figure 36. Microstructure of hub in wheel made of commercial A356 alloy heat treated by T_6 Optical

Until now a few studies have been done on the behavior of hydrogen in aluminum in electrolytic process [78]. In electrolytic pot the surface of aluminum melt is usually crusted over by fused cryolite, acting as an insulator to isolate the liquid from the atmosphere and protecting melt from hydrogen pick-up. In this case the hydrogen level is very low. Prodhan [47] studied the behavior of hydrogen in Al-Si alloy during solidification under electric field, indicating that the hydrogen level can be decreased from 2.5ppm to 1.7ppm. This is within the acceptable limit of pore-free castings [47]. It would be expected that in electrolytic process the hydrogen in Al-Si alloy melt can be removed under electric field. Undoubtedly, the decreased hydrogen level in DEASAs strongly weakens the trend to porosity, enhancing the impact resistance and fatigue strength. It is evident that the electrolytic process would be a powerful mean to reduce the hydrogen level in alloy.

6. Conclusion

1. DEASAs either hypoeutectic or eutectic display self-modified structure in ingot with excellent structural inheritance upon remelting. DEASA is self-undercooled alloy.
2. Hypereutectic DEASAs with silicon in range from 13% to 17% exhibit completely self-modified eutectic structure, but are subjected to fully fading upon remelting due to disappearance of undercooling in freezing.
3. DEASAs have high volume of α-Al dendrite that is associated with the high undercooling in freezing.
4. DEASAs are insensitive to cooling rate in freezing.
5. Iron-bearing precipitate in DEASAs appears in small curved shape as iron level increases to near 0.5%.
6. 30% DEASA ingot in metallic charge with added Sr-level of 0.002-0.003% is necessary to produce high quality Al-Si casting with self-modified structure.
7. Automobile wheel made of DEASA display high impact resistance and fracture strength that is associated to small amount of Sr-modifier added into melt, low hydrogen concentration and small curved shape of iron-rich compound.
8. Electrolysis is a potential measure to produce high quality Al-Si casting.

Author details

Ruyao Wang and Weihua Lu
Institute of Material Science and Engineering, Donghua University,
Shanghai, P.R.China

7. References

[1] Hayward C.R(1955) An Outline of Metallurgical Practice. Third Edition. Toronto: D.Van Nostrand Company, Inc.285-286p.
[2] Frilley J (1911) Revue de Metallurgie. j. 8(7): 518-523
[3] Bullough V.L (1973) US Patent No.3 765 878 Oct.16.
[4] Qiu Z.X, Zhang Z.L, Grjotheim K, Kvande H (1987) Aluminium.j. 6312:1247-1250.
[5] Orman Z (1976) Rudy Met. Niezelez. j. 21(5): 162-164.
[6] Weslan D (1984) US Patent No.4 425 308. Jan.10, 1984.
[7] Keller R, Weld B.J, Tabereau A.T (1990) In: Light metals 1990. TMS. pp.333-340.

[8] Moxnes B, Gikling H, Kvande H, Rolseth S, Straumsheim K(2003) In: Crepeau P , editor. Light Metals 2003. TMS. pp.329-334.
[9] Tabereaux A.T, McMinn C.J(1978) In: Light Metals 1978 .TMS.pp.209-222.
[10] McMinn C. J, Tabereaux A.T (1976) US Patent No 3 980 537 Sept.14, 1976.
[11] Yang K.Q, Gu Q.S, Tian G.Y, Li Q.C(1994) Chinese Patent. ZL94 11 6235.4, March 4th 1994.(In Chinese)
[12] Yang K.Q, Yang S (1997) Foundry. j. No.1: 44-46. (In Chinese)
[13] Yang D.X,.Wang R.Y(1994) Hot Working Technology. j. No.2: 19-21.(In Chinese)
[14] Wang R.Y, Lu W.H (2001) Light Metal Age. j. 59(5/6):.6-10.
[15] Wang R.Y,.Lu W.H ,Hogan L.M (2003) Mat.Sci.Eng. j.A348: 289-298.
[16] Bloomfield L.A, Freeman R.R, Brown W.L (1954) Phys. Rev. Letter. j. 54:2246-2249.
[17] Mitsuo S (1997) Liquid Metals. London: Academic Press. pp.41, 235-236.
[18] Iida T, Guthrie R.I.L (1988) The Physical Properties of Liquid Metals. Oxford: Clarendon Press. pp.18-46.
[19] Gui M.C, Li Q.C, Jun J(1995) Special Casting and Nonferrous Alloys.j.15(1):5-8.(In Chinese)
[20] Eskin D.G (1996) Z. Metallkunde.j. 89(4):295-301.
[21] Popel P.S, Tchikova O.A., Brodova J.G, Makeev V.V (1992) Nonferrous Metals. J. No.9: 53-56(In Russian).
[22] Bian X.F, Wang W.M (2000) Materials letters .j. 44(1): 54-58.
[23] Brodova J.G, Popel P.S,.Eskin G.I (2002) Liquid Metal Processing. Application to Aluminum Alloy Production. London: Taylor and Francis. pp.85-145.
[24] Wang, L.D. Zhu D.Y, Wei Z.I, Chen Y.L, Huang L.G, Li Q.J, Wang Y.S (2011) Advance of Materials Research. J.146-147:79-89.
[25] Nikitin V.I (1991) Foundry Production.j. No.4:.4-5(In Russian)
[26] Li P.J,. Xiong Y.H, Zhang Y.F, Zeng D.B(2003) Trans.Nonferrous Met. Soc.China.j. 13 (2):.329-334.
[27] Singh M. Kumar R (1973) J. Mater. Sci. j. 8: 317-323.
[28] Zhang L,.Bian X.F, Ma J.J(1995) Foundry. j. No.10:7-12(In Chinese).
[29] Wang W.M, Bian X.F, Wang H.R, Wang Z, Lin Z.G, Liu J.M(2001) J. Materials Research. J. 16(12): 592-3598.
[30] Xu C.L, Jing Q.C(2006) Mat.Sci.Eng.j. A437 (2): 451-455
[31] Gui M.C, Li Q.C, Jia J(1995) Special Casting and Nonferrous Alloys. j. 15 (1):.5-8 (In Chinese)
[32] Zhang R,.Zhang L.M, Yang Z.H, Liu L(2011) Materials Sci. Forum. J. 654-656:.1412-1415.
[33] Ri E.K, Ri K.S, Khimukhin S.N, Kalugin M.Y, Statsenko D.P, Kryuchkov I.V (2011) Foundry Production .j. No.7: 10-12 (In Russian).
[34] Awano Y, Shimizu Y (1987) Imono.J.59(4): 233-238; 59(7):.415-420 (In Japanese).
[35] Li P.J,NikitinV.I, Kandalova E.G, Nikitin K.V(2002) Mater.Sci.Eng.j.A332:371-372.
[36] Müller K(1998) Metall. J.52 (1-2):29-35.
[37] Gubinko A.Y (1991)Foundry Production. No.4: 19-20 (In Russian).

[38] Zhao Z.L, Wang J.L, Lu L (2011) Materials and Manufacturing Processing.j. 26(2):249-254.

[39] Prodhan A, Sanyal D (1998) Materials Science Letter.j. 16(11):.958-961.

[40] Prodhan A, Sivaramakrishnan C.S, Chakrabariti A. K.(2001)Met Mat. Trans B. j.32(2):.372-3780.

[41] Yu S.R,.Zheng Y.H, Feng H, Cai L.G(2010) Hot working technology.j.39(9):47-50 (In Chinese).

[42] Conral H (2000) Mat. Sci. Eng. J.A287(8): 205-212.

[43] Li H, Bian X.F, Liu X.F, Ma J.J (1996) Special Casting and Nonferrous Alloys. J.16(3): 8-10 (In Chinese).

[44] Huang L.G, Gao Z.Y, Zhang Z.M (2009) Foundry Technology.j.30(5):650-652 (In Chinese).

[45] Temchenko S.L, Zadoroschnai N.A (2005) Foundry Production.j. No.9: 12-13.(In Russian)

[46] B.A.Rabkin, S.L.Temchenko(2003) Foundry Production.j.No.10:.17-19(In Russian).

[47] Prodhan A (2009) AFS Trans. J.117: 63-77

[48] Loper C.R, Lu D.Y, Kang C.S (1985) AFS Trans. j. 93:533-543.

[49] Wang R.Y, Lu W.H (2007) AFS Trans. J.115: 241-248.

[50] Thall B.M, Chalmer B (1950) J.Inst. Metals. j. 77:79-97.

[51] Kim C. B, Heine R.W (1963-1964) J. Inst. Met. j. 92:367-376.

[52] Jenkinson D.C, Hogan L.M (1975) J. Crystal Growth. j. 28:171-187.

[53] Fredriksson H (1991) Scand. J. Metallurgy.J.20: 43-49.

[54] Talaat El-Benawy, Fredriksson H (2000) Trans. J. of Japanese Institute of Light Metals. j. 41:507-515.

[55] Stuhldreier G, Stoffregen K.W (1981) Giesserei. j. 68: 404-409.

[56] Stoffregen K.W (1985) Giesserei.j. 72: 545-549.

[57] Mondolfo L.F(1979) Aluminum Alloys, Structures and Properties.London: Butterworths.pp.513-515; 534-537; 566-575;592-594;604;614-615;617-618.

[58] Apelian D, K.Sigworth G, Whaler K.R (1984) AFS Trans.j. 92: 297-370.

[59] Wang R.Y, Lu W.H.Hogan L.M (1995) Mat. Sci.Tec. j. 11(5):.441-449.

[60] Wang R.Y, Lu W.H, Hogan L.M(1999) J. Crystal Growth .j. 207:43-54 .

[61] Wang R.Y, Lu W.H, Hogan L.M(1997) Trans. Metall. Mater.j.28A:1233-1243.

[62] Jackson K.A(1958) Mechnism of growth, In: Liquid Metals and Solidification, ASM, Cleveland, OH.pp.174-186.

[63] Gilmer G.H,.Leaming H.J,.Jackson K.A (1974) Liquid Metals and Solidification. Proc.4th Int. Conf. on Crystal Growth, Amsterdam, North-holland. pp.495.

[64] Kang H.S.,.Yoon W.Y,.Kim K.H, Kim M.H. Yoon E.P(2004) Mater. Sci. Forum. j. 449-452: 169-172.

[65] Oswalt K.J, Misra M.S(1980) AFS Trans. J. 88: 845-862.

[66] Mi J.W , Cheng J.N,.Yu Y.M (1990) In: C.Q.Chen, F.A.Starke.editors. Proceedings of the second international conference on aluminum alloys. Beijing: pp.566-570.

[67] Dobatkin V.I, Eskin G.I.(1990) In: C.Q.Chen, F.A.Starke.editors. Proceedings of the second international conference on aluminum alloys. Beijing: pp278-282

[68] Li S.S, Zhu Y.F, Zeng D.B (1999) Foundry. j. No.8:53-58 (In Chinese).

[69] Lui X.F, Bian X.F,.Ma J.J et al(1994) Foundry.j. No.10:18-23 (In Chinese)

[70] Lu W.H., Wang R.Y(1995) Special Casting and Nonferrous Alloys.j.15(2):1-5. (In Chinese).

[71] Gruzleski J.E., Closset B.M (1990) The Treatment of Liquid Aluminum-Silicon Alloys. AFS Inc. Des Plaines, IL, US. pp.223; chapter 4.

[72] Anyalebechi P.N(2003) In: P N. Crepeau. Editor. Light Metals 2003. TMS. pp.971-981.

[73] Eady J.A,.Smith D.M (1986) Materials Forum.j. 9(4):.217-223.

[74] Surappa M.K, Blank E.J, Jacquet.C (1986) Script Metallurgica.j. 20(9):.1281-1286.

[75] Caceeres C.H (1995) Script Metallurgica et Materialia.j. 32(11):1851-1886.

[76] Mascre C, Lefebvre M (1959) Fonderie. j. No.166:484-497.

[77] Arbenz H(1962) Giesserei.j. 49:105-110.

[78] Zhang M.J, Qiu Z.X, Di H.L (1987) Nonferrous Metals. No.1:27-31;No.2:29-34 (In Chinese).

Chapter 7

Electrolytic Enrichment of Tritium in Water Using SPE Film

Takeshi Muranaka and Nagayoshi Shima

Additional information is available at the end of the chapter

http://dx.doi.org/10.5772/52705

1. Introduction

Just after Voltaic cell was invented at the beginning of eighteenth century, the cell was applied to electrolysis in water and it was found that water is composed of hydrogen and oxygen. Since this discovery stimulated the scientific understanding on the behavior of ions in solution with atomic level, the electrolytic technology spreaded to the various fields such as refining of metals, metal plating, generation of alkaline substances and removal of trace toxic materials in water and soil. And this technology nowadays becomes an important field in electrochemistry.

Since tritium is generated at the stratosphere by the nuclear reaction of nitrogen and neutron and decays with radioactive half-life of 12.3 years releasing beta-ray, it always exists in a constant amount in nature. On the other hand, large amounts of tritium were produced by the nuclear bomb tests in the atmosphere until the early 1960s and it fell out all over the world, although the concentration of the fallout tritium is lowered to the background level because it passed for several decay times of half-life. Tritium is also generated from atomic power plants and atomic fuel reprocessing plants currently and tritium contaminated waste vapor is released to the air and the wasted water into seawater now in the world. Then the flow of the released waste has to be monitored to know the diffusion process in the environment and to keep the safety in mankind and ecosystem.

Liquid scintillation counting system is used to count the released electrons by the beta-decay. Lower limit of tritium concentration in a low background liquid scintillation counting system is around 0.4 Bq/L for counting time of 1000 minutes using 100 mL vial (Hagiwara et al., 2012). As tritium concentration in environment is 1 Bq/l or lower and especially lower than 0.5 Bq/L in seawater in Japan (Yamashita & Muranaka, 2011), it is difficult to measure the concentration even using such a low background liquid scintillation counting system. To solve the problem tritium concentration in sample water is enriched by water electrolysis before counting. Since the tritium concentration in the enriched water exceeds the lower limit of the counting system, the concentration can be measured with the sufficient accuracy even for seawater.

Various methods are proposed to obtain the enrichment factor in the tritium electrolysis in sample water. Among the suggested methods a method using apparatus constant presented by Kakiuchi, M. et al. (Kakiuychi et al., 1991) is noteworthy because the constant is not so fluctuated comparing with a separation factor (Inoue & Miyamoto, 1987). Details are described in the next paragraph.

Tritium enrichment apparatus by electrolysis is roughly classified into two categories. One is the method using alkaline electrolyte and another is that using solid polymer electrolytic film (SPE film). Electrolysis using SPE film was established firstly by Saito, M. et al. (Saito et al., 1996) and a large current apparatus was developed commercially (Saito, 1996). SPE film is used as a medium to carry hydrogen ions from anode electrode to cathode one in electrolytic cell instead of OH^- in alkaline electrolysis. Since this method has merits that alkalizing and neutralizing process of electrolytic solution is not necessary and there is no limit for electrolytic enrichment factor, this method is used in many groups recently.

Since it is important to know the tritium concentration in seawater at the area in Aomori prefecture, Japan because tritium contaminated waste is released into coastal seawater after the reprocessing is opened at Rokkasho, Aomori, Japan. We therefore tried the experiment of electrolytic enrichment in water with a handmade electrolytic device used SPE film from the mid age of 1990's (Muranaka & Honda, 1996).

Although Saito, M. et al. made an electrolytic apparatus which can release hydrogen and oxygen gas separately to remove the accident of the explosion by the mixing of the two gases, we designed an electrolytic device which can be put into the sample water cell as in the alkaline electrolysis (Muranaka & Honda, 1996; Muranaka et al., 1997). Method and apparatus for electrolytic enrichment are described precisely in section three and characteristics of the designed devices are described in the section four.

Tritium concentrations in seawater samples collected along the Pacific coast in Aomori prefecture, Japan were below 0.3 Bq/L except for the case in which the sample collection and the release of tritium contaminated wastewater into coastal sea from the atomic fuel reprocessing plant are matched in time (Muranaka & Shima, 2011). Tritium concentrations in land water in Aomori prefecture are around 0.5 Bq/L, very low comparing with those collected twenty years ago in the same locations. These results are described in section five in detail.

2. Electrolytic enrichment of tritium in water

2.1. Principle of electrolytic enrichment

The basic formulas for electrolysis are shown in equation (1)~ (3). (1) is the reaction in anode electrode, (2) is that in cathode one and (3) shows the whole reaction for electrolysis.

Tritium, deuterium and light hydrogen exist in water as THO, DHO and H_2O, respectively. It is confirmed experimentally that H_2O is easily electrolyzed than THO and DHO. Tritium and deuterium in water thus are enriched after the continued electrolysis. Schematic diagram of electrolytic enrichment using SPE film is depicted in Fig.1. In anode electrode sample water is decomposed into oxygen and hydrogen ion and the generated H^+ ion passes through the SPE film. This H^+ ion is combined to the electron arrived at cathode electrode through the external electric circuit and is released as hydrogen gas.

$$H_2O \rightarrow O_2/2 + 2H^+ + 2e^- \tag{1}$$

$$2H^+ + 2e^- \rightarrow H_2 \tag{2}$$

$$H_2O \rightarrow O_2/2 + H_2 \tag{3}$$

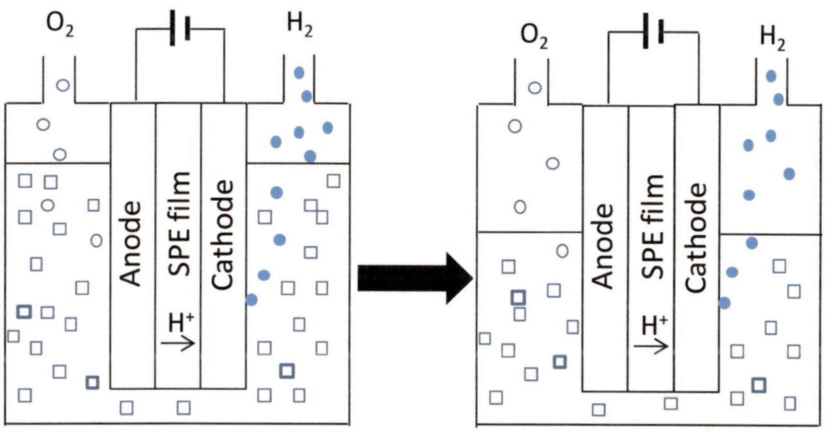

Figure 1. Electrolytic enrichment of tritium in water.
Opened squares in the electrolytic cell represent H_2O and those with comparatively thick border show THO or DHO. Opened and filled circles mean oxygen gas and hydrogen gas, respectively. Electrolysis proceeds from left hand figure to right hand one, that is, the number of THO or DHO molecules are not so reduced although those of H_2O molecules are gradually decreased in the process of electrolysis.

2.2. Determination of tritium enrichment factor

2.2.1. A method using tritium recovery factor

Tritium recovery factor which is defined by (4) is used to determine tritium enrichment factor in electrolytic enrichment.

$$R=(T_f\ V_f)/(T_i\ V_i) \tag{4}$$

Where T_i and T_f represent tritium concentration before and after the electrolytic enrichment, and V_i and V_f indicate the volume of sample water before and after the enrichment. Tritium concentration before enrichment can be represented by (5) which is transformed from (4).

$$T_i = T_f / \left((V_i / V_f)R\right) \quad (5)$$

Since V_i/V_f in (5) means the reduction factor in sample volume, tritium concentration before enrichment can be obtained dividing tritium concentration after the enrichment by the product of volume reduction factor and tritium recovery factor in the electrolytic enrichment. This product therefore means the tritium enrichment factor. In the constant volume reduction factor, tritium concentration before the enrichment can be deduced from only the tritium concentration after the enrichment if tritium recovery factor in the enrichment is provided using standard sample water. To obtain correct volume reduction factor volume of sample water before and after the enrichment must be measured precisely. The volume or the weight of sample water after the enrichment is especially necessary to be measured correctly to obtain the tritium concentration in sample water accurately, because the volume after the enrichment is lower than that before enrichment and the error rate becomes larger.

2.2.2. *A method using apparatus constant*

An apparatus constant is defined by (6)(Kakiuchi et al.,1991).

$$k = \ln(T_f / T_i) / \ln(D_f / D_i) \quad (6)$$

Tritium concentration can be given by (7) transforming the expression of (6).

$$T_i = T_f / (D_f / D_i)^k \quad (7)$$

Deuterium concentration before and after enrichment and apparatus constant are contained instead of the product of volume reduction factor and tritium recovery factor as the tritium enrichment factor in the expression of (7). If apparatus constant for electrolytic cell is determined, tritium concentration can be deduced from tritium concentration after the enrichment and deuterium concentration before and after the enrichment.

The relation between tritium recovery factor R and tritium separation factor β is shown in (8)（Inoue & Miyamoto, 1987）. In (9) the relation between apparatus constant and tritium separation factor β and deuterium separation factor α is represented(Kakiuchi et al.,1991).

$$R = (V_f / V_i)^{1/\beta} \quad (8)$$

$$k = \alpha(\beta-1) / \beta(\alpha-1) \quad (9)$$

Tritium separation factor β is known to be varied by some experimental factors such as electrode materials, surface condition of the electrode, current density in electrolytic enrichment, temperature in the water and so on (Satake & Takeuchi, 1987; Inoue & Nam, 1994). But it is confirmed that apparatus constant is not so vary, although α and β are varied by the run condition of the repeated electrolysis (Kakiuchi et al.,1991).

It is estimated that the variation of α and β is canceled in the apparatus constant represented in (9). Therefore the use of apparatus constant and deuterium concentration before and after the enrichment will be better than that of volume reduction factor and tritium recovery factor to obtain tritium enrichment factor correctly.

3. Method and apparatus for electrolytic enrichment

3.1. An electrolytic enrichment of tritium using alkaline electrolyte

In the conventional method for electrolytic enrichment of tritium a pair of metal electrode is set in an electrolytic cell and sodium peroxide (Na_2O_2) is added to the sample water in the cell to give it the electric conductivity. As the structure of this type of electrolytic cell is so simple that two or more cells can be connected in series and electrolytic enrichment in multiple cells is possible to be electrolyzed at the same time. Schematic diagram of two electrolytic cells in series is shown in Fig.2.

On the other hand this method has a disadvantage of a tedious process after the electrolysis that electrolytic material must be neutralized and the precipitation material has also to be removed from the solution. Another demerit of this method is that electrolytic enrichment factor is limited by the solubility of the electrolyte.

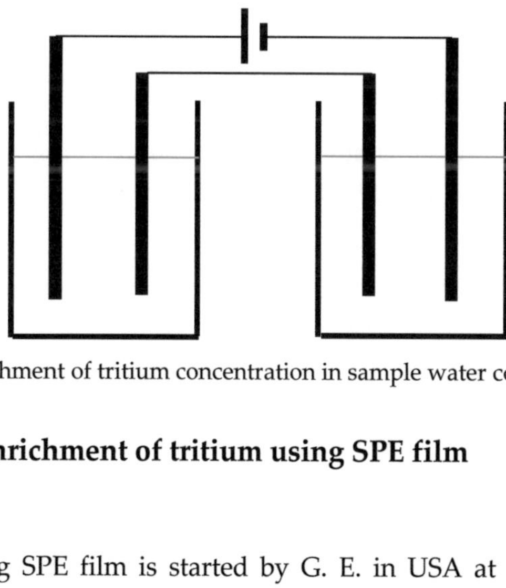

Figure 2. Electrolytic enrichment of tritium concentration in sample water connecting each cell in series.

3.2. An electrolytic enrichment of tritium using SPE film

3.2.1. SPE film

Water electrolysis using SPE film is started by G. E. in USA at 1970s applying fuel cell technology. The used SPE film is a kind of cation exchangeable membrane (NAFION® 117, Dupont).

This membrane is composed of carbon-fluorine backbone chains with many perfluoro side chains containing sulfonic acid groups (SO_3^-) and hydrogen plus ion can wade through the groups as depicted in Fig.3. The hydrogen ion reached to the cathode through the membrane film receives an electron from the external electric circuit and a pair of hydrogen atoms are connected to be hydrogen gas.

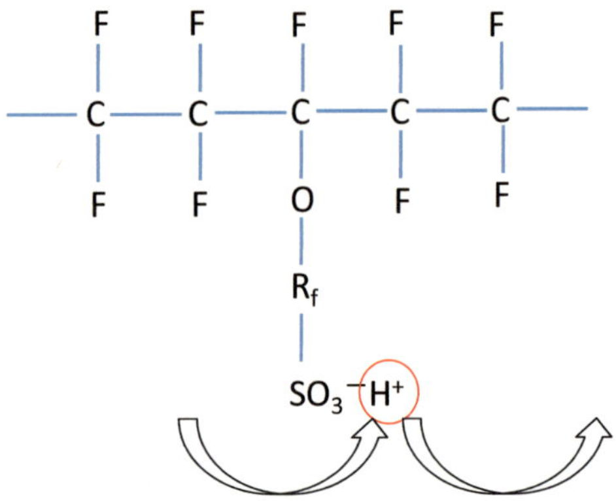

Figure 3. The structure of SPE film and the transfer of hydrogen ion from a sulfonic acid group to the next one.

In this method sample water can be enriched by electrolysis without adding any electrolyte to the water and then the tedious process after the electrolysis is not necessary and the enrichment factor is not limited by the solubility of the electrolyte. In addition to these merits electrolytic current can be increased removing generated gases quickly from the surface of both electrodes.

3.2.2. Electrolytic apparatus using SPE film

3.2.2.1. Oxygen-hydrogen separating apparatus

Oxygen- hydrogen separating apparatus using SPE film was originally devised by Saito et al. (Saito et al., 1996). Schematic diagram is shown in Fig.4.

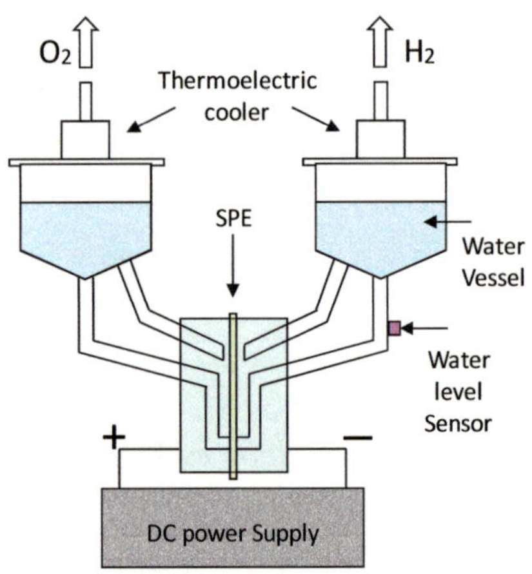

Figure 4. Schematic diagram of electrolytic cell devised by Saito et al..

In this type of electlytic device SPE film is sandwiched by porous electrodes, water vessels being connected at the bottom are placed in both side of electrodes, oxygen and hydrogen gases are released from each side of water vessel through the thermoelectric cooler to avoid the evaporation loss of water. Water level sensor is also provided not to exceed the lower limit of the water volume. This type of electrolyzer which can operate by a large electrolytic current of 50A was released commercially from Permelec Electrode LTD.

3.2.2.2. Oxygen-hydrogen non-separating apparatus

We designed an electrolytic device using SPE film like a conventional cell using alkaline electrolyte, because this type is simple in structure, can be connected in series and is easy to be cooled by a chiller.

1. An electrolytic device using corrugated electrodes

At the beginning an electrolytic cell having corrugated electrodes was designed. The composition of the device and the photograph of the electrode are represented in Fig.5 and in Fig.6, respectively. Generated oxygen and hydrogen gases are released from not in contact space between the electrode and SPE film in the vertical direction (Muranaka & Honda, 1996; Muranaka et al., 1997; Shima, 2007) .

The size of SPE film is 4.2cm square, used electrode is 4cm×4cm in size, 0.2mm in thickness and Teflon plate and acrylic plate are 6cm square. Slots in the Teflon plate can be grooved to fit the folded dents in the electrode. Acrylic plate placed in the outermost is used to keep the position of the each part including gold lead wire of 0.6mm in diameter tightening with stainless steel bolts. Anode electrode is made of platinum plate to avoid the dissolution during the electrolysis. The electrolytic cell has a capacity of 1.9L, is made of polypropylene, 28cm in height, 10cm in diameter at upper side and 8.5cm at the bottom. This electrolytic cell is set in the cooling water bath and is cooled at two degrees centigrade or lower in the bath.

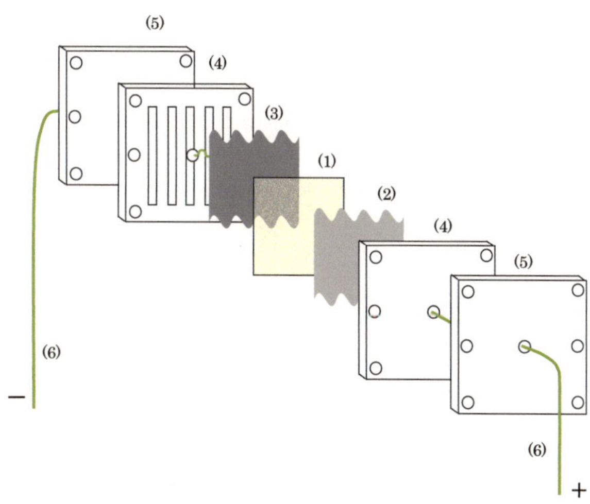

(1)SPE film(NAFION®117), (2)corrugated anode (platinum plate), (3)corrugated cathode(nickel plate), (4)Teflon plate, (5)acrylic plate, (6)gold lead wire.

Figure 5. Schematic diagram of the initial type electrolyzer.

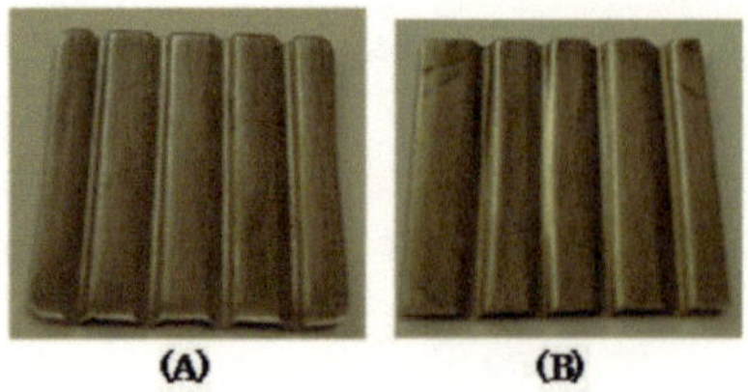

(A) and (B) represent an anode made of platinum plate and a cathode of nickel plate, respectively.

Figure 6. Photographs of the electrodes used for the initial type electrolyzer.

2. A newly designed electrolytic device using porous electrode

The drawback of the device using corrugated electrodes is that electrolysis is not carried out at not in contact area between the electrode and SPE film, and generated gas cannot be released smoothly at in the contact area. Although the lead wire contacts to the electrode with a spiral form to increase the contact area, electrolytic current could not be increased because of the heat loss by the electric current between the lead wire and the electrode not in contact to the wire directly. A new electrolytic device was designed to improve these faults. Schematic diagram of the electrolyzer is shown in Fig.7. (A) is the figure for the elements, (B) is the cross section and (C) is the assembly drawing of the electrplyzer (Shima & Muranaka, 2007b; Shima, 2007; Muranaka & Shima, 2008).

SPE film(1) in (A) and (B) is Nafion117. Anode (2) is made of porous titanium metal covering with rare metals and their oxides. Cathode (4) is made of porous stainless steel. Dimensionally stable electrode (DSE) manufactured by Permelec Electrode was used as the porous electrode. These electrodes asked were 4cm square and 2.6mm in thickness. Platinum mesh (3) was inserted between the SPE film and the anode DSE so that the cell can be easily dismantled without the detachment of the catalyst from the base metal electrode. The porous electrode (5) used to collect electrolytic current is made of gold, 4cm square and has many small holes of 1.0mm in diameter. These two plates allowed for homogenous current flow through SPE film and for release of electrolytic gases. Current lead wire is touched to the folded edges of the current collector to maintain the electric contact between the two elements. A spacer (6) was placed between the supports (7) to maintain constant separation between the elements of the electrolytic device. The spacer serves to restrict the thickness of the SPE film as it absorbs water from sample water and swells. These supports have many small holes of 1.5mm in diameter distributed coaxially to the holes of the current collectors to release the generated gases smoothly.

In Fig.8 electrolytic device positioned in the cell is shown (Shima & Muranaka, 2007b). Electrolytic device is placed in a container so that it could be tilted slightly in order to release oxygen gas generated beneath the anode electrode. Hydrogen gas released from the cathode electrode can therefore rise vertically without any obstacles. Generated oxygen and hydrogen gases are released in a mixed condition like an alkaline electrolysis. Although this method is felt to be dangerous at a glance, the burst timing of the bubbles containing hydrogen gas is different from each other and the released hydrogen is diffused into the air immediately. Since the released hydrogen gas mass from the bubble is therefore departed from the explosion limit each other, it will be safe up to some electrolytic current in such a open system.

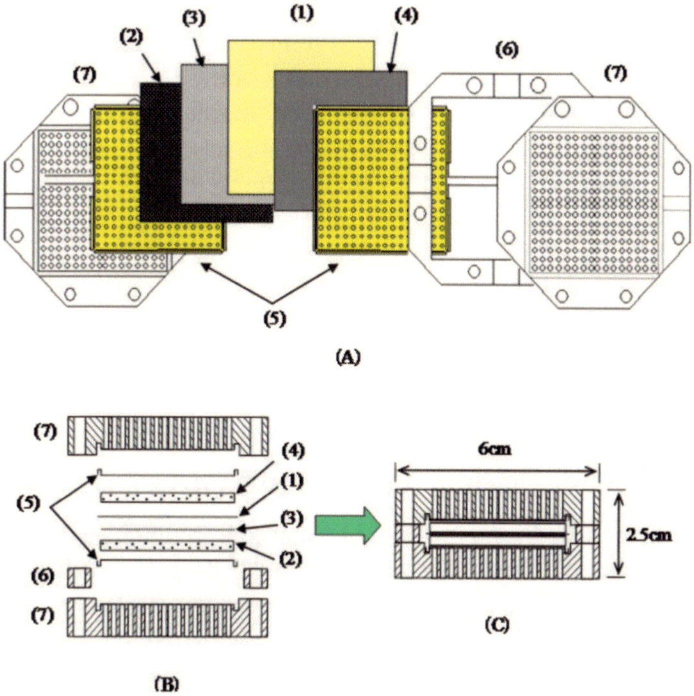

Figure 7. Schematic diagram of the electrolyzer.
(A) is the figure for the element, (B) is the cross section and (C) is the assembly drawing of the electrolyzer. (1) SPE film(Nafion 117), (2) porous anode DSE, (3) Pt mesh, (4) porous cathode DSE, (5) electrolytic current collectors, (6) spacer, and (7) support (Shima & Muranaka, 2007b; Shima, 2007; Muranaka & Shima, 2008). [(A) partly modified and (B) were reprinted with permission from Japan Radioisotope Association and the American Nuclear Society.]

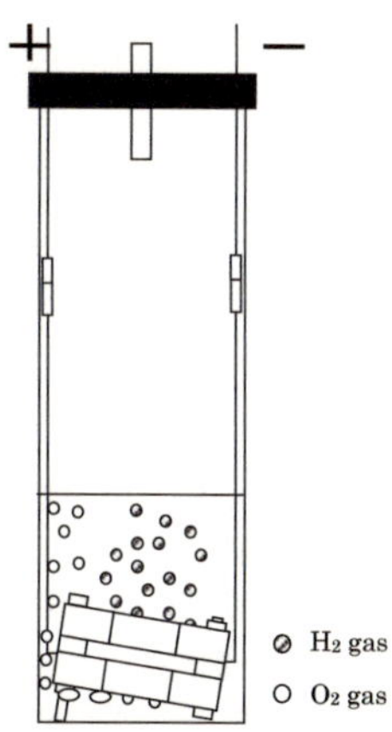

Figure 8. Schematic diagram of the electrolytic cell containing the newly designed electrolytic device (Shima & Muranaka, 2007b). [Reprinted with permission from JRA]

4. Characteristics of the electrolytic device

Tritium in water samples were measured by a low background liquid scintillation counting system (LB-II, Aloka). Enriched water of 40mL and a scintillator solution (Aquasol-2, Packard) of 60mL were mixed in Teflon vials. Tritium concentrations were estimated by beta ray counting, with each sample counted ten times over a period of 50 minutes for four cycles. Total counting time for each sample was therefore 2000 minutes. To prevent erroneous results due to electrostatic charges on the surface of sample vial, we removed the first three of the ten repeated counts and estimated tritium concentrations using the residual counting data. The hydrogen stable isotope ratio was measured by an isotope ratio mass spectrometer (Delta plus, Thermo Fisher Scientific) connected to a pretreatment device (H/Device, Thermo Fisher Scientific). Water droplets (1.2μL) were converted to hydrogen gas in this device according to the chemical reaction (10)(Sato et al., 2005).

$$2Cr + 3H_2O \xrightarrow{800^\circ C} 3H_2 + Cr_2O_3 \tag{10}$$

4.1. Characteristics of the electrolytic device using corrugated electrodes

Water samples were enriched using the same tritium concentrations in two cells in series to demonstrate the apparatus constant derived from cell B could be used to estimate the tritium concentration in cell A. The flow chart is represented in Fig.9.

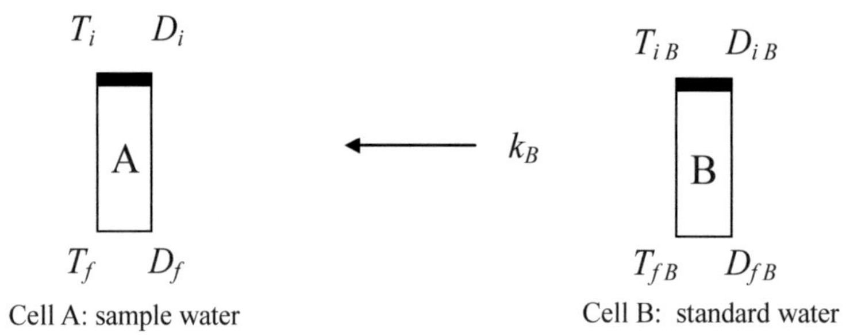

$$T_i = \frac{T_f}{(D_f / D_i)^{k_B}} \longleftarrow k_B = \frac{\ln(T_{fB} / T_{iB})}{\ln(D_{fB} / D_{iB})}$$

T_i : Tritium concentration in sample water before enrichment
T_f : Tritium concentration in sample water after enrichment
D_i : Deuterium concentration in sample water before enrichment
D_f : Deuterium concentration in sample water after enrichment
T_{iB} : Tritium concentration in standard water before enrichment
T_{fB} : Tritium concentration in standard water after enrichment
D_{iB} : Deuterium concentration in standard water before enrichment
D_{fB} : Deuterium concentration in standard water after enrichment

Figure 9. Flow chart of calculating tritium concentration in cell A connected to standard cell B in series using an apparatus constant in the neighbor cell of B.

A tritium contained solution was prepared each time with a concentration of about 8 Bq/L, inputted to both cells of A and B and electrolyzed from the initial weight of 300g to the final one of 60g with the electrolytic current of 3A which corrrespons to the currrent density of about 0.2A/cm^2 in the device. Experimental results are presented in Table 1 (Muranaka et al., 2005). Tritium concentrations in four prepared samples were initially measured to be in the range from 7.56 to 8.12 Bq/L. In Table 1 T_{iB} and D_{iB} in cell B are considered to be equal to T_i and D_i in cell A, respectively because the same prepared solution is divided into two cells of A and B. But T_{fB} and D_{fB} in cell B are considered not to be equal to T_f and D_f in cell A, respectively because the electrolytic condition is not same during the long time electrolytic run although these two cells are connected in series. Apparatus constants calculated from data in Table 1 are shown in Table 2. Tritium and deuterium concentrations after the enrichment in cell A are shown in Table 3. And in Table 4 calculated tritium concentrations using the apparatus constant derived each run in Table 2 and those using an averaged apparatus constant are represented as "Estimated (1)" and as "Estimated (2)", respectively together with the directly measured values. In the case using each apparatus constant differences between the directly measured and estimated results are in the range of 0.7 to 2.5%, with the average difference of 1.5%. Conversely, if we use the averaged apparatus constant of 1.070 from Table 2, the difference increased to the range of 0.7 to 7.5% with the average difference being 3.8%. From this result it is confirmed that the apparatus constant obtained from each electrolytic run is preferable to the averaged apparatus constant of the four results in this experiment.

No.	T_{iB}[Bq/L]	D_{iB}[%]	T_{fB}[Bq/L]	D_{fB}[%]
1	7.56	0.0147	26.61	0.0468
2	7.69	0.0148	27.30	0.0459
3	8.09	0.0147	25.75	0.0442
4	8.12	0.0148	25.81	0.0459

Table 1. Tritium and deuterium concentration in electrolytic enrichment test. Suffix B means standard cell B corresponding to Fig.9. T_{iB} and D_{iB} in cell B are considered to be equals to T_i and D_i in cell A, respectively as explained in the text (Muranaka et al., 2005). [Reprinted with permission from ANS]

No.	T_{fB}/T_{iB}	D_{fB}/D_{iB}	Ratio $(T_{fB}/T_{iB})/(D_{fB}/D_{iB})$	Apparatus constant k_B
1	3.52	3.18	1.106	1.087
2	3.55	3.10	1.145	1.119
3	3.18	3.01	1.059	1.052
4	3.18	3.10	1.025	1.022
Average	3.36	3.10	1.083	1.070
$\pm\sigma$	±0.21	±0.07	±0.05	±0.04
C.V.[%]	6.3	2.3	4.9	3.9

Table 2. Enrichment ratios and derived apparatus constants (Muranaka et al., 2005). [Reprinted with permission from ANS]

No.	T_f[Bq/L]	D_f[%]
1	26.00	0.0469
2	27.77	0.0459
3	26.55	0.0450
4	26.20	0.0469

Table 3. Tritium and deuterium concentrations after the enrichment in cell A.

No.	Direct value [Bq/L]	Estimated (1) [Bq/L]	Difference [%]	Estimated (2) [Bq/L]	Difference [%]
1	7.56	7.37	2.5	7.51	0.7
2	7.69	7.83	1.8	8.27	7.5
3	8.09	8.18	1.1	8.02	0.9
4	8.12	8.06	0.7	7.63	6.0
		Average	1.5	Average	3.8

Table 4. Difference between directly measured and the estimated tritium concentrations in the experiment (Muranaka et al., 2005). [Reprinted with permission from ANS]

4.2. Characteristics of a newly designed electrolytic device used porous electrodes

Electrolytic voltages for three kinds of electrolytic devices described in the paragraph three are shown in Fig.10 (Muranaka & Shima, 2008). (A) is the variation of an electrolytic voltage for a commercially available apparatus, (B) represents that for a newly designed device and (C) is that for the conventional device using corrugated electrodes. Electrolytic current is 6A for (A) and (B) which corresponds to the current density of about 0.3A/cm^2 and 0.4A/cm^2 for the device of (A) and (B) , respectively and the current is 3A (0.2A/cm^2) for (C). Although the structure in the devices of (A) and (B) is different, the electrolytic voltage in both devices are lower than that in the device of (C). Decrease of the voltage will therefore depend on the used materials for the electrodes. The electrolytic time will be shortened because electrolytic current can be increased with small electric power in both devices of (A) and (B). The voltage for the device (B) is somewhat higher than that for (A). One of the origin is the platinum mesh inserted between the anode and SPE film for the device (B) to separate them after the electrolysis easily.

Tritium recovery factors which are defined by the formula of (4) are represented for the device of (A) and (B) in Table 5 (Muranaka & Shima, 2008). Electrolytic current is 6A for all experiments. Electrolytic enrichment was repeated to study the stability of both cells. R3, R5 and R10 in the table mean that the volume reduction factors are three, five and ten, respectively. Coefficient of variation (C.V.) is in the range of 0.5 to 2.5% and the difference in the stability for both devices is not so large. Tritium recovery factor of five times in the device (B) is larger than that in the device (A). This will be caused from the temperature in the sample solution. Sample water in the device (A) is cooled by air cooling, on the other hand that in the device (B) is cooled by water bath.

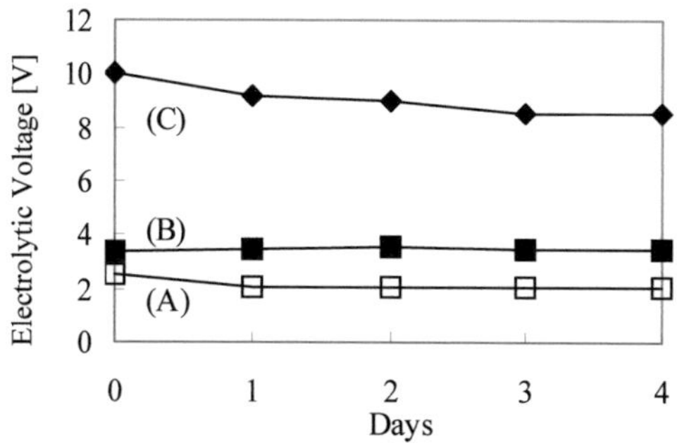

(A) is a commercially available apparatus(TRIPURE XZ027, PERMELEC ELECTRODE LTD), (B) is the newly designed electrolytic device, and (C) is a conventional enriching device equipping with corrugated metal electrodes. The electrolytic current is 6A for (A) and (B) which corresponds to the current density of about 0.3A/cm^2 and 0.4A/cm^2 ,respectively, and the current is 3A (0.2A/cm^2) for (C) (Shima & Muranaka, 2007b, Muranaka & Shima, 2008). [Reprinted with permission from JRA and ANS]

Figure 10. Electrolytic voltages during electrolysis.

No.	(A)	(B)		
	R5	R3	R5	R10
1	0.658	0.902	0.860	0.774
2	0.681	0.910	0.831	0.782
3	0.676	0.913	0.810	0.783
4	0.676	0.910	0.843	0.768
Average	0.673±0.010	0.909±0.005	0.836±0.021	0.777±0.007
C.V.(%)	1.5	0.6	2.5	0.9

Table 5. Comparison of tritium recovery factors obtained using (A) the commercially available apparatus, and (B) the newly designed electrolyzer. R3, R5 and R10 mean that volume reduction factor in the electrolysis are 3, 5 and 10, respectively. C.V. means the coefficient of variation (Muranaka & Shima, 2008). [Reprinted with permission from ANS]

Apparatus constant was studied by Kakiuchi, M. (Kakiuchi, 1999) and tritium and deuterium separation factor were measured by Momoshima, N. et al. (Momoshima, et al., 2005) on the characteristics of the commercially available device.

4.3. Two-stage electrolysis

Electrolytic enrichment is possible by the electrolytic current up to 50A for commercially available apparatus. This ability leads to the shortening of the enrichment time. On the other hand, tritium recovery factor for the newly designed device is larger than that for the large current electrolytic apparatus and the decomposition of the device is easier in the designed device. Therefore we adopted a two-stage electrolysis using the large current electrolytic apparatus in the first stage and the designed device is used in the second stage (Shima & Muranaka, 2007a). In Fig.11 the flow chart is represented for this electrolytic method.

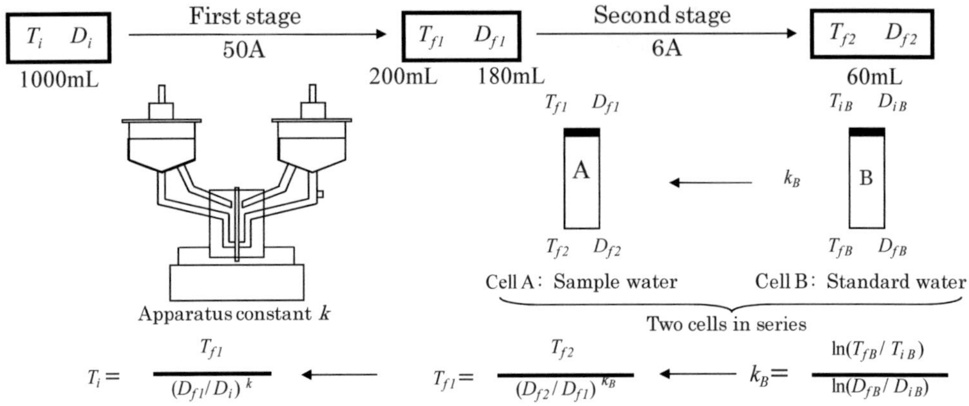

Figure 11. A flow chart of the two-stage electrolytic enrichment. T and D stand for tritium and deuterium concentration, respectively. Suffixes 1 and 2 indicate the stage of the enrichment, while suffixes $_i$ and $_f$ mean before and after enrichment, respectively. Suffix B indicates cell B which contains standard water .

Comparison of tritium recovery factors between one- and two-stage electrolysis is show in Table 6. In the two-stage electrolysis sample water was enriched from 1000mL to 200mL by the commercially available apparatus and from 180mL to 60mL by the designed device as depicted in Fig.11. The volume reduction factor is 5 times in the first stage, three times in the second and the total volume reduction factor is therefore fifteen times. On the other hand, as sample water is enriched in the commercially available apparatus only from 900mL to 150mL by the electrolytic current of 50A and after that it is enriched to 60mL by the current of 20A, the total electrolytic reduction factor is same to the two-stage electrolysis mentioned before.

From Table 6 tritium recovery factor using such two-stage electrolysis is somewhat larger than that using one-stage enrichment. This is due to the higher tritium recovery factor of the second stage electrolysis using designed device showed in Table 5. Since this designed device is easy to decompose, it can be possible to reduce the tritium memory in the previous electrolysis exchanging SPE film in the designed device with a new one after finishing one run. As two-stage electrolysis combined both merits of the electlyzers is useful, this system was adopted in the electrolytic enrichment of seawater described in the next section.

No.	one-stage enrichment	two-stage enrichment
1	0.583	0.659
2	0.593	0.649
3	0.602	0.662
Average	0.593±0.010	0.657±0.007

Table 6. Comparison of tritium recovery factors between one- and two-stage electrolysis.

5. Tritium concentration of land water and coastal seawater collected in Aomori prefecture, Japan

In 1985 the governor in Aomori prefecture, Japan accepted to construct an atomic fuel reprocessing facility at Rokkasho village in this prefecture. This fact motivated us to investigate tritium background concentration in environmental water in this area because tritium would be released to environment after the plant is completed. Low background liquid scintillation counting system was introduced in 1988 in our laboratory and the study began to start.

5.1. Tritium concentration in land water

5.1.1. Tritium concentration in land water twenty years ago

Tritium concentration in precipitation from 1990 to 1993 was measured at that time without pretreatment of electrolytic enrichment. Sampled water of 40mL was mixed to the scintillation liquid of 60mL (Aquasol2, Packard) in a Teflon vial of 100mL and the beta ray released from the sample solution was counted by the liquid scintillation counter (LB II, Aroka). Counting for fifty minutes was repeated four times for one vial and proceeded to the next vial. After all the placed vials on the conveyor belt were finished to count, these counting were repeated for seven cycles. Therefore total counting time for a vial is 1400 minutes. The calculation of tritium concentration was depended on the document issued by Science and Technology Agency at that time (Science and Technology Agency, Japan, 1977).

The concentration error corresponding with statistic error of 1σ is about ±0.3[Bq/L]. The detection limit is represented by (11) (Hagiwara et al., 2012).

$$A = (100/E)\times(1000/V)\times(K/2)\times[K/T_s + ((K/T_s)^2 + 4N_b(1/T_s + 1/T_b))^{0.5}] \qquad (11)$$

Where A means the detection limit with the unit of [Bq/L], K shows the width of the standard deviation (K equals three is used usually), T_s is the measuring time for sample water with the unit of second, T_b is that for background water with the same unit and N_b means the counting efficiency for background sample with the unit of cps.

Sampling was conducted at Hachinohe Institute of Technology, Japan. Precipitation was collected once a day after the precipitation.

The results were represented in Fig.12 to show the variation throughout the year (Muranaka & Honda, 1997). The detection limit was estimated about 0.6[Bq/L] using (11) for the counting condition and the tritium concentration below this value is therefore unreliable.

From the results it was confirmed that tritium concentrations are higher in the season from April to June than those in the period from August to September. This will be explained as follows. The moisture containing somewhat higher tritium concentration is mainly carried from the continent across the Sea of Japan in this season. On the other hand, precipitations in the period from August to September mainly carruied by Typhoon or Tropical cyclone generated at Pacific Ocean where tritium concentration is lower.

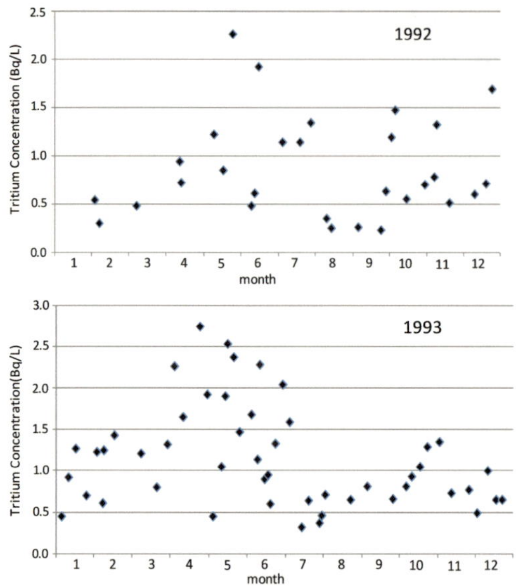

Figure 12. Tritium concentration in the precipitation collected at Hachinohe institute of technology in 1992 and 1993.

Sampling locations of lakes, marshes and rivers in Aomori prefecture, Japan are depicted in Fig.13. Water volume of 1L was sampled at each site once a year on a continued sunny period in autumn.

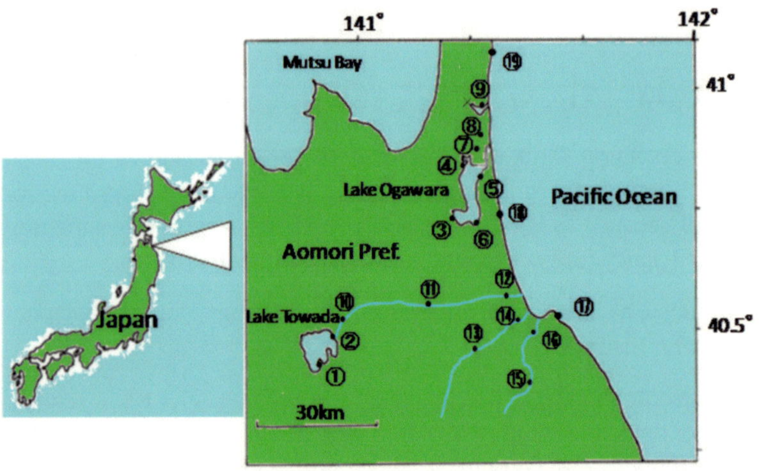

①Lake Towada, Yasumiya, ②Lake Towada, nenokuchi, ③Lake Ogawara, touhoku, ④Lake Ogawara, Rokkasho, ⑤ Lake Ogawara, Yawata, ⑥Lake Ogawara, Misawa, ⑦Tamogi marsh, ⑧Ichiyanagi Marsh, ⑨Obuchi marsh, ⑩Oirase riber, Ishigedo, ⑪Oirase riber, Miyuki bridge, ⑫Oirase riber, Koun bridge, ⑬Mabechi riber, Hukuda bridge, ⑭Mabechi riber, Uruichi, ⑮Niida riber, Simamori, ⑯Niida riber, Ishidearai. ⑰Shirahama beach in Hachinohe city, ⑱Misawa fishing port and ⑲Tomari beach in Rokkasho village. Water samples were not collected in these three points of ⑰～⑲ twenty years ago. Seawater in these sites was collected since 2006 when sample water is able to be enriched by electrolysis. The cross symbol indicates the nuclear reprocessing plant.

Figure 13. Sampling locations of environmental water in Aomori prefecture, Japan

Tritium concentrations are represented in Fig.14 (Muranaka & Honda, 1997). They could be classified into four divisions. A shows the tritium concentration in Lake Towada. Since the groundwater which contains fallout tritium wells up from the bottom in Lake Towada, tritium concentration will be higher than other divisions. B is the division from Lake Ogawara which is a brackish lake. The tritium concentration is lower due to the mixing entering river water and inflow of seawater. C is the water collected at Obuchi marsh which is also brackish. But the mixing amount of seawater is so large that tritium concentration will become lower. Since river water in the division D is the mixed water of groundwater and precipitation, the tritium concentration is lower than the groundwater only like in Lake Towada. But the tritium concentration in D is higher than those in water sampled from brackish lakes.

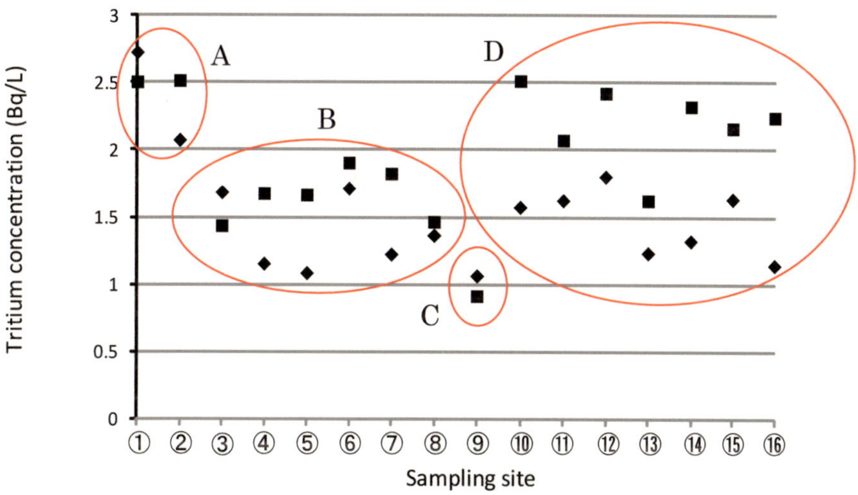

Numbers in horizontal axis indicate sampling sites in the same way denoted in Fig.13. The notation of ◆ and ■ mean tritium concentrations in the sample collected in 1992 and 1993, respectively.

Figure 14. Tritium concentration in environmental water collected in 1992 and 1993 at Aomori area, Japan.

5.1.2. Recent tritium concentration in land water

Recent tritium concentration in land water has decreased gradually due to the lowering of the influence by the fallout tritium generated until the early 1960s. Land water samples were collected in 2010 at the sevral points same to those sampled twenty years ago. The volume of sample water is reduced from 800mL to 200mL by the electrolytic current of 50A and from 200mL to 80mL by 20A using the commercially available apparatus only with the volume reduction factor of ten times because tritium concentration is not so lower like seawater. The volume reduction factor is ten times corresponds to the tritium enriched factor of about 7.2 times. The results are presented in Fig.15 with the same divisions to Fig.14. Tritium concentration is in the range from 0.4Bq/L to 0.55Bq/L except for the sample collected at Obuchi marsh. Tritium concentration is decreased comparing with those measured about twenty years ago. It is pointed out that tritium concentration in the sample collected at Lake Towada decreased and it becomes closer to that sampled in Lake Ogawara recently.

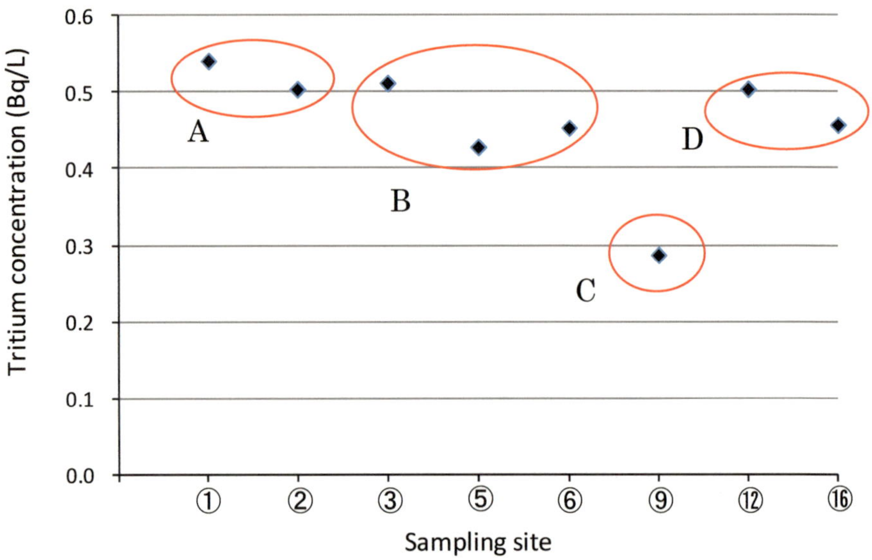

Figure 15. Tritium concentration in environmental water collected in 2010 at Aomori area. Numbers in the horizontal axis are same as those in Fig.13.

5.2. Tritium concentration in coastal seawater

Sampling points were selected along the coastal beach of the Pacific Ocean in Aomori prefecture including two southern sites, a northern site and a nearby site from the plant which is noted by the cross symbol in Fig.13. These sampling sites are indicated as the sites of ⑰～⑲ and ⑨. Samples were collected two or three times once a year. After sampled water was distilled under reduced pressure to remove contained salts, it was enriched by the two-stage electrolysis described in the secion four.

Tritium concentration for the samples collected from Shirahama beach are shown in Fig.16 (Muranaka & Shima, 2011). Shirahama beach locates fifty five kirometers far from the nuclear reprocessing plant in the southern direction. (a) and (b) in Fig.16 indicate beginning (March, 2006) and the stop time (December, 2008) of the active test in the plant, respectively. Tritium concentration is less than 0.4 Bq/L outside of the period for the active test. But tritium concentrations are sometimes increased during the test. It was confirmed that the most prominent increase in tritium concentration observed on January 2008 is due to the tritium-contaminated waste water released from the plant (Muranaka & Shima, 2011). These temporary increase were observed at other three sampling sites.

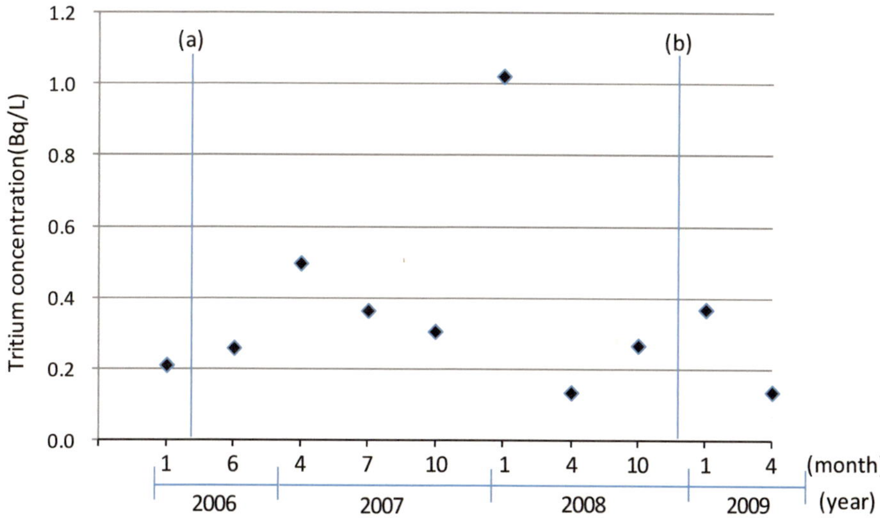

(a) and (b) in the figure represent beginning (March, 2006) and the stop time (December, 2008) of the active test in the plant, respectively.

Figure 16. Tritium concentration in seawater collected at Shirahama beach in Hachinohe, Aomori, Japan (Muranaka & Shima, 2011). [Reprinted with permission from ANS]

6. Conclusion

Electrolytic enrichment using SPE film is an indispensable technology to study tritium concentration in environment nowadays. These devices are generally classified into two types. One is an oxygen-hydrogen separating apparatus and another is oxygen-hydrogen non-separating type. The former is commercially available and is useful to electrolyze a sample quickly with large electrolytic current. On the other hand the latter device also has some merits. It has heigher recovery factor than the former device when the electrolyric cell is cooled by a chilly water bath and can prevent a memory effect from the previous run by exchanging SPE film with a new one. The reduced weight by electrolysis in the cell is able to measure by an electrolytic balance precisely. This last merit is especially useful in the case when the electrolytic volume after the enrichment is little due to the large volume reduction factor or the sample volume is originally little like water sample contained in plant. Since each electrolytic device has its own merit, two-stage electrolysis will be one of a practical method for the sample which has lower tritium concentration such as seawater. Among described analysis in environmental water samples in Aomori area coastal seawater was enriched by the two-stage electrolysis and it was confirmed that the tritium-contaminated waste water released from atomic fuel reprocessing plant at Rokkasho impacted to the tritium concentration in coastal seawater more than fifty kirometer far from the plant. Oxygen hydrogen non-separating electrolyzer has a potential to be used connecting in series to enrich tritium concentrations of several samples at once like a conventional alkaline method.

Author details

Takeshi Muranaka *
Hachinohe Institute of Technology, Japan

Nagayoshi Shima
Hachinohe Institute of Technology, Japan
Entex co, Japan

Acknowledgement

One of the authors, Muranaka T. wishes to thank a guraduated student, Mr. Yamashita, J. and under graduated studens for taking part in this study.

7. References

Ogiwara, K. et al. (2012). Current Status and History for the Lowering of Background of Liquid Scintillation Counters, *RADIOISOTOPES*, Vol.61, No.2 (February 2012)pp.79-85, ISSN0033-8303(in Japanese)

Inoue, Y. & Miyamoto, K. (1987). Statistical Evaluation of Variations in Electrolytic Enrichment Factors of Tritium, *International Journal of Radiation, Applications and Instrumentation. Part A. Applied Radiation and Isotopes,* Vol.38, No.12, pp.1013-1018

Inoue, Y. and Dang Thi Phuong Nam. (1994). Dependence of Electrochemical Separation Factor of Tritium on Temperature and Current Density, *Abstracts of the 31st Annual Meeting on Radioisotopes in the Physical Sciences and Industries*, p.31, Tokyo, Japan, July11-13, 1994 (in Japanese)

Kakiuchi, M. et al. (1991). Natural Tritium Measurements: Determination of the Electrolytic Enrichment Factor of Tritium, *International Journal of Radiation, Applications and Instrumentation. Part A. Applied Radiation and Isotopes,* Vol.42, No.8, pp.741-748

Kakiuchi, M. (1999). Correlation between Enrichment Factor of Deuterium and That of Tritium by Electrolysis Using Solid Polymer Electrolyte, *RADIOISOTOPES*, Vol.48, No.2, (February 1999), pp.79-86, ISSN 0033-8303 (in Japanese)

Momoshima, N., Nagano, Y. and Toyoshima, T. (2005). Electrolytic Enrichment of Tritium with Solid Polymer Electrolyte for Application to Environmental Measurements, *Fusion Science and Technology,* Vol.48, No.1(July/August2005), pp.520-523, ISSN: 1536-1055, *Proceedings of the Seventh International Conference on Tritium Science and Technology, Baden-Baden, Germany, September* 12-17, 2004

Muranaka, T. & Honda, K. (1996). Tritium Concentration in Environmental Water Samples Collected from the Area of Aomori Prefecture, *Proceedings of the 5th Low Level Counting Conference using Liquid Scintillation Analysis,* pp.18-25, Yokohama, Japan, June20-21, 1996

---

* Corresponding Author

Muranaka, T. and Honda, K. (1997). Tritium Concentration Measurements of Environmental Water Samples in Aomori Prefecture, *The Bulletin of Hachinohe Institute of Technology*, No.16, (Febrary1997), pp.169-175, ISSN 0287-1866, (in Japanese)

Muranaka, T. et al. (1997). Tritium Enrichment Device by Electrolysis Using Solid Polymer Electrolytic Film, *The Bulletin of Hachinohe Institute of Technology*, No.16, (February 1997), pp.177-182, ISSN 0287-1866 , (in Japanese)

Muranaka, T., Shima, N. and Sato, H. (2005). A Study to Estimate Tritium Concentrations of 1 Bq/L or Lower in Water Samples, *Fusion Science and Technology*, Vol.48, No.1(July/August2005), pp.516-519, ISSN: 1536-1055, *Proceedings of the Seventh International Conference on Tritium Science and Technology, Baden-Baden, Germany, September 12-17, 2004*

Muranaka, T. & Shima, N. (2008). Improved Electrolyzer For Enrichment of Tritium Concentrations in Environmental Water Samples, *Fusion Science and Technology*, Vol.54, No.1 (July2008), pp.297-300, ISSN: 1546-1055, *Proceedings of the Eighth International Conference on Tritium Science and Technology, Rochester, New York, USA, September 16-21, 2007*

Muranaka, T. & Shima, N. (2011). Variation of Tritium Concentration in Coastal Seawater Collected Along the Pacific Coast in Aomori Prefecture, *Fusion Science and Technology*, Vol.60, No.4, (November2011), pp.1264-1267, ISSN:1536-1055, *Proceedings of the Ninth International Conference on Tritium Science and Technology*, Nara, Japan, October *24-29, 2010*

Saito, M. et al. (1996). Tritium Enrichment by Electrolysis Using Solid Polymer Electrolyte, *Radioisotopes*, Vol.45, No.5, (May 1996), pp.258-292, ISSN 0033-8303 (in Japanese)

Saito, M. (1996). Automatic stop type SPE tritium enrichment apparatus, *Proceedings of the 5th Low Level Counting Conference using Liquid Scintillation Analysis*, pp.102-110, Yokohama, Japan, June20-21, 1996

Satake, H. and Takeuchi, S. (1987). Electrolytic enrichment of tritium with Ni-Ni and Fe-Ni electrodes, *Annual report of Tritium Research Center, Toyama University, Japan*, Vol.7, pp.73-80, ISSN:0287-1408 (in Japanese)

Sato, H., Muranaka, T. & Shima, N. (2005). Characteristics of Stable Hydrogen Isotope Ratio in Precipitation in the Hachinohe Area, Japan, *Radioisotopes*, Vol.54, No.7, pp.229-232, ISSN0033-8303

Science and Technology Agency, Japan (1977). Tritium analysis (in Japanese)

Shima, N. (2007). A study of the electrolytic enrichment prior to measure tritium concentration in environmental water, Doctorial thesis, Hachinohe Institute of Technology (in Japanese)

Shima, N. & Muranaka, T. (2007a). Two-stage electrolysis to enrich tritium in environmental water, *Proceedings of the International Symposium on Environmental Modeling and Radioecology*, pp.247-250, Rokkasho, Aomori, Japan, October 18-20, 2006

Shima, N. & Muranaka, T. (2007b). Characteristics of a Newly Designed Electrolyser to

Enrich Tritium in Environmental Water, *Radioisotopes*, Vol.56, No.8, (August 2007), pp.455-461, ISSN 0033-8303 (in Japanese)

Yamashita, J. and Muranaka, T. (2011). Measurement of Tritium Concentrations in Seawater Collected along the Pacific Coast in Aomori Prefecture after Enrichment by Electrolysis, *The Bulletin of Laboratory for Energy, Environment and Systems Hachinohe Institute of Technology,* Vol.9, (March2011), pp.9-14, ISSN 1347-658X, (in Japanese)

Chapter 8

Analysis of Kinetics Parameters Controlling Atomistic Reaction Process of a Quasi-Reversible Electrode System

Yuji Imashimizu

Additional information is available at the end of the chapter

http://dx.doi.org/10.5772/51896

1. Introduction

For understanding the mechanism of electrolysis it is important to estimate kinetics parameters controlling the atomistic reaction process of metal electrode that is polarized in an electrolyte solution, but it seems not to have been performed satisfactorily. The reason for this is attributed to the fact that because actual electrode reactions proceed quasi-reversibly via consecutive two processes which consist of surface reaction and volume diffusion of ions involved in the reaction, the expression for its current density/overpotential relationship have become complex and not been presented explicitly. This is also related to the subjects of studies concerning the process of deposition or dissolution of atoms in crystal growth or its dissolution.

It is well known that the etch pits having a crystallographic symmetry are formed at dislocation sites of the low indices surfaces of a crystal which was etched under a specified condition (e.g. Gilman et al., 1958; Young, Jr., 1961). The dislocation etch pit is thought to be formed via a nucleation and growth process of two-dimensional pits at the dislocation site or via a spiral dissolution of the surface step which is caused by screw dislocation (Burton et al., 1951; Cabrela and Levine, 1956). Therefore elucidation of its formation mechanism is important for understanding of the surface step motion which is thought to play major role in the dissolution process of a crystal, and dissolution kinetics of crystals in the etch pit formation has been investigated and discussed by some researchers (e.g. Ives and Hirth, 1960; Schaarwächter, 1965; Jasper and Schaarwächter ,1966; Van Der Hoek et al., 1983) so far.

However the research concerning parameters controlling surface step motion in the dissolution of crystals has not been satisfactorily performed. Especially it has not been examined quantitatively except for a few studies (e.g. Onuma, 1991). This is principally due to the reason that because the dissolution of a crystal proceeds generally via a dissolution reaction of surface atom and diffusion process of the dissolved atom (ion) into interior of solution, it is difficult to experimentally inspect the dissolution kinetics of surface step which depends on both processes. Since the dissolution rate of a metal crystal which is anodically dissolved under polarization in an electrolyte solution can be investigated by measurement of current density, dissolution mechanism of metal crystals has been researched electrochemically (e.g. Despic and Bockris, 1960; Lee and Nobe, 1986). However because of the same reason as the above mention, discussions on the results have become complex and not always contributed to understanding of surface step motion.

Recently, however, it has been proposed by the author that an expression to analyze the relationship between anodic current density and overpotential of a quasi-reversible electrode system including both the consecutive reaction processes is derived explicitly on the basis of an appropriate assumption (Imashimizu, 2010, 2011). According to the analysis, if the anodic and cathodic diffusion-limited current densities are measured for a given quasi-reversible electrode system, we can experimentally determine the kinetics parameters controlling dissolution process of crystals of the metal electrode, by assuming expressions for the activation and concentration overpotentials which are driving forces of surface reaction process and volume diffusion process respectively.

Thus the dissolution rates at dislocation-free and edge dislocation sites of (111) surface when a copper crystal was anodically dissolved in an electrolyte solution are investigated and discussed based on the above thinking, in this chapter. The relationships between anodic current density and overpotential are analyzed and discussed electrochemically by using the method developed for anodic dissolution processes of quasi-reversible electrode as described above. Activation enthalpy, transfer coefficient and surface concentrations of the ions involved in the dissolution process are experimentally estimated, and kinetics parameters controlling anodic reaction of the copper crystal/electrolyte system are quantitatively examined. An expression for the vertical dissolution rate at dislocation site is proposed based on a nucleation model of two-dimensional pit, and the critical free energy change at nucleation is quantitatively examined.

2. Experimental procedures for study of dissolution kinetics of copper crystals

2.1. Preparation of specimens

Single crystals of copper with [111] direction about 10 mm in diameter were prepared from the starting material of re-electrolyzed copper of 99.996 % purity by using the pulling method. They were divided into the cylindrical crystals approximately 15 mm length by a strain-free cutting. A terminal for detection of electric current and potential was soldered to an end surface of the cylindrical crystals. Another end surface was chemically polished so that the deviation of surface orientation from [111] direction is within 8.7×10^{-3} rad, and was further electrolytically polished in a high concentrated phosphoric acid solution. The crystal specimen was embedded in a Teflon holder with paraffin so that its polished surface is exposed. Then it was supplied for the electrolysis experiment after the boundary portion between the paraffin and the periphery of polished surface was covered with a vinyl seal having a hole 6 mm in diameter (Watanabé et al., 2003).

2.2. Apparatus for potentiostatic electrolysis

Schematic diagram of the electrolytic cell for this experiment is shown in Fig.1 (Watanabé et al., 2003). The crystal specimen was immersed in the electrolyte solution which consists of 5 kmol m^{-3} NaCl, 0.25 kmol m^{-3} NaBr and 10^{-4} kmol m^{-3} CuCl (Jasper and Schaarwächter, 1966) so that (111) surface of copper is located at approximately 5mm below the surface of electrolyte solution, and was held at a specified temperature. Then, the crystal specimen was set under a constant overpotential, and (111) surface of the crystal was anodically dissolved for a prescribed time, while anodic current density/time curve was recorded. The potentiostatic electrolysis experiments were performed at a range of lower overpotential and at a range of higher overpotential. After that the structure of dissolved surfaces were observed by use of the optical microscope system equipped with lens for interferometry.

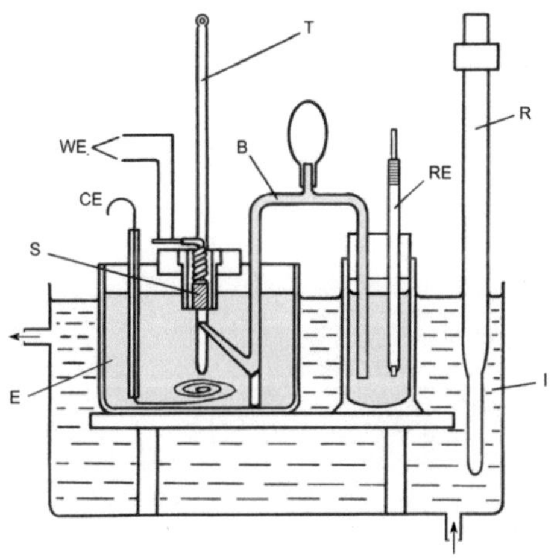

Figure 1. Schematic diagram of electrolytic cell. S: sample; E: electrolyte; WE: terminal for potential and current; RE: saturated calomel electrode; CE: counter electrode of platinum wire; B: salt bridge; I: thermobath; R: Regulator; and T: thermometer.

2.3. Features of anodic current density/time curves and structure of dissolved surface

Figure 2 shows typical anodic current density/time curves which were recorded while the copper crystal was anodically dissolved for 360 s or 600 s at the respective overpotentials. Anodic current density under any condition decreases steeply immediately after start of electrolysis and reaches a nearly constant current density i_s when it was carried out at an overpotential lower than about 125 mV as shown by the curve of 87 mV in Fig.2. Figures 3 (a), (b) and (c) are the optical micrographs of the (111) surfaces which were dissolved for 600 s at overpotentials in a range of 60 mV to 125 mV being held at 298K. The surfaces are rather smooth though etch pits tend to be formed as overpotential increases.

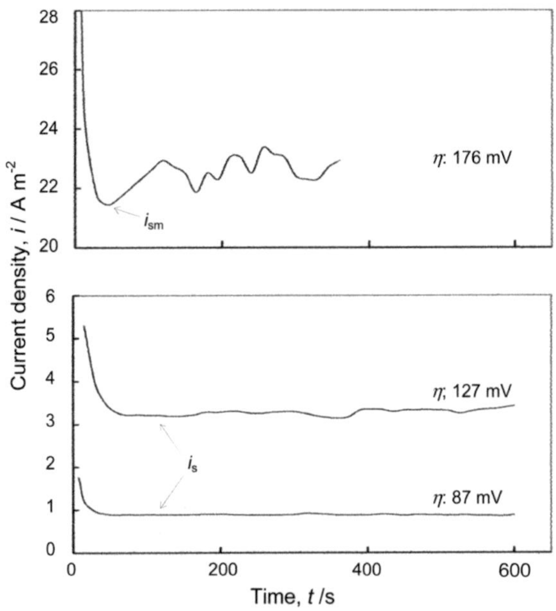

Figure 2. Anodic current density/time curves under potentiostatic electrolysis at 298K.

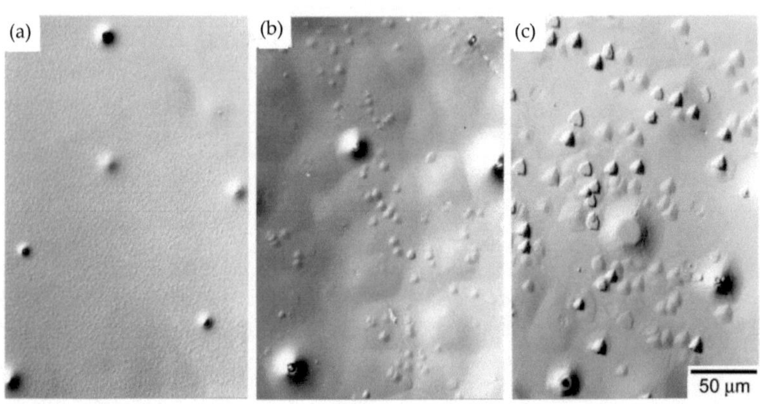

Figure 3. Optical micrographs of the surfaces which were anodically dissolved at lower overpotentials of (a): 58 mV, (b): 88 mV, and (c): 108 mV.

On the other hand, the current density reaches a minimum current density i_{sm} that is pointed by arrow after the initial steep decrease when an overpotential higher than about 125 mV was applied. Then it tends to increase gradually along with fluctuating and take a higher steady value as shown by the curve of 176 mV. Figures 4 (a), (b), and (c) are optical micrographs of the surfaces which were dissolved for 300 s at 156, 166, and 176 mV respectively being held at 298K. One can see that etch pits are significantly formed.

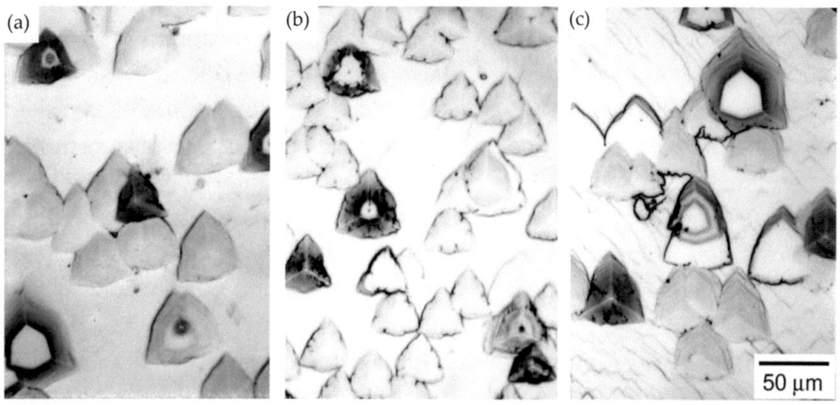

Figure 4. Optical micrographs of the surface that were anodically dissolved under potentiostatic electrolysis at higher overpotentials of (a): 156 mV; (b): 166 mV; (c): 176 mV.

2.4. Measurement of anodic current density

2.4.1. Steady anodic current densities at lower overpotentials

The initial steep decrease of current density is principally due to the fact that a diffusion layer of the dissolved atoms (ions) forms in the neighborhood of crystal surface in process of time so as to decrease the undersaturation which is driving force for the dissolution. Therefore an approximately constant current density after its initial steep decrease is thought to be a steady current density i_s which flows accompanying with the consecutive two dissolution processes consisting of surface reaction and volume diffusion of dissolved atoms. This shows that the copper crystal/electrolyte system is a quasi-reversible electrode. Also the steady anodic current densities at lower overpotentials are thought to have only a little influence of formation of dislocation etch pits. Thus we assume that i_s is related to the vertical dissolution rate v_s at dislocation-free site of surface, which is given by the following expression (Schaarwächter and Lücke, 1967):

$$v_s = \frac{\Omega}{ne} i_s \tag{1}$$

where e [C] is elementary charge (electronic charge), n [1] the charge number transferred at reaction and Ω [m^3] the atomic volume.

In this experiment, the potentiostatic electrolysis at overpotentials in a range from about 60 mV to 125 mV were carried out for 600 s at each temperature of 268, 283, 298 and 308 K and the relationships between the steady anodic current densities i_s and applied overpotentials η were investigated.

2.4.2. Minimum anodic current densities at higher overpotentials

On the other hand, the current density reaches a minimum after an initial steep decrease as shown in the curve of 176 mV in Fig.2 when an overpotential higher than about 125mV was applied. Then it tends to increase gradually together with fluctuating and take a higher steady value as described in Section 2.3. This is thought to be due to the fact that etch pits remarkably formed at dislocation sites and grew along with time under higher overpotential as shown in Fig.4 (Schaarwächter and Lücke, 1967; Imashimizu and Watanabé, 1983). That is, it is because nucleation and growth of etch pits at dislocation site resulted in an increase of the anodic current density which represents an average dissolution rate of whole surface exposed to electrolyte solution as the areas occupied by etch pits increase. Based on the above knowledge, we assume that the initial minimum current density i_{sm} under potentiostatic electrolysis at higher overpotentials is approximately equal to a current density that is equivalent to the dissolution rate of dislocation-free site of surface because the contribution to anodic current density of dislocation etch pit formation is thought to be a little in the initial stage of electrolysis. That is, an average value of i_{sm} was assumed to give the vertical dissolution rate at dislocation-free site of surface approximately as represented by the relation:

$$v_s \approx v_{sm} = \frac{\Omega}{ne} i_{sm} \tag{2}$$

Thus the electrolysis experiment was carried out for a prescribed time from 60 s to 360 s at each overpotential of 156, 166, 176 and 186 mV keeping the temperature at 298 K and at each temperature of 268, 283, 298 and 308 K under an overpotential of 176 mV. The initial minimum current densities i_{sm} were obtained from the anodic current density/time curves measured under every condition.

2.5. Measurement of polarization curve and estimation of the diffusion-limited current densities

It needs to estimate activation overpotential η_a and concentration overpotential η_c for analyzing the relationship between anodic current density and applied overpotential as described in Section 1. Thus the polarization curves in a range of overpotential of about -400 mV to 400 mV were measured three times at each temperature of 298K and 308K by the potential step method. Then the anodic and cathodic diffusion-limited current densities were estimated.

2.6. Direct measurement of dissolution rates of surface

2.6.1. Vertical dissolution rate of surface

After the (111) surface of a copper crystal specimen was anodically dissolved at every condition of specified overpotrntials and temperatures as described in Section 2.4.2, it was observed by use of the optical microscope equipped with objective lenses for two-beam interferometry and multiple interferometry.

Figure 5 (a) shows the micrograph of a part of boundary region between the crystal surface exposed to the electrolyte solution and the peripheral portion covered with vinyl seal, which was photographed with two-beam interferometry mode. The vertical dissolution amounts s of surface shown by the illustration was measured from a deviation of the interference stripes caused by the step which was formed at that boundary region after dissolved. The vertical dissolution amounts s of surface under each condition was plotted against dissolution time t. The increasing rate $\dot{s}$ of s with t was obtained from the gradient of each linear relationship, and the vertical dissolution rate of surface under every condition was estimated by the $\dot{s}$.

2.6.2. Dissolution rates at dislocation site of surface

Figures 5 (b) and (c) show a pair of micrographs of identical dislocation etch pits formed on dissolved surface which were photographed with optical mode and multiple interferometry mode. In this work, the depth d of the dark (deep) pits that were formed at positive edge dislocation sites (see Appendix A1) were measured by drawing the vertical cross sections of the pits that is shown by the illustration with use of the micrograph pairs such as Figs.5 (b) and (c). Also the width w (average distance from center to the three sides of pit) of those dark pits that is shown by the illustration were measured on the micrograph such as Fig.5 (b). Measurements of the depth and width of pit were performed about more than 20 dark pits formed on the surface dissolved under every condition, and the respective average values d and w were obtained. The depth d and the width w of dark pits were plotted against dissolution time t.

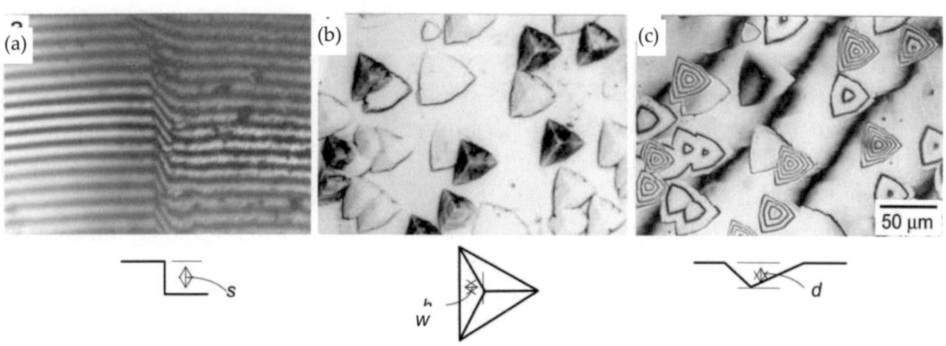

Figure 5. (a): Two-beam interferometry micrograph at the boundary between dissolved and undissolved surfaces; (b): Optical micrograph; (c): Multiple interferometry micrograph of the same view as b.

The increasing rate $\dot{d}$ of d with t was obtained from the gradient of each linear relationship, and the vertical dissolution rate v_{ed} at edge dislocation site was estimated by

$$v_{ed} = \dot{d} + v_s, \tag{3}$$

where v_s means the vertical dissolution rate at dislocation-free site of the surface. Also the increasing rate $\dot{w}$ of w with t was obtained from the gradient of each linear relationship, and the lateral dissolution rate v_w at edge dislocation site was estimated from the relation:

$$v_{\mathrm{w}} = \frac{w}{d} v_{\mathrm{ed}} \approx \frac{\dot{w}}{\dot{d}}\left(\dot{d} + v_{\mathrm{s}}\right). \tag{4}$$

2.7. Analysis of relationship between current density and overpotential

Under potentiostatic electrolysis of the copper/electrolyte system in the present experiment, the copper crystal is thought to be dissolved accompanying an anodic current according to a simple electrode reaction expressed by the following equation (Lal and Thirsk, 1953; Jasper and Scaarwächter, 1966):

$$\mathrm{Cu} + 2\mathrm{Cl}^- \rightarrow \mathrm{CuCl_2}^- + \mathrm{e}, \tag{5}$$

where the contribution to current density of reaction of Br^- ion involved in dissolution process as inhibitor is assumed to be disregarded. The anodic current density i_s flowing steadily at an applied overpotential η is generally expressed by a relation:

$$i_{\mathrm{s}} = i_0 \left\{ \left(\frac{C_{\mathrm{Cl^-}}}{C^0_{\mathrm{Cl^-}}} \right)^2 \exp\left(\frac{\alpha n e \eta}{kT} \right) - \frac{C_{\mathrm{CuCl_2^-}}}{C^0_{\mathrm{CuCl_2^-}}} \exp\left(-\frac{(1-\alpha) n e \eta}{kT} \right) \right\}, \tag{6}$$

where exchange current density i_0 is represented by

$$i_0 = ne\beta k_{\mathrm{s}} {C^0_{\mathrm{Cl^-}}}^{2(1-\alpha)} {C^0_{\mathrm{CuCl_2^-}}}^{\alpha} \nu \exp\left(-\frac{\Delta H_0}{kT} \right) \tag{7}$$

(Tamamushi, 1967; Maeda 1961). k_s is surface density of kink that is active site at dissolution of surface atom, α the transfer coefficient, ΔH_0 the activation energy (enthalpy) at dissolution of an atom, ν the atomic frequency, and β a supplementary factor of rate constant of electrode reaction. Also, C_{Cl^-} and $C_{CuCl_2^-}$ are the surface concentrations of Cl^- and $CuCl_2^-$ ions involved in a steady anodic dissolution, and $C^0_{Cl^-}$ and $C^0_{CuCl_2^-}$ the ones in equilibrium state. They are represented as a relative surface density as follows.

If the electrolyte solution contacting with crystal surface contained X ions of m kmol m^{-3}, the surface concentration C_X [1] can be expressed by the following relation:

$$C_{\mathrm{X}} = \frac{m \times 10^3 N_{\mathrm{A}} \left[\mathrm{m}^{-3}\right] \times \xi \left[\mathrm{m}\right]}{(bb^*)^{-1} \left[\mathrm{m}^{-2}\right]}, \tag{8}$$

where N_A is the Avogadro constant, bb^* the area occupied by an atom and ξ the thickness of electrolyte solution layer contacting with the crystal surface (Imashimizu, 2011).

The anodic dissolution of copper crystal in this experiment is thought to proceed quasi-reversibly with a surface reaction and volume diffusion of dissolution atom as described in Section 2.4.1. So we assume that the activation overpotential η_a and the concentration overpotential η_c are written by

$$\eta_a = \eta - \eta_c \text{ and } \eta_c = \frac{kT}{ne}\ln\left\{\frac{C_{CuCl_2^-}/C^0_{CuCl_2^-}}{\left(C_{Cl^-}/C^0_{Cl^-}\right)^2}\right\}, \tag{9}$$

where the folloing relations:

$$\frac{C_{Cl^-}}{C^0_{Cl^-}} = 1 - \frac{i_s}{i_{lCl^-}}, \quad \frac{C_{CuCl_2^-}}{C^0_{CuCl_2^-}} = 1 - \frac{i_s}{i_{lCuCl_2^-}} \tag{10}$$

are given, if i_{lCl^-} and i_{lCuCl2^-} are the anodic and cathodic diffusion-limited current densities of the electrode reaction respectively (Tamamushi, 1967). Thus activation overpotential η_a and concentration overpotential η_c are assumed to be given by Eqs.(9) and (10), when the anodic dissolution of copper crystal proceeds steadily at an applied overpotential η by a quasi-reversible electrode reaction of Eq.(5). Also surface undersaturation σ is defined by

$$\sigma = 1 - \exp\left(-\frac{ne\eta_a}{kT}\right) = 1 - \exp\left(-\frac{ne\eta}{kT}\right)\left\{\frac{C_{CuCl_2^-}/C^0_{CuCl2^-}}{\left(C_{Cl^-}/C^0_{Cl^-}\right)^2}\right\}. \tag{11}$$

Then, the Eq.(6) is reduced to

$$i_s = ne\beta k_s {C^0_{Cl^-}}^{2(1-\alpha)} {C^0_{CuCl_2^-}}^{\alpha} \left(\frac{C_{Cl^-}}{C^0_{Cl^-}}\right)^2 \sigma\nu\exp\left(-\frac{\Delta H_0 - \alpha ne\eta}{kT}\right) \tag{12}$$

by using Eqs. (7), (9) and (11). Also Eq.(12) leads to the following relation:

$$i_s\left(\frac{C_{Cl^-}}{C^0_{Cl^-}}\right)^{-2}\sigma^{-1} = i_0(T)\exp\left(\frac{\alpha ne\eta}{kT}\right). \tag{13}$$

Thus if the anodic and cathodic diffusion-limited current densities i_{lCl^-} and i_{lCuCl2^-} are obtained, the experimental relationship of i_s/η would be represented with use of Eqs.(10) and (11) by Eq.(13). Then α and $i_0(T)$ would be estimated from the gradient and the constant term of the linear relationship of $\ln\{i_s(C_{Cl^-}/C^0_{Cl^-})^{-2}\sigma^{-1}\}$ vs. $ne\eta/kT$. Also ΔH_0 would be estimated from the gradient of the linear relationship of $\ln\{i_0(T)\}$ vs. $1/T$.

On the other hand, concerning the complex term consisting of surface concentrations of Cl^- and $CuCl_2^-$ ions,

$$C_{\mathrm{Cl^-}}{}^{2(1-\alpha)} C_{\mathrm{CuCl_2^-}}{}^{\alpha} = C^0{}_{\mathrm{Cl^-}}{}^{2(1-\alpha)} C^0{}_{\mathrm{CuCl_2^-}}{}^{\alpha} \left(\frac{C_{\mathrm{Cl^-}}}{C^0{}_{\mathrm{Cl^-}}} \right)^2 \exp\left(\frac{\alpha n e \eta_{\mathrm{c}}}{kT} \right) \tag{14}$$

is lead from Eq.(9). Therefore applying Eq.(14) to Eq.(12) lead to

$$i_{\mathrm{s}} = n e \beta k_{\mathrm{s}} C_{\mathrm{Cl^-}}{}^{2(1-\alpha)} C_{\mathrm{CuCl_2^-}}{}^{\alpha} \sigma \nu \exp\left(-\frac{\Delta H}{kT} \right), \tag{15}$$

where ΔH is given by the relation:

$$\Delta H = \Delta H_0 - \alpha n e \eta_{\mathrm{a}}. \tag{16}$$

We can see that Eqs.(15) and (16) are formulae for the steady current density expressed with use of the parameters β, k_s, C_{Cl-}, C_{CuCl2-}, σ, α and ΔH involved in the surface reaction process when the anodic dissolution progresses steadily by a quasi-reversible electrode reaction.

Thus undersaturation σ, transfer coefficient α and activation enthalpy ΔH_0 for the anodic dissolution reaction of copper crystal/electrolyte system will be estimated from experimental results, and a supplementary factor β and kink density k_s will be examined by a model of crystal dissolution in this study.

3. Experimental results

3.1. Polarization curves and undersaturation in anodic dissolution

The polarization characteristic of the copper crystal/electrolyte system at 298K is shown in Fig. 6. The anodic and cathodic diffusion-limited current densities i_{lCl-} and $i_{lCuCl2-}$ shown in the diagram were obtained by averaging the values measured three times. Table 1 shows those diffusion-limited current densities obtained from the polarization characteristics measured at 298 and 308 K by a similar method.

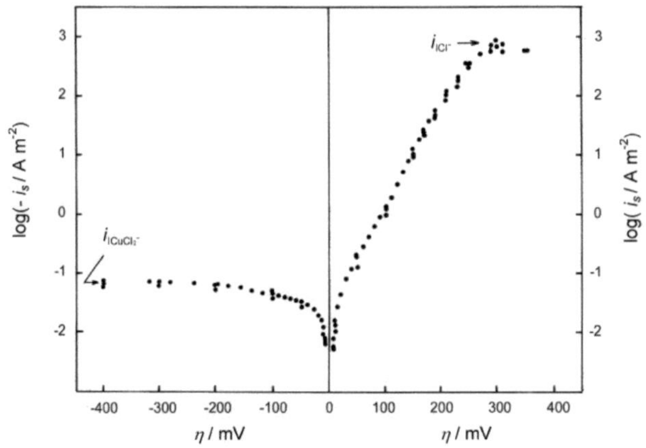

Figure 6. Polarization curve of the copper crystal/elecrolyte system. i_{lCl-} and $i_{lCuCl2-}$ mean the anodic and cathodic difffusion-limited current densities.

T /K	i_{lCl^-}/A m^{-2}	$i_{lCuCl_2^-}$/ A m^{-2}
298	827	-0.0732
308	1072	-0.156

Table 1. Measurements of the anodic and cathodic diffusion-limited current densities i_{lCl^-} and $i_{lCuCl_2^-}$.

The undersaturation σ were estimated from experimental polarization characteristics such as Fig.6 with use of Eqs.(10) and (11). The diagram that plotted σ against $ne\eta/kT$ in a range of $(ne\eta/kT)$ about 0 to 6 is shown in Fig.7. The black dots in the diagram show the values of σ which are calculated from the $(i_s/i_{lCl^-})/(ne\eta/kT)$ relationship that was derived by substituting the experimental values i_0, α, i_{lCl} and i_{lCuCl2} into Eq.(6).

The experimental relationships of $\sigma/(ne\eta/kT)$ at 298K and 308K approximately consist with each other, and also with the calculated relationship. However, the experimental curves of $\sigma/(ne\eta/kT)$ deviate from the calculated curve in a range of $(ne\eta/kT)$ larger than about 5. This is because the experimental current density includes an increase of current density attributed to significant formation of etch pits at higher overpotentials than about 125 mV. We assumed that the $\sigma/(ne\eta/kT)$ relationship does not almost depend on temperature from the result of Fig.7.

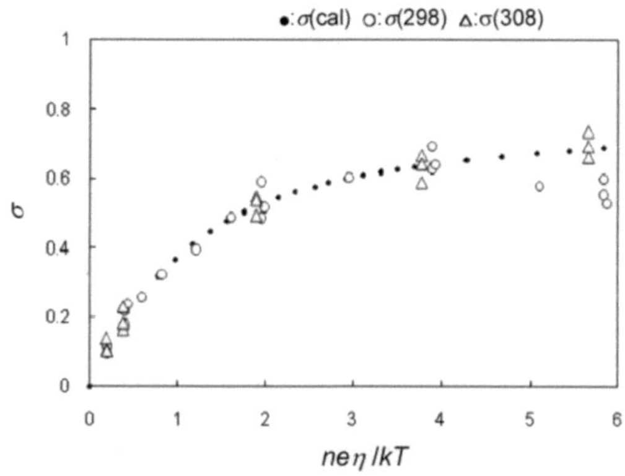

Figure 7. Plots of undersaturation σ against normalized overpotential $ne\eta/kT$. $\sigma(298)$ and $\sigma(308)$ designate experimental values at 298K and 308K respectively. σ(cal) is the calculated one.

3.2. Estimations of parameters controlling exchange current density

Figure 8 is the diagram that plotted the steady current densities against overpotentials lower than 127 mV which were measured at 268K, 283K, 298K and 308K. Figure 9(a) is the diagram that plotted $\ln(i_s\sigma^{-1})$ obtained from Fig. 8 against $ne\eta/kT$ at every temperature taking account of $(C_{Cl^-}/C^0_{Cl^-})^{-2} \approx 1$. The linear relationships at every temperature in the diagram are drawn so that they have a same gradient given by averaging. The transfer coefficient α was estimated from the gradient of their linear relationships. Then also the exchange current densities $i_0(T)$ at each temperature were estimated from the constant terms of them. Figure 9(b) is the diagram that plotted $\ln\{i_0(T)\}$ against $1/T$. The activation enthalpy for the anodic dissolution reaction of copper crystals was estimated from the gradient of the linear relationship shown in Fig. 9(b).

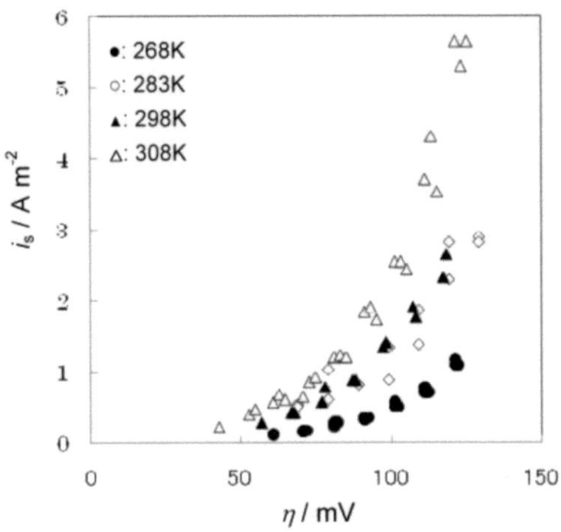

Figure 8. Plots of steady current densities against overpotentials lower than about 125 mV

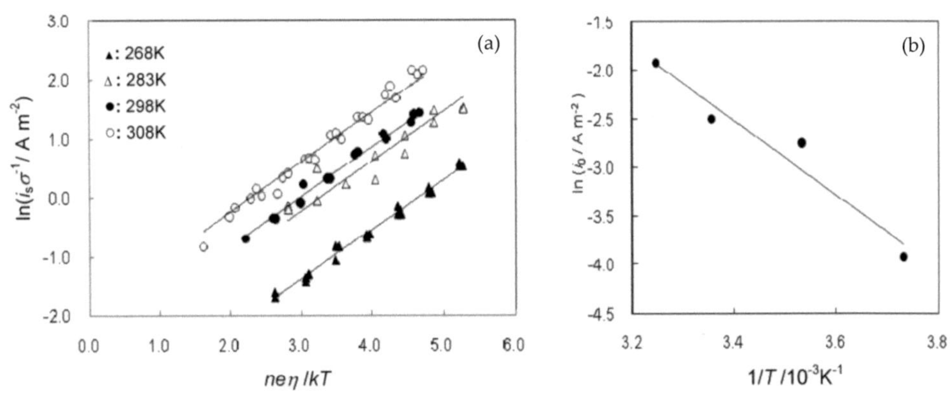

Figure 9. (a): Plots of $\ln(i_s\sigma^{-1})$ against $ne\eta/kT$; (b): Plot of $\ln\{i_0(T)\}$ against $1/T$.

Then, because surface concentrations $C^0_{Cl^-}$and C^0_{CuCl2} of Cl^- and $CuCl_2^-$ ions are calculated by Eq.(8) when (111) surface of a copper crystal is in equilibrium with the electrolyte solution consisting of 5 kmol m^{-3} NaCl and 10^{-4} kmol m^{-3} CuCl, the complex term $C^0_{Cl^-}{}^{2(1-\alpha)}C^0_{CuCl2^-}{}^{\alpha}$ in Eq.(7) giving exchange current density can be evaluated by using the transfer coefficient α estimated above.

Thus the estimations of parameters controlling exchange current density are summarized in Table 2. The value of βk_s was evaluated by substituting i_0, ΔH_0 and $C^0_{Cl^-}{}^{2(1-\alpha)}C^0_{CuCl2^-}{}^{\alpha}$ into Eq.(7), where the atomic frequency $\nu = 6.21\times10^{12}$ [s^{-1}], elementary electric charge, $e = 1.602\times 10^{-19}$[C] and $n = 1$ were assumed.

α	ΔH_0 / eV	i_0 /10^{-2}A m^{-2}	$C^0_{Cl^-}{}^{2(1-\alpha)}C^0_{CuCl2^-}{}^{\alpha}$	βk_s /10^{16} m^{-2}
0.84	0.33	8.2*	2.36×10^{-6}	1.32

* the value at 298K

Table 2. Estimation of transfer coefficient α, activation enthalpy ΔH_0, exchange current density i_0 at 298K, and a factor βk_s affecting reaction rate constant.

3.3. Anodic dissolution rates at higher overpotentials

3.3.1. *Estimation of vertical dissolution rate of surface from anodic current density*

Figures 10 (a) and (b) are examples of the anodic current density/time curves which were recorded when the copper crystal was dissolved for 240 s at higher overpotential. The vertical dissolution rate v_{sm} at dislocation-free site of surface under every condition was determined from an average of the initial minimum current densities i_{sm} pointed by arrow of *i*/*t* curves (measured for five different dissolution time in a range of 60 s to 360 s under each condition) shown in Fig.10 as described in Sections 2.4.2.

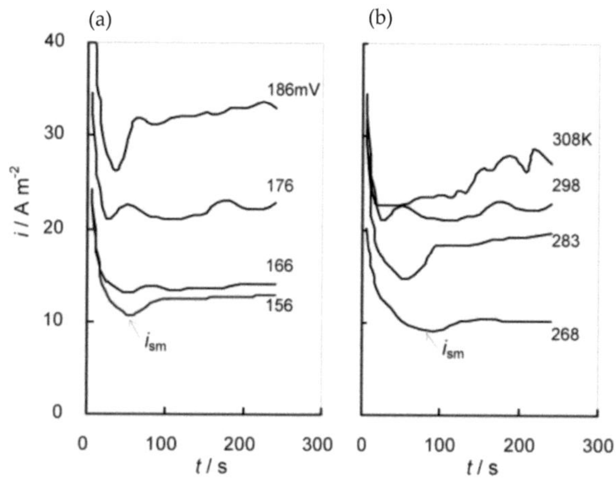

Figure 10. Examples of anodic current density/time curves. (a): The effect of overpotential; (b): The effect of temperature. Initial minimum i_{sm} was obtained in every curve.

3.3.2. *Estimation of vertical dissolution rate of surface by direct measurement*

Figures 11 (a) and (b) are the diagrams that plotted vertical dissolution amounts *s* of surface against dissolution time *t* as described in Sections 2.6.1. The vertical dissolution rates $\dot{s}$ of surface were estimated from the gradient of the linear relationship of *s*/*t* shown in Fig.11.

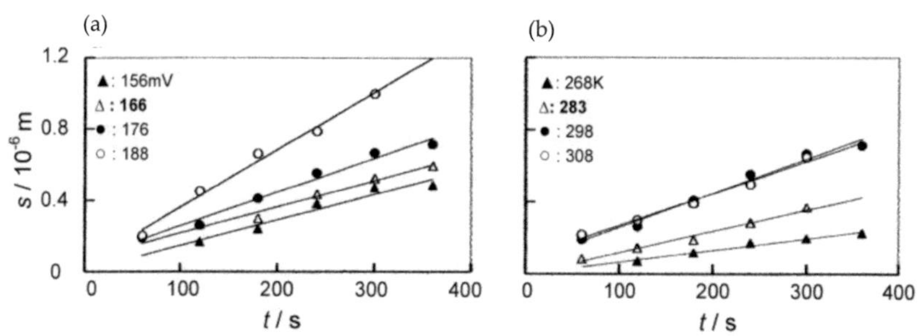

Figure 11. Vertical dissolution amounts of surface vs. dissolution time. (a): Effect of overpotential at 298K; (b): Effect of temperature at 176mV

3.3.3. Estimation of dissolution rates at edge dislocation site by direct measurement

Figures 12 (a) and (b) are the diagram that plotted the depth d of the dark etch pits which were formed at positive edge dislocation sites on the surface dissolved under each condition against dissolution time t, as described in Sections 2.6.2. Also Figs.13 (a) and (b) are the diagrams that plotted similarly the width w of the same dark etch pits as the above mention against dissolution time t. The increasing rate $\dot{d}$ of depth d of etch pit with t and the increasing rate $\dot{w}$ of width w with t were obtained from the gradient of each linear relationship shown in those results.

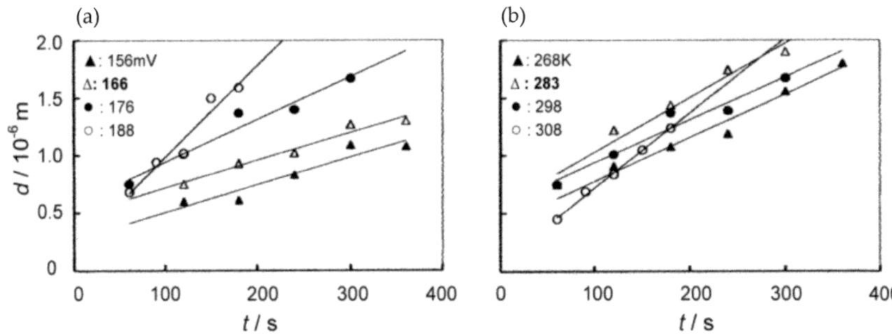

Figure 12. Depth of dark etch pit vs. dissolution time. (a): Effect of overpotential at 298K; (b): Effect of temperature at 176mV

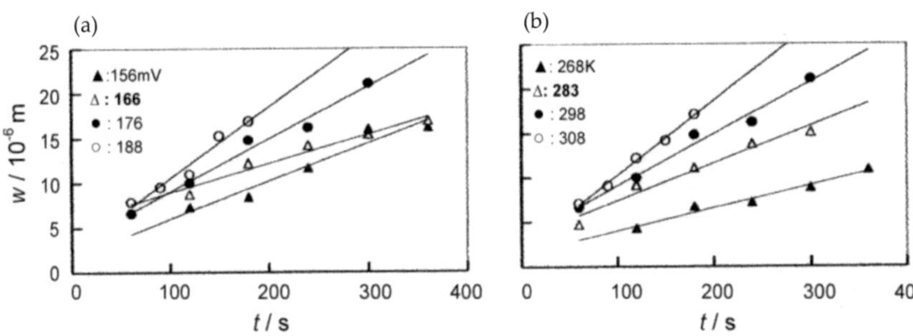

Figure 13. Width of dark etch pit vs. dissolution time. (a): Effect of overpotential at 298K; (b): Effect of temperature at 176mV.

3.4. Effects on the dissolution rates of overpotential and temperature

Figures 14 (a) and (b) are the diagrams that plotted the logarithm values of the dissolution rates. It can be seen that the value of log v_{sm} approximately consists with that of log $\dot{s}$. This shows that both v_{sm} and $\dot{s}$ represent the dissolution rate v_s of dislocation-free site of surface approximately. However, the value of v_{sm} seems to be more exact than that of $\dot{s}$, because while the former is a quantity related to total dissolution amounts of whole surface, the latter is that related to the dissolved amounts of a part of surface near to boundary between the portion exposed to electrolyte solution and the portion covered by vinyl seal. Thus the dissolution rate v_s of dislocation-free surface was assumed to be given not by $\dot{s}$ but v_{sm}. Then v_{ed} and v_w in Fig.14 show the values estimated from Eq.(3) and Eq.(4) in which v_s was substituted by v_{sm}.

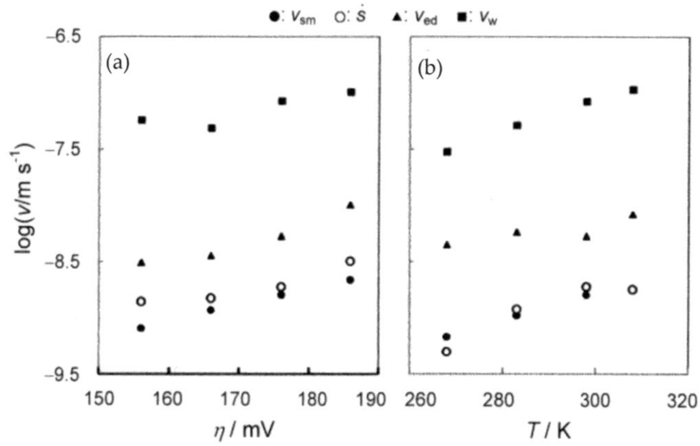

Figure 14. (a): Plots on a logarithmic scale of dissolution rates v_{sm}, $\dot{s}$, v_{ed}, and v_w against overpotential η; (b): Similar plots of v_{sm}, $\dot{s}$, v_{ed}, and v_w against temperature T

It can be seen that both log v_{sm} and log v_w tend to increase rather homogeneously with an increase of η, from Fig.14 (a). However, the tendency of log v_{ed} are somewhat different and in accelerative. Also, it can be seen from Fig.14 (b) that though log v_{sm} and log v_w tend to similarly increase with an increase of T, the tendency of log v_{ed} are somewhat little, compared to the former two. This is seen from the fact that the increasing rate ($\Delta\log v_{ed}/\Delta T = 5.4\times10^{-3}$) of the latter is less than that ($\Delta\log v_{sm}/\Delta T = 1.1\times10^{-2}$, $\Delta\log v_w/\Delta T = 1.4\times10^{-2}$) of the former two.

4. Discussion

4.1. Atomistic dissolution model of crystal surface

4.1.1. Vertical dissolution rate at dislocation-free site of surface

Concerning the dissolution of a crystal, the atomistic model illustrated in schematic diagram of Fig.15 has been proposed (Burton et al., 1951; Schaarwächter, 1965). The dissolution of crystals proceeds via a lateral retreat motion of surface step of an atomic height that is induced by dissolving of surface atom from the kink sites into the solution. The vertical dissolution rate v_s of surface is given by lateral retreat rate v_h and surface density $\tan\theta$ of surface step, which is expressed by the following equation:

$$v_s = v_h \tan\theta = v_h \frac{a}{\lambda}, \tag{17}$$

where θ is an average inclination of crystal surface to a low index face, a an atomic height of surface step, and λ the mean distance between adjacent surface steps. The lateral retreat rate v_h of surface step is expressed by

$$v_h = b^* k^* \sigma_s \nu \exp\left(-\frac{\Delta H_s}{kT}\right), \tag{18}$$

where ΔH_s is the activation enthalpy for dissolution of an atom at kink site of surface step, ν the atomic frequency, k^* the retreat rate constant of surface step, and b^* the unit retreat distance. σ_s is surface undersaturation, which is written as

$$\sigma_s = 1 - \exp\left(-\frac{\Delta\mu}{kT}\right), \tag{19}$$

where $\Delta\mu$ is the chemical potential difference of dissolution atom between two phases of a crystal/ solution system (Schaarwächter, 1965).

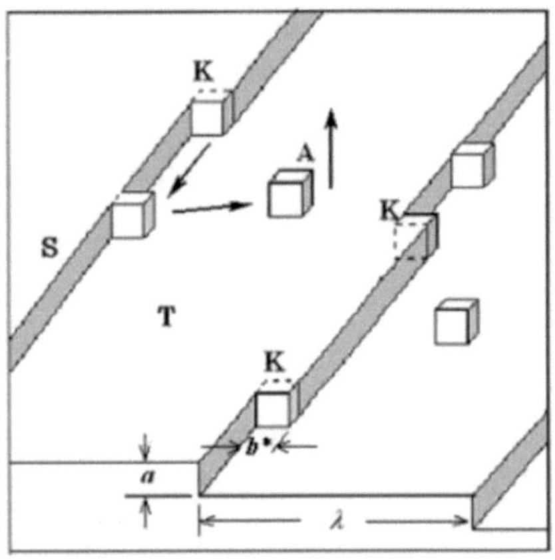

Figure 15. Atomistic model for dissolution process of a crystal surface. Atom dissolves from kink site of surface step into solution. K: Kink; S: Surface step; T: Terrace; A: Ad-atom

4.1.2. Lateral dissolution rate at edge dislocation site

The dislocation etch pit is thought to be formed via a successive nucleation and growth processes of two-dimensional pits at the dislocation site (Schaarwächter, 1965) or via a spiral dissolution of the surface step which is caused by screw dislocation (Cabrera and Levine, 1956). We discuss the dissolution rate at edge dislocation site of (111) surface of copper crystals, based on a nucleation and growth model of two-dimensional pits (Schaarwächter, 1965) that is illustrated in Fig 16, in the following.

Since the lateral dissolution rate v_w is thought to represent horizontal growth rate of two-dimensional pit nucleated at edge dislocation site of surface, it may be corresponding to the lateral retreat rate v_h of surface step along (111) face. Thus we assume that v_h is given by v_w as shown in the following relation:

$$v_h \approx v_w. \tag{20}$$

4.1.3. Vertical dissolution rate at edge dislocation site

On the other hand the vertical dissolution rate at positive edge dislocation site would be examined by the nucleation rate of two-dimensional pit at dislocation site as follows.

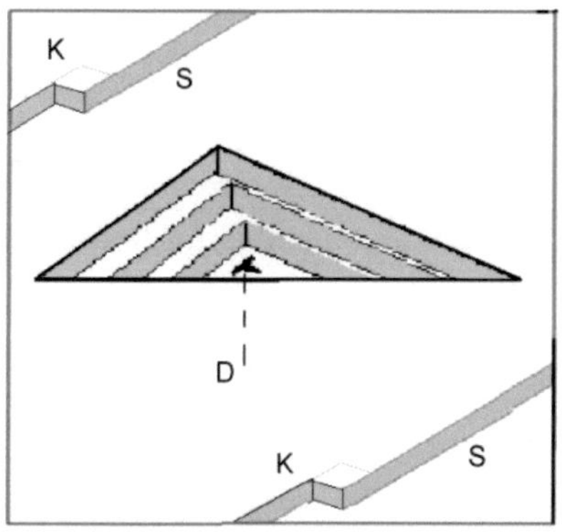

Figure 16. Illustration for dislocation etch pit formation by successive nucleation and growth of two-dimensional pits. D: Dislocation; S: Surface step; K: Kink.

According to the classical nucleation theory, if ΔG_{ed}* is the critical free energy change at nucleation of a two-dimensional pit at edge dislocation site, a steady state nucleation rate I of two-dimensional pit would be expressed by

$$I = Zr\exp\left(-\frac{\Delta G_{ed}{}^{*}}{kT}\right), \tag{21}$$

where r is a separation rate of an atom from an active site of the two-dimensional pit into the solution and Z the Zeldovich factor (Toschev, 1973). Since the separation rate r is assumed to be a similar quantity to the dissolution rate of an atom from kink site of surface, it depends on the surface concentrations of Cl^- and $CuCl_2^-$ ions as known from the Eq. (15) in Section 2.7, and is expressed by

$$r \propto C_{Cl^-}{}^{2(1-\alpha)} C_{CuCl_2^-}{}^{\alpha} \nu \exp\left(-\frac{\Delta H_0}{kT}\right). \tag{22}$$

Accordingly, the vertical dissolution rate v_{ed} at edge dislocation site of surface is expressed by

$$v_{ed} = aK_s C_{Cl^-}{}^{2(1-\alpha)} C_{CuCl_2^-}{}^{\alpha} \nu \exp\left(-\frac{\Delta G_{ed}{}^{*} + \Delta H}{kT}\right), \tag{23}$$

where a is the depth of "two-dimensional pit and K_s is an undetermined constant including Zeldovich factor and others (see Appendix A2.).

According to the nucleation theory of dissolution of crystals, ΔG_{ed}* is small compared to ΔG_s* which is the critical free energy change at nucleation of a two-dimensional pit at dislocation-free site of surface, because of strain energy of dislocation core. It is expressed by

$$\Delta G_{ed}{}^{*} = p\Delta G_s{}^{*} = p\frac{\pi a \Omega \gamma^2}{\Delta \mu} \tag{24}$$

and

$$p = \left(1 - \frac{\alpha_C q}{4\pi} \frac{Gb}{\gamma}\right)^2 \leq 1, \tag{25}$$

where γ is the interfacial free energy of the crystal and solution at step of the two-dimensional pit, G the shear modulus and q and α_c the constants (Schaarwächter, 1965).

4.2. Relations between vertical dissolution rate of surface and anodic current density

4.2.1. Expression for dissolution rate of dislocation-free site of surface

When the copper crystal is anodically dissolved by the simple electrode reaction of Eq.(5) the vertical dissolution rate v_s of dislocation-free surface at lower overpotentials and the v_{sm} at higher overpotentials would be estimated by Eq.(1) and Eq.(2) respectively as described in Section 2.4. Thus it is experimentally estimated with use of Eqs. (1), (2), and (15) by the following expression:

$$v_{sm} \approx v_s = \Omega \beta k_s C_{Cl^-}{}^{2(1-\alpha)} C_{CuCl_2^-}{}^{\alpha} \sigma \nu \exp\left(-\frac{\Delta H}{kT}\right). \tag{26}$$

According to the dissolution model of crystals, the dissolution rate at dislocation-free site of surface is expressed from Eqs. (17) and (18) by

$$v_s = \frac{a}{\lambda} b^* k^* \sigma_s \nu \exp\left(-\frac{\Delta H_s}{kT}\right). \tag{27}$$

Therefore, the following relations are obtained from Eqs.(16), (26) and (27) concerning the rate constant of the lateral retreat rate of surface step and activation enthalpy for the dissolution.

$$k^* = \beta \frac{b}{x_0} C_{Cl^-}{}^{2(1-\alpha)} C_{CuCl_2^-}{}^{\alpha} = \beta \frac{b}{x_0} C^0{}_{Cl^-}{}^{2(1-\alpha)} C^0{}_{CuCl_2^-}{}^{\alpha} \left(\frac{C_{Cl^-}}{C^0{}_{Cl^-}}\right)^2 \exp\left(\frac{\alpha n e \eta_c}{kT}\right) \tag{28}$$

and

$$\Delta H_s = \Delta H = \Delta H_0 - \alpha n e \eta_a \tag{29}$$

Also from Eqs. (11) and (19)

$$\Delta\mu = ne\eta_a \tag{30}$$

is obtained. It can be seen that the rate constant k^* of lateral retreat motion of surface step is electrochemically expressed by Eq.(28) and that it increases with an increase of concentration overpotential η_c.

4.2.2. Estimation of kinetics parameters controlling the dissolution rate

As mentioned above the dissolution rate v_{sm} at dislocation-free site of surface under higher overpotentials is expressed by an approximate equation:

$$v_{sm} \approx \Omega \beta k_s {C^0}_{Cl^-}{}^{2(1-\alpha)} {C^0}_{CuCl_2^-}{}^{\alpha} \sigma \nu \exp\left(-\frac{\Delta H_0 - \alpha n e \eta}{kT}\right), \tag{31}$$

from Eqs. (14) and (26), where we assumed $i_s << i_{lCl^-}$, that is,

$$\frac{C_{Cl^-}}{{C^0}_{Cl^-}} = 1 - \frac{i_s}{i_{lCl^-}} \approx 1. \tag{32}$$

Thus concerning the dissolution rate at dislocation-free site of surface which have a constant kink density k_s, a following approximate expression is lead from Eq.(31) (Imashimizu, 2011).

$$\ln v_{sm} \approx \ln(\nu\, \Omega \beta k_s {C^0}_{Cl^-}{}^{2(1-\alpha)} {C^0}_{CuCl_2^-}{}^{\alpha}) + \ln(\sigma) - \frac{\Delta H_0 - \alpha n e \eta}{kT} \tag{33}$$

Figures 17 (a) and (b) are the diagrams that plotted the dissolution rate v_{sm} shown in Figs.14 (a) and (b) on a natural logarithmic scale against η (T = 298K) and $1/T$ (η =176mV) respectively. It can be seen that the values of $(\alpha ne/kT)$ and $((\Delta H_0 - \alpha ne\eta)/k)$ are estimated by comparing the Eq.(33) with gradients of the linear relationships drawn in Figs. 17 (a) and (b), because the overpotential and temperature dependences of σ in the range of 156 mV to 186 mV are assumed to be a little. Thus, α and ΔH_0 were also obtained from overpotential dependence of vertical dissolution rate v_{sm} of surface at higher overpotentials and temperature dependence of that. The estimations are shown in Table 3, showing α and ΔH_0 are in good agreement with those values in Table 2.

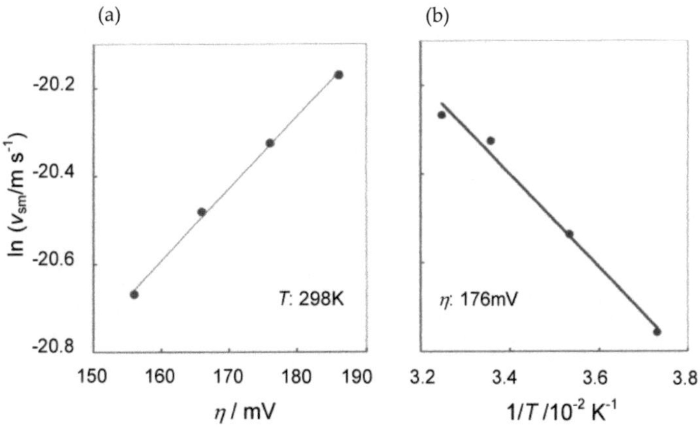

Figure 17. Vertical dissolution rate v_{sm} at dislocation-free site of surface on a natural logarithm scale. (a): The plot against overpotential η; (b): The plots against the inverse $1/T$ of temperature.

According to atomistic dissolution model of a crystal surface illustrated in Fig.15, the relation of $\theta = \tan^{-1}(a/\lambda) = \tan^{-1}(v_s/v_h)$ is lead from Eq. (17), which represents the inclination angle of surface to (111) face. Since it is approximately given by $\theta \approx \theta^* = \tan^{-1}(v_{sm}/v_w)$ with use of Eqs.(20) and (26), the values of θ^* obtained from Fig.14 were plotted against η and T in Figs.18 (a) and (b). It can be seen that the tendencies of change in θ^* against η and T are not clear and not reasonable. The average value of θ^*_{av} is 2.1×10^{-2} rad, which is a little large compared to a deviation 8.7×10^{-3} rad from [111] direction that was aimed when we prepared the surface of specimen as described in Section 2.1. This is probably attributed to the fact that actual surface exposed to electrolyte solution was slightly spherical as a whole and was having microscopic swells. That is, the variation of their values seems to be due to experimental error. Thus the vertical dissolution rate at dislocation-free site of surface is assumed to be given by retreat rate of the surface steps which preexists on the prepared surface, which gives following relation:

$$\beta k_s = \beta \frac{1}{\lambda x_0} \approx \beta \frac{b}{x_0} \frac{\tan\theta^*}{ab} \tag{34}$$

Accordingly $\beta(b/x_0)$ is calculated from βk_s in Table 2 by using Eq. (34), which is shown in Table 3 where assumed $\theta^* = \theta^*_{av}$ (0.021 rad).

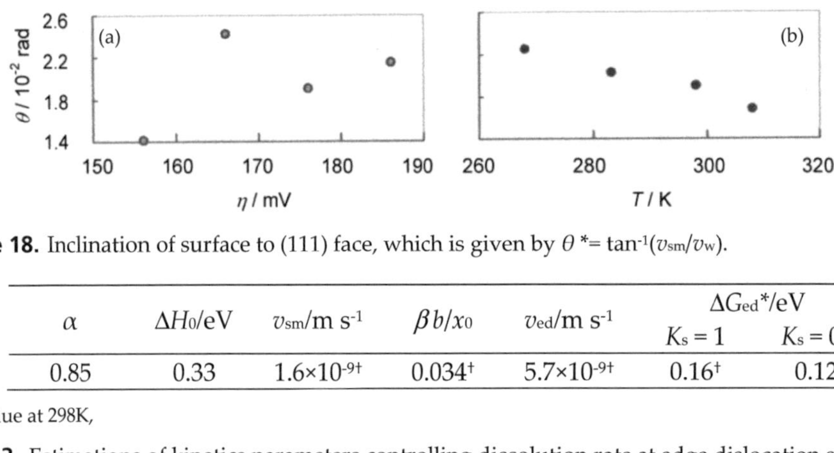

Figure 18. Inclination of surface to (111) face, which is given by $\theta^* = \tan^{-1}(v_{sm}/v_w)$.

α	ΔH_0/eV	v_{sm}/m s^{-1}	$\beta b/x_0$	v_{ed}/m s^{-1}	ΔG_{ed}^*/eV $K_s = 1$	ΔG_{ed}^*/eV $K_s = 0.2$
0.85	0.33	$1.6\times10^{-9\dagger}$	$0.034^\dagger$	$5.7\times10^{-9\dagger}$	$0.16^\dagger$	$0.12^\dagger$

†the value at 298K,

Table 3. Estimations of kinetics parameters controlling dissolution rate at edge dislocation site of surface of copper crystals.

4.3. Vertical dissolution rate at dislocation site

4.3.1. Estimation of the critical free energy change for nucleation of two-dimensional pit

As mentioned in Section 4.1.2 the dissolution rate v_{ed} at edge dislocation site is expressed by Eq. (23), but if Eq. (14) is applied it is reduced to

$$v_{ed} = aK_s {C^0_{Cl^-}}^{2(1-\alpha)} {C^0_{CuCl_2^-}}^{\alpha} \left(\left(\frac{C_{Cl^-}}{C^0_{Cl^-}} \right)^2 \right) \nu \exp\left(-\frac{\Delta G_{ed}^* + \Delta H_0 - \alpha n e \eta}{kT} \right). \tag{35}$$

Accordingly, if we assume $C_{Cl^-}/C^0_{Cl^-} \approx 1$, ΔG_{ed}^* is given by

$$\Delta G_{ed}^* \approx -kT\ln\left(\frac{v_{ed}}{aK_s C^0_{Cl^-}{}^{2(1-\alpha)} C^0_{CuCl_2^-}{}^{\alpha} \nu}\right) - \Delta H_0 + \alpha ne\eta. \quad (36)$$

Thus ΔG_{ed}^* under each condition was estimated by Eq.(36) with use of experimental value of v_{ed} as well as estimations of α and ΔH_0 which were obtained in Section 4.2.2. The ΔG_{ed}^* estimated with use of two assumed values of undetermined constant K_s for a specified condition (η = 176 mV and T = 298K) are shown together with the values α and ΔH_0 in Table 3, where $a = 2.09\times10^{-10}$ m and $\nu = 6.21\times10^{12}$ s^{-1} were used.

According to the precedent theoretical study (Schaarwächter 1965), in which the conditions for the formation of visible etch pit at dislocation site were investigated on the basis of a proposed nucleation model, the critical free energy change is estimated to be 0.115 eV. The present estimation of ΔG_{ed}^* approximately consists with that value as shown in Table 3, though the exact value of K_s can not be evaluated in this study. This is seemed to be reasonable as described in Appendix A2.

On the other hand, however, it was admitted that the value of ΔG_{ed}^* varies with overpotential and temperature as mentioned below.

4.3.2. Overpotential and temperature dependences of ΔG_{ed}^*

Figures 19 (a) and (b) are the diagrams that plotted the square root of ΔG_{ed}^* estimated assuming K_s = 1 by Eq. (36) against η and T respectively. It can be seen that $\Delta G_{ed}^{*1/2}$ is not constant but changes in different manners with increases in η and T. The reason for this is probably that $\Delta G_{ed}^{*1/2}$ is proportional to the interfacial energy γ as known from Eqs. (24) and (25).

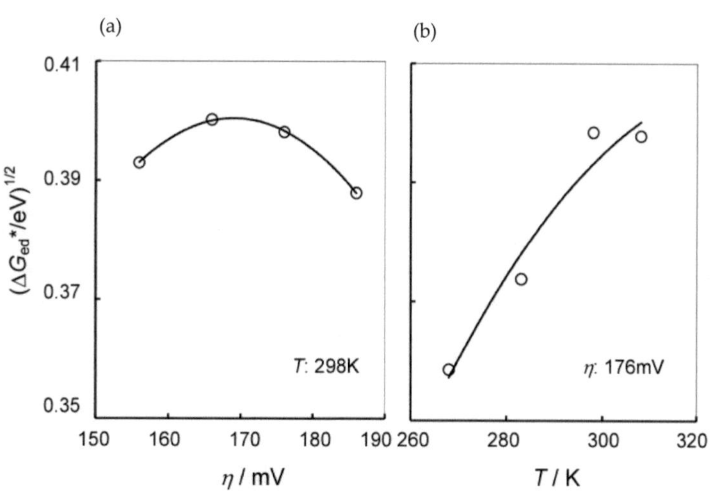

Figure 19. Square root of the critical free energy change for the formation of a two-dimensional pit. (a): The overpotential dependence; (b): The temperature dependence.

It is known that the interfacial energy varies with electrode potential according to so-called electrocapillary curve (Tamamushi, 1967). Therefore, the change in $\Delta G_{ed}^{*1/2}$ with η is surmised to be due to the potential dependence of γ, because the overpotential dependences of the undersaturation σ and therefore that of $\Delta\mu = ne\eta_a = -kT\ln(1-\sigma)$ in an overpotential range of 156 to 186 mV are assumed to be a little as described in Section 4.2.2. This is supported by the fact that Fig.19 (a) indicates a quadratic dependence similar to the electrocapillary curve. Also, it is inferred from Fig.19(b) and Eq.(24) that γ should increase with an increase in T, because $\Delta\mu$ tend to increase with increase in T. This is probably attributed to a decrease in specific adsorption of anion accompanied by an increase of interfacial energy with rising of temperature.

The overpotential dependence of log v_{ed} is in accelerative, and somewhat different from that of both log v_{sm} and log v_w. Also the increasing rate of log v_{ed} with increase in temperature is smaller than that of both log v_{sm} and log v_w as shown in Figs.14 (a) and (b). The reason for this seems to be attributed to the overpotential and temperature dependences of the interfacial energy of the electrode surface as mentioned above.

5. Conclusions

Following conclusions were obtained from the results and discussion:

1. The transfer coefficient, activation enthalpy and surface concentrations of the ions which control the dissolution reaction were estimated from measurements of the relationships between steady anodic current densities and applied overpotentials when copper crystals are dissolved in an electrolyte solution under potentiostatic electrolysis.
2. The values of a supplementary factor and kink density affecting rate constant of dissolution reaction were examined.
3. The dissolution rate at edge dislocation site of (111) surface of copper was discussed quantitatively by a nucleation model of two-dimensional pit based on the classical nucleation theory.
4. The present estimation of the critical free energy change ΔG_{ed}^* for nucleation of a two-dimensional pit at edge dislocation site reasonably consisted with the evaluation by the precedent study.
5. The overpotential and temperature dependences of dissolution rate at edge dislocation site were somewhat different from those dependences of dissolution rate at dislocation-free site. The reason for this is probably that ΔG_{ed}^* changes according to the overpotential and temperature dependences of interfacial energy.

6. Appendix

A1. Kinds of dislocation etch pits and their characters

The surface of copper specimen on which some small glass spheres 300 μm in diameter were dropped beforehand was anodically etched by the present method. Fig.20 (a) is an optical micrograph of dissolved surface in which Rosseta pattern composed of dark and light etch pits was formed at the portion that was hit by a small glass sphere. This proves that dark and light etch pits are formed at the sites of positive and negative edge dislocations respectively because the six arms of Rosseta pattern are composed of rows of a pair of positive and negative edge dislocations.

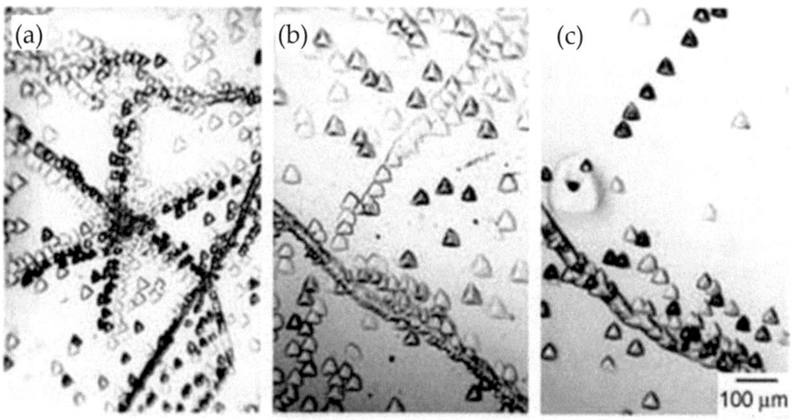

Figure 20. Optical micrographs for identifications of dark and light pits. The surfaces dissolved by the present method; (a): Rosetta pattern composed of etch pits; (b): a distribution of etch pits. (c): etch pits formed by a chemical etchant in the same portion as that observed in b.

In another experiment, the surface of prepared copper specimen was anodically etched first by the present method, and a distribution of etch pits were observed by the optical microscope. Subsequently after electropolished the etched surface of specimen, the surface was etched for 10 s by a modified Young's etchant prepared by Marukawa (Marukawa, 1967), and the same portion as the previous portion was observed. Figs.20 (b) and (c) are a pair of optical micrographs of the surfaces etched by such two methods. It has been reported by Marukawa that the dark (deep) and light (shallow) pits are formed at screw dislocations and edge dislocations on the surface etched by the modified Young's etchant respectively. Accordingly it can be seen that the light etch pits are formed at the sites of screw dislocations on the surface that was anodically etched by the present method, by comparing the kinds of etch pits which are observed in these micrographs. Thus Table 4 is obtained concerning dislocation characters related to dark and light etch pits.

Etching	Edge dislcation (positive)	(negative)	Screw dislcation
Chemical†	Light	Light	Dark
Electrolytic††	Dark	Light	Light

† by modified Young's etchant , †† present method

Table 4. Relations between the dislocation characters and the kinds of etch pits which are formed by two etching methods

In this work, the depth and width of the dark (deep) pits were measured to investigate the dissolution amounts at positive edge dislocation sites.

A2. Estimation of undetermined constant K_s

As described in the Section 4.1.3, if the separation rate r of an atom at nucleation of two-dimensional pit is a quantity similar to the dissolution rate of an atom from kink site of surface, it would need to take account of supplementary factor β affecting the exchange current density as a parameter involved in the separation rate r. Then the dissolution rate v_{ed} at edge dislocation site derived from the nucleation rate Eq. (21) is represented afresh by

$$v_{ed} \approx a\beta Z C_{Cl^-}{}^{2(1-\alpha)} C_{CuCl_2^-}{}^{\alpha} \nu \exp\left(-\frac{\Delta G_{ed}{}^* + \Delta H}{kT}\right). \tag{37}$$

Thus, we assume that the undetermined constant K_s is approximately given by a relation:

$$K_s = \beta Z. \tag{38}$$

We have assumed in the Section 2.7 that the exchange current density i_0 is given by Eq. (7) for simplification, but to be exact i_0 should be expressed with use of the activities of the ions involved in the electrode reaction instead of the concentrations. Also, transmission coefficient should be taken account of as pre-exponential factors in Eq. (7). Therefore it is generally hard to estimate β including some unknown factors. However, concerning β of the present electrode reaction, $\beta\ (b/x_0) = 0.034$ was estimated experimentally as shown in Table 3. Also it can be seen from an observation of etch pit by optical microscope that surface steps have a structure along a crystallographic direction of the crystal. Accordingly if (b/x_0) is assumed to be a quantity of 0.02 to 0.2, it would give an estimation of $\beta = 0.17$~1.7.

On the other hand, if we assume the free energy change $\Delta G_{ed}\ (j)$ for formation of a two-dimensional pit consisting of j vacancies at edge dislocation site, it is written as

$$\Delta G_{ed}(j) = 2\gamma(\pi a \Omega)^{1/2} j^{1/2} - \frac{\Delta\mu}{p'} j \tag{39}$$

Then the critical size j^* of two-dimensional pit and the critical free energy change $\Delta G_{ed}^*(j^*)$ are given by

$$j^* = p'^2 \frac{\pi a \Omega \gamma^2}{\Delta\mu^2} \quad \text{and} \quad \Delta G_{ed}{}^*(j^*) = p' \frac{\pi a \Omega \gamma^2}{\Delta\mu} \tag{40}$$

respectively. It can be seen that $\Delta G_{ed}^*(j^*)$ is expressed by the same relation as Eq. (24), and that the factor p' has the same contents with Eq. (25), that is, $p' = p$. Then, Zeldovich factor is expressed from the definition (Toschev, 1973) by

$$Z = \sqrt{-\frac{1}{2\pi kT}\left(\frac{\partial^2 \Delta G_{ed}(j)}{\partial j^2}\right)_{j^*}} = \sqrt{\frac{1}{4\pi kT}\left(\frac{\Delta\mu^2}{p^2 \Delta G_{ed}{}^*}\right)}. \tag{41}$$

Accordingly, $Z = 0.76$ is estimated, if $p = 0.18$ (Schaarwächter 1965), $\Delta\mu = 0.027$ eV($\sigma = 0.65$) (Imashimizu, 2011), $\Delta G_{ed}^* = 0.12$ eV (Table 3) and $kT = 0.0257$ eV ($T = 298$K) are used.

Thus $K_s = 0.13\sim1.3$ is estimated from Eq. (38), which suggests the reasonability of the assumed value of K_s shown in Table 3.

Author details

Yuji Imashimizu
Mineral Industry Museum, Faculty of Engineering and Resource Science, Akita University

Acknowledgments

The author thanks Emeritus Prof. Dr. J. Watanabé of Akita University for affording an opportunity to accomplish this study. He also wishes to express his thanks to Messrs.T. Wakayama and Y. Hirai, who carried out much of experimental work.

7. References

Burton, W. Cabrera, N. & Frank, F. (1951). The Growth of Crystals and the Equilibrium Structure of their Surfaces, *Philosophical Transactions of the Royal Society*, A 243, pp.299-358.

Cabrera, N. & Levine, M. (1956), On the Dislocation Theory of Evaporation of Crystals, *Philosophical Magazine*, Vol. 1, pp. 450-458.

Despic, A. & Bockris, J. (1960), Kinetics of the Deposition and Dissolution of Silver, *The Journal of Chemical Physics*, Vol. 32, No.2, pp. 389-402.

Gilman, J., Johnston, W. & Sears, G. (1958), Dislocation Etch Pit Formation in Lithium Fluoride, *Journal of Applied Physics* Vol.29, No. 5, pp.747- 754

Imashimizu, Y. & Watanabé, J. (1983), Dissolution Rate at Dislocations on a (111) Surface of Copper Crystal under Potentiostatic Electrolysis, *Transactions of The Japan Institute of Metals*, Vol. 24, No.12, pp.791-798.

Imashimizu,Y. (2010), Dissolution Rate of the (111) Surface of Copper Crystals under Potentiostatic Electrolysis, *Journal of the JRICu*, Vol. 49, No.1, ISSN 1347-7234, pp. 258-263. (in Japanese).

Imashimizu, Y. (2011), Dissolution kinetics at edge dislocation site of (111) surface of copper crystals, *Journal of Crystal Growth*, Vol. 318, pp.125-130.

Ives, M. & Hirth, J. (1960), Dissolution Kinetics at Dislocation Etch Pits in Single Crystals of Lithium Fluoride, *The Journal of Chemical Physics*, Vol.33, No. 2, pp.517-525.

Jasper, L & Schaarwächter, W. (1966), Ätzgrubenbildung an Versetzungen in Kupfereinkristallen durch potentiostatische Elektrolyse, *Zeitschrift für Metallkunde*, Bd. 57, H. 9, pp.661-668.

Lal, H. & Thirsk, H. (1953), The Anodic Behaviour of Copper in Neutral and Alkaline Chloride Solutions, *Journal of Chemical Society*, pp. 2638-2644.

Lee, H. & Ken Nobe (1986), Kinetics and Mechanism of Cu Electrodissolution in Chloride Media, *Journal of Electrochemical Society*, Vol. 133, No. 10, pp.2035-2043.

Maeda, M. (1961), *Denkyoku No Kagaku*, Gihodo, ASIN: B000JAMTNW, Tokyo, Japan (in Japanese)
Marukawa, K. (1967), Dark and Light Pits on (111) Surface of Copper, *Japanese Journal of applied Physics*, Vol. 6, No. 8. pp.944-949.
Onuma, K., Tsukamoto, K. & Sunagawa, I. (1991), Dissolution kinetics of K-alum crystals as judged from the measurements of surface undersaturations, *Journal of Crystal Growth*, Vol. 110, pp.724-732.
Schaarwächter, W. (1965), Zum Mechanismus der Versetzungsätzung, I. Die Bildung zweidimensionaler Lochkeime an den Enden von Versetzungslinien, *physica status solidi*, Bd. 12, pp. 375-382.
Schaarwächter, W. (1965), Zum Mechanismus der Versetzungsätzung II Entstehungsbedingungen für Atzgruben, *physica status solidi*, Bd. 12, pp. 865- 876.
Schaarwächter , W. & Lücke, K. (1967), Der Einfluβ der versetzungsstruktur auf die Auflösung von Kristallen, *Zeitschrift für Physikalische Chemie Neue Folge*, Bd. 53, pp. 367-386.
Tamamushi, R. (1967), *Denkikagaku*, Tokyo Kagaku Doujin, NCID: BN01795418, Tokyo, Japan (in Japanese)
Toschev, S. (1973), Homogeneous nucleation, In: *Crystal Growth: An Introduction*, P. Hartman, (Ed.), 1-49. ISBN North-Holland 0 7204 1821 6 / ISBN American Elsevier 0444 10463 1, Amsterdam
Van Der Hoek, B.;Van Enckevort, W. & Van Der Linden, W. (1983), Dissolution Kinetics and Etch Pit Studies of Potassium Aluminium Sulphate, *Journal of Crystal Growth*, Vol.61, pp.181-193.
Watanabé, J.; Imashimizu, Y. & Sugawara, S. (2003), Dissolution Kinetics at Dislocation Site of Copper Crystal, *Journal of the Society of Materials Engineering for Resources of Japan*, Vol. 16, No. 1, pp.13-20. (in Japanese).
Young, Jr., F. (1961), Etch Pits at Dislocations in Copper, *Journal of Applied Physics*, Vol. 32, No. 2, pp. 192-201.

Chapter 9

Scale-Up of Electrochemical Reactors

A. H. Sulaymon and A. H. Abbar

Additional information is available at the end of the chapter

http://dx. doi. org/10. 5772/48728

1. Introduction

Electrochemical technology can provide valuable cost efficient and environmentally friendly contributions to industrial process development with a minimum of waste production and toxic material. Examples are the implementation of electrochemical effluent treatment, for example, the removal of heavy metal ions from solutions, destruction of organic pollutants, or abatement of gases. Further progress has been made in inorganic and organic electro synthesis, fuel cell technology, primary and secondary batteries, for example, metal-hydride and lithium-ion batteries. Examples of innovative industrial processes are the membrane process in the chloralkali industry and the implementation of the gas-diffusion electrode (GDE) in hydrochloric acid electrolysis with oxygen reduction instead of hydrogen evolution at the cathode [1]. The main advantages of electrochemical processes are:

- *Versatility*: Direct or indirect oxidation and reduction, phase separation, concentration or dilution, biocide functionality, applicability to a variety of media and pollutants in gases, liquids, and solids, and treatment of small to large volumes from micro liters up to millions of liters.
- *Energy efficiency*: Lower temperature requirements than their non electrochemical counterparts, for example, anodic destruction of organic pollutants instead of thermal incineration; power losses caused by inhomogeneous current distribution, voltage drop, and side reactions being minimized by optimization of electrode structure and cell design.
- *Amenability to automation*: The system inherent variables of electrochemical processes, for example, electrode potential and cell current, are particularly suitable for facilitating process automation and control.
- *Cost effectiveness*: Cell constructions and peripheral equipment are generally simple and, if properly designed, also inexpensive. The backbone of any electrochemical technology is the electrochemical reactor, therefore the perfect design and scale-up plays an important role in successful of this electrochemical technology [2].

2. The principal of similarity

Dimensional analysis as a basic concept underlying the theory of transport processes and chemical reactor is familiar to every chemical engineer in analyzing laboratory data for reacting systems, the various rate constants, transfer coefficients, transport properties and reactor dimension must be combined in such a way that dimensional consistency is maintained. Most engineers are familiar with the Buckingham theorem which may be considered a formal restatement of the requirement of dimensional consistency [3]

3. Scale-up philosophy

An electrochemical reaction can be conducted in batch or continuous (mixed/plug flow) mode. Further, the reactors can be operated with or without recycle. Further classifications are possible on the basis of flow arrangement (parallel/series flow) or electrical connections (monopolar/bipolar). The electrodes may be flat (2-dimensional solid electrode) or porous (3-dimensional) electrode. Three-dimensional electrodes are used when high surface area is desired to compensate for the inherent low current density of the process. The electrodes can be configured to be horizontally placed or vertically placed [4]. The philosophy of scaling-up chemical process units requires the values of corresponding dimensionless groups of the two units are similar [5]. Several similarity criteria have been defined to guide the engineer to scale-up a reactor [6]. The criteria normally employed in thermo chemical reactors are those of geometric, kinematic and thermal similarity between the reactors. In the case of electrochemical reactors an additional criterion necessary to define the scale-up is that of current/potential similarity. These four criteria are discussed below:

3.1. Geometric similarity

Geometric similarity is achieved by fixing the dimensional ratios of the corresponding reactors. However, for electrochemical reactors, this criterion cannot normally be met, as increasing the inter-electrode gap would give a high voltage drop and increased energy costs. Further, in 3D electrodes an increased electrode thickness may cause a decrease in the average electric potential and/or promote secondary electrode reactions [7]. Therefore, geometric similarity is usually sacrificed in favor of current/potential similarity in electrochemical reactors. Scale-up in electrochemical reactors is achieved by using multiple cells and reactor units [8].

3.2. Kinematic similarity

Kinematic similarity is concerned with the flow velocities within a system. In any continuous reactor, the gas and liquid flow loads, or more generally the Reynolds number govern the pressure drop, fluid hold-up and mass transfer capacity in the system [9]. Therefore it was desirable to maintain similar gas and liquid flow velocities through the corresponding reactors.

3.3. Thermal similarity

Thermal similarity implies matching the temperatures in corresponding portions of the reactors under comparison. This condition may be approached by temperature control through internal heat transfer surfaces and/or heat exchange with recycling reactants. Thermal similarity is difficult to maintain in the scale-up of electrochemical reactors due to the effect of Joule heating within inter-electrode dimensions of the order of millimeters. The method of providing cooling channels between cells (as in conventional fuel cell stacks) is not practical in other types of electrochemical reactor and in any case would defeat the purpose of the reactor design.

3.4. Current/potential similarity

Electrochemical reactors, unlike their thermo chemical or chemical counterparts, require electrical similarity and this is usually the most important criterion in the scale-up of such reactors. Electrical similarity exists between two units when corresponding electrode potential and current density differences bear a constant ratio [8]. This criterion necessitates a constant inter- electrode gap on scale-up. One factor normally employed to quantify the effect is the Wagner number (Wa), which may be defined as:

$$W_a = \left(\frac{k}{L}\right)\left(\frac{dV}{di}\right) \tag{1}$$

Where (k) is the electrolyte conductivity, V the electrodepotential, i the current density and L the characteristic length. For electrical similarity, the Wagner number in the two reactors should have the same value at all points being compared. In 3D electrodes, an inverse Wagner number qualitatively describes the current distribution in the system [10]. These sets are the rules to get uniform current distribution for scale-up. i. e. for reactors operating under kinetic control, better uniformity is obtained for a higher slope of polarization curve, larger conductivity of electrolyte (k in equation (1)), smaller characteristic length (L in the equation (1)) and lower average current density. The constraints of current/potential similarity require that the scale-up of electrochemical reactors to industrial capacity is usually achieved by:

i. Fixing the inter electrode gap while increasing the superficial area of individual cells
ii. Stacking individual cells in monopolar or bipolar multi-cell reactors (e. g. containing up to 200 cells)

4. Current and potential distribution in electrochemical reactor

4.1. Current and potential distribution in planar electrode

Many industrial electrochemical processes use channel flow between two plane, parallel electrode as shown in Fig. (1)

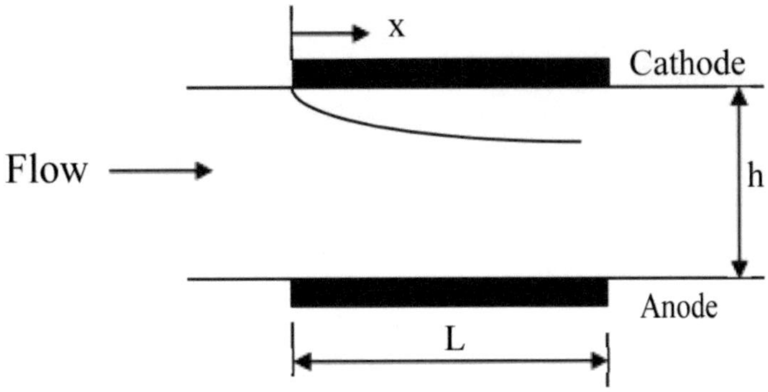

Figure 1. Parallel plate electrodes

The current distribution on such electrode is dependent on type of polarization occurring on the electrode surface; therefore current distribution can be classified as primary, secondary and limiting current distribution [11].

4.1.1. Primary current distribution

The primary current and potential distribution apply when the surface overpotentail can be neglected and the solution adjacent to an electrode can be taken equipotentail surface. Calculation of primary current distribution and resistance represents a first step toward analyzing and optimizing an electrochemical system. The cell resistance calculated can be coupled with calculation including mass-transfer and kinetic effects to optimize approximately a given cell configuration.

Calculation the primary current and potential distribution involves solution of Laplace's equation [$\nabla\phi = 0$]. Solution methods are analytically and numerically, the analytical methods involves the method of image [12],separation of variables [13], superposition [14,15], and Schwarz-Christoffel transformation [16]. The Schwarz-Christoffel transformation is a powerful tools for solution of Laplace's equation in systems with planar electrodes. This method was used by Moulten [17] which gave a classical solution for the primary current distribution for two electrode placed arbitrarily on the boundary of a rectangle, in their analysis they considered a special case of planer cells in which two plane electrodes placed opposite each other in walls of flow channel (Fig. (1)). The potential distribution is shown in Fig. (2) for L=2h. The current lines are represented by solid curves and equipotentail surface by dash curves. The two sets curves should be perpendicular to each other every where in the solution.

Moulten [17] represented the current distribution by the following equation:

$$\frac{i}{i_{av}} = \frac{v \cosh v / K(\tanh_2 v)}{\left|\sinh v - \sinh(2xv/L)\right|} \tag{2}$$

Where v = l/2h ,x measured from the center of electrode ,(K)is the complete elliptic integral of the first kind tabulated in reference [18]. Primary current distribution is determined by geometric factors alone, thus ,only the geometric ratios of cell are a parameter. Wagner number expresses the ratio of the polarization resistance at the interface over the ohmic resistance in the electrolyte approaches to zero in this case.

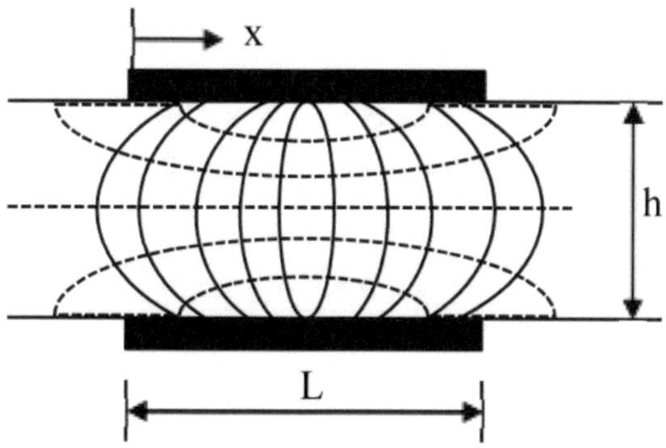

Figure 2. Current and potential lines in parallel plates electrode

4.1.2. Secondary current distribution

When slow electrode kinetics are taken into consideration, the electrolytic solution near the electrode is no longer an equipotentail surface, and the result of calculation is secondary current and potential distribution. Secondary current distribution predominates if the kinetic resistance is higher than the ohmic resistance. The general effect of electrode polarization is to make the secondary current potential nearly uniform than the primary current distribution and an infinite current density at the edge of electrode is eliminated. This can be regarded as the result of imposing an additional resistances at electrode interface [19]. Calculation of secondary current distribution was achieved analytically and numerically by several investigators [20]. The prediction of current distribution using numerical methods is an essential step in the rational design and scale-up of electrochemical reactors and in engineering analysis of electrochemical processes.

A numerical method for predicting current density distribution in multi- ion electrolytes was developed, assuming steady-state, 2D, dilute solution theory and constant properties. The parallel-plate electrochemical reactor (PPER) geometry was used. The calculation of current density for the PPER had been a frequent subject of study but reported mathematical models exist only for limited cases. For example, the models of Parrish and Newman [21] and Caban and Chapman [22] were based on the thin diffusion layer approach, while by Pickett [23] on the mass transfer control assumption, and those of White, Bain and Raible [24] and Nguyen, Walton and White [25] ignore axial diffusion and axial migration and do not account for high velocities

4.1.3. Tertiary current distribution

The combined effects of activation and concentration polarization give rise to changes in primary current distribution resulting in what is known as tertiary current distribution. In such cases the potential of the solution does not obey the Laplace's equation due to concentration variation. When the concentration at surface of electrode approaches to zero, the limiting current condition occurs and distribution of current is limited by the mass transfer rate through the diffusion layer [26]:

$$i/i_{av} = 2/3[x/L]_{-1/3} \tag{3}$$

Figure (3) shows the three types of current and potential distribution

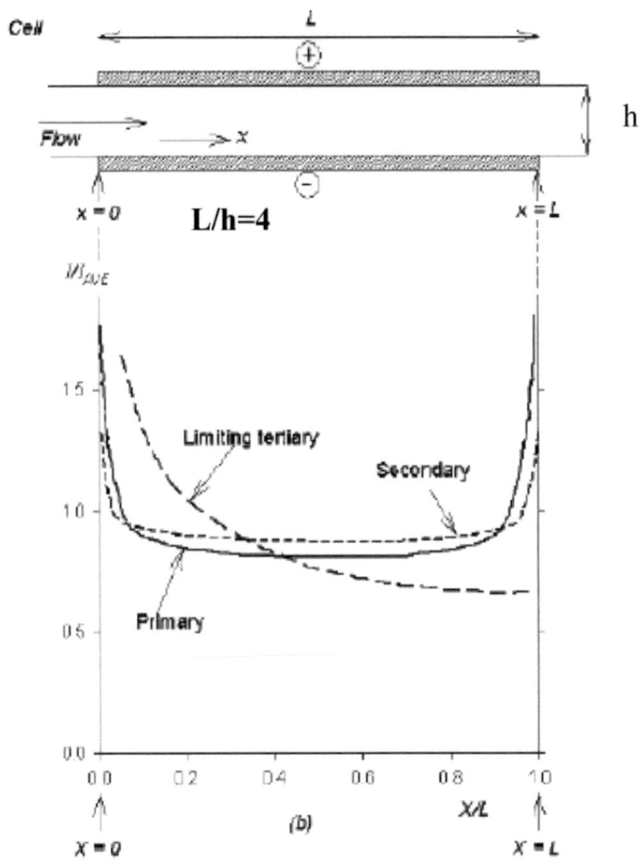

Figure 3. Current distribution in parallel plate electrode [2]

4.2. Current and potential distribution in cylindrical electrodes

Electrodes have been rotated at least since 1905 to provide some quantitative control of solution convection. Hydrodynamic theory for such electrode is generally considered to originate with Levich [27]. A number of rotating electrode geometries have been explored , namely the rotating cylinder electrode(RCE), rotating cone electrode, rotating hemispherical electrode, have been used, but only the RCE has grown to be generally accepted and increasingly widely used. Electrochemical reactors based on the RCE have particularly used in metal ion removal from dilute aqueous solutions, where the metal can be deposited on the surface of an inner rotating cathode. The enhanced mass transport to such RCE cathodes has already been considered and a number of industrial devices have utilized this (and related) technology:

i. The Eco-Cell [28, 29] and the Eco-Cascade cell [30]
ii. The MVH cell [31]
iii. The Turbocel [32]
iv. A cell from Enthone±OMI [33]
v. A rotating cathode band cell from Heraeus El-ektrochemie GmbH [34].

The design and application of RCE reactors in metal ion removal have been extensively reviewed [35, 36, 37]. The primary and secondary current distribution were studies by many investigators for example, the primary and secondary current distribution for deposition of copper–nickel alloys from a citrate electrolyte was studied by Madore and Landolt [38,39]where an empirical equation for the primary current distribution on rotating cylinder electrode has been reported as follows:

$$\frac{j_x}{j_{ave}}\frac{0.535-0.458\times\left(x/h\right)}{\left[0.0233+\left(x/h\right)^2\right]^{0.5}}+8.52\times10^{-5}\times\exp\left[7.17\times\left(\frac{x}{h}\right)\right] \quad (4)$$

Modeling of RCE cells has been focused on both the primary and secondary current distribution with very few studies of the tertiary current distribution.

4.3. Current and potential distribution in fixed bed electrodes

Packed bed electrodes can be used for electrochemical recovery of heavy metals from a variety of industrial and laboratory model solutions (Bennion and Newman [40]; Doherty et al. [41]; El-Deab et al. , [42]; Gaunand et al. , [43]; Lanza and Bertazzoli, [44]; Podlaha and Fenton, [45] ; Ponce de León and Pletcher, [46]; Saleh, [47]; Soltan et al. , [48].

The study of the behavior of fixed bed electrode falls within the scope of electrochemical engineering considered to be the application of the principal of analysis and design of the chemical engineering discipline to electrochemical processes. Fixed bed electrode does not normally operate with a uniform reaction rate and potential, because of ohmic voltage losses within their structure and consequently the specific surface area is not used to full. From a practical point view, it is desirable to utilize most of the internal surface area of the electrode. The estimation of the utilized specific surface area is therefore of great importance in the design and scale-up of these electrodes and one approach is by analogy to the chemical engineering analysis of heterogeneous reactions such as gas–solid catalytic reaction and gas–liquid absorption with chemical reaction, extended to the electrochemical case. Therefore, an effectiveness or effectiveness factor is introduced into account for the fraction of bed thickness electrochemically reactive. This effectiveness (ε) is defined as follows:

$$\varepsilon = \frac{\text{observed electrolytic current for the desired reaction}}{\text{current obtained with an electrode whose over potential is the same at every point for that reaction}} \quad (5)$$

As a rough guide to the operation of fixed bed electrodes, the effectiveness should have a minimum value of 0. 5 if they are to be considered worthwhile alternative to plate electrode cell [49].

Two principle configuration of fixed bed electrodes have been developed ,the flow through porous electrodes(FTPE) where fluid and current flow are parallel, and flow-by porous electrode (FBPE) where fluid flow perpendicularly to current [50]. The current and potential distribution in the first configuration is one-dimensional (the electrode potential varies in the direction of current flow) while the second type involving a two –dimensional problem(the electrode potential varies in the direction parallel to or perpendicular to current flow),which is inherently more complex than the first type.

In the first configuration, a dilemma can arise in choosing a sufficient electrode length to ensure a high conversion factor of the reacting species but avoiding at the same time, a too large potential distribution and consequently bad process selectivity. This is an obstacle for any tentative industrial application of the flow through porous electrode [51]. Alternatively in the second configuration the existence of two degree of freedom (thickness and length of electrode) make it possible to obtain simultaneously a uniform potential distribution and adequate residence time. Therefore this configuration is more adapted to an industrial use [52]. On the other hand, due to fact that first configuration obeys one-dimensional model ,it is very valuable in the theoretical formulations of current and potential distribution in porous electrode [53]. In addition Fedikiw [54] found that if the aspect ratio (length/thickness) of flow-by electrode is large, it is reasonable to assume that the potential field is governed by one dimensional Laplace's equation and consequently the flow through type can be considered to simulated the behavior of a horizontal slab of the flow-by electrode.

4.3.1. One-dimensional model

The complex structure of porous electrode is almost or always reduced for purpose of analysis to a one-dimensional representation. A one-dimensional model of porous electrode is most frequently used; this choice avoids considerable mathematical difficulty and at the same time provides solutions which are in reasonable agreement with experiment [55].

Coeuret et al. [56–58] were the first to study and analyse the current and potential distribution in flow through fixed bed by using a mathematical approach similar to that proposed by the chemical engineering discipline for heterogeneous reaction. They represented the current and potential distribution by the following dimensionless relationship:

$$\frac{\eta_{(X)} - \eta_{(o)}}{\eta_{(L)} - \eta_{(o)}} = \frac{\cosh\left[K_n\left(X/L\right)\right]}{\cosh\left(K_n\right) - 1} \quad (6)$$

$$K_n = \sqrt{\frac{n+1}{2}\frac{i_o zFSL^2}{RT\gamma}\left(\frac{\alpha}{\eta_{(L)}}\right)^{1-n}} \tag{7}$$

Where *Kn* is an effectiveness criterion for fixed bed electrode operating under Tafel polarization regime. Its value determines the effectiveness factor by the following relation:

$$\varepsilon_n = \frac{\tanh(K_n)}{K_n} \tag{8}$$

Where (εn)is the effectiveness factor for packed bed electrode under Tafel region.

It was found in scale-up of electrochemical reactor [59],the effectiveness factor (εn) for scale-up depends on bed thickness (*L*) and the potential at the back of electrode ($\eta(L)$). The conclusion was that effectiveness factor increases as bed thickness (*L*) decreasing and potential ($\eta(L)$) increasing.

Therefore a higher fraction of bed thickness reactive can be obtained either at lower thickness or higher value of potential. Practically, it is preferred to use higher thickness of bed to ensure a higher current supplied or production rate. On the other hand, it is necessary to utilize the maximum portion of surface area with maximum effective bed thickness. It is found that bed thickness not higher than 0. 6 cm must be taken as a maximum limit in the scale-up of the system under study and this system should not be operated beyond this value of bed thickness because it will be not utilized in the reaction zone and causing higher capital cost.

In recent years there are many researches have been done in scale –up of electrochemical reactor [60-62],the aim of these studies and the previous one are how to scaling up the electrochemical reactors to industrial case maintaining the same potential and current distribution. These factors have a vital role in developing the electrochemical system at different fields especially in waste water treatments.

4.3.2. Tow dimensional model

On the contrary to the one-dimensional model a few works have been concerned with the fundamental study of fixed bed electrode obeying two dimensional model. Alkire et al. [63] were the pioneers in the analysis of two dimensional model , where a finite difference method was adopted for Laplace differential equation solution. Their study was extended to the limiting current condition in subsequent work [64].

5. Mass transfer criteria for scale-up

5.1. Design equation for scale up of electrochemical reactor

The electrochemical cell can be controlled by mass transfer at the electrode surface. In the electrochemical cell, for example, a metal- ion concentration at the cathode surface decreased by electrolysis. The mass flux of a metal-ion generated by a special concentration gradient can be described by Fick's law.

$$\dot{n} = -D\,\mathrm{grad}\,c \tag{9}$$

Equation (9) can be expressed in one-dimensional form which can be applied to diffusive mass transport for large flat electrodes:

$$\dot{n} = -D\left(\frac{\partial c}{\partial y}\right)_{y=0} \tag{10}$$

The current density of an electrochemical cell removing a metal-ion by electrochemical conversion couples the rate of electrochemical conversion with diffusive mass transport at and toward the electrode surface:

$$i = \dot{n}_0 \nu_e F = D\left(\frac{\partial c}{\partial y}\right)_{y=0} \nu_e F \tag{11}$$

The index 0 refers to y=0 which means at the electrode surface. The mass transfer rate and associated current densities are given by the product of the mass transfer coefficient km and the concentration difference:

$$i = n_0 \nu_e F = k_m\left(c_\infty - c_0\right)\nu_e F = k_m \Delta c \nu_e F \tag{12}$$

Setting C0=0 defines the mass transfer limited current density ilim:

$$i_{\lim} = k_m c_\infty \nu_e F \tag{13}$$

An explicit equation for the mass transfer coefficient, km, can be found for laminar flow. Under turbulent flow one can only measure the mass transfer coefficient by measuring the mass transport limited current densities. However, this is a tedious affair as mass transfer is often influenced by a great number of variables.

Dimensional analysis allows one to reduce the number of variables which have to be taken into account for mass transfer determination by introducing dimensionless groups. For mass transfer under forced convection, there are at least three dimensionless groups. Those are the Sherwood number, Sh, which contains the mass transfer coefficient, the Reynolds number, Re, which contains the flow velocity and defines the flow condition (laminar/turbulent) and the Schmidt number, Sc, which characterizes the diffusive and viscous properties of the respective fluid and describes the relative extension of the fluid-dynamic and concentration boundary layer.

$$Re = \frac{wL}{\nu} \qquad Sh = \frac{k_m L}{D} \qquad Sc = \frac{\nu}{D}$$

The experimental determination of Sh is quite easy in an electrochemical reactor. Measuring the limiting current and using equation (13), one obtains Sh from ilim.

$$Sh = \frac{i_{\lim} L}{c_{\infty} v_e FD} \tag{14}$$

In general, the dependence of Sh on Re and Sc can be presented in the form of a power series:

$$Sh = a\, Re^{b}\, Sc^{1/3} \left(\frac{D_{C/A}}{L} \right)^{c} \left(\frac{D_{W/C}}{L} \right)^{d} \tag{15}$$

The design equation for the reactor scale-up is deduced by using dimensionless terms. Daewon et. al. [65] found that the dependence of Sh on Re, Sc, characteristic lengths, DC/A/L and DW/C/L could be described in the following form:

$$Sh = 1.24\, Re^{0.12}\, Sc^{1/3} \left(\frac{D_{C/A}}{L} \right)^{-0.87} \left(\frac{D_{W/C}}{L} \right)^{-0.42} \tag{16}$$

Therefore the characteristic length plays an important role in scale-up of the electrochemical reactor in addition to Reynolds and Sherwood number.

Recently the performance of a novel pilot plant scale ,fixed bed flow through cell ,consisting of a cathode formed by a bundle of stainless steal tubes have been investigated [66]. Two mass transfer correlation which represented the flow in bandle of tubes have been obtained for two tube diameters(0. 6cm and 1. 0 cm):

$$\mathrm{Sh} = 0.411\ \mathrm{Re}_{0.871} \left[0.6\mathrm{cm\ outer\ diameter\ tubs} \right] \tag{17}$$

$$\mathrm{Sh} = 0.295\ \mathrm{Re}_{0.84} \left[0.6\mathrm{cm\ outer\ diameter\ tubs} \right] \tag{18}$$

6. Conclusion

The design and scale-up of the electrochemical reactor play an important role in the development of industrial electrochemical processes. Therefore studying the controlling factors on scale-up make the operation of the system more efficient and economic on the commercialization stage. Current and potential distributions are the most significant parameters characterizing the operation of the electrochemical cell. The current density on the electrodes is directly proportional to the reaction rate and its distribution critically affects the electrochemical process. In parallel plate and rotating cylinder electrodes, primary and secondary current and potential distribution are very important, the primary current distribution apply when the surface overpotentail can be neglected and the solution adjacent to an electrode can be taken to be equipotentail surface, while secondary current distribution predominates if the kinetic resistance is higher than the ohmic resistance. In porous electrodes, an effectiveness or effectiveness factor should be taken into account which refers to the fraction of bed thickness electrochemically reactive. The effectiveness should have a minimum value of 0. 5 if they are to be considered worthwhile alternative to plate electrode cell. Studding above factors in addition to the mass transfer correlation are an important parameter for scale-up the electrochemical reactor and should be considered in any future study in any electrochemical reactor before implant to industrial application.

Nomenclature

A_e	electrode area (cm2)
C	metal concentration (mol cm-3)
D	diffusion coefficient (cm2 s-1)
$D_{C/A}$	gap between cathode and anode (cm)
$D_{W/C}$	gap between reactor wall and cathode(cm)
F	Faraday constant (96,487 C mol-1.)
i	current density(A cm-2)
i_o	exchange current density (A cm-2)
K_1	effectiveness criterion for linear polarization
K_n	effectiveness criterion for Tafel polarization
k_m	mass transfer coefficient(cms-1)
L	thickness of screen or length of electrode (cm)
M_i	molecular weight of chemical species i
n	mass flux(gcm-2s-1)
n_0	mass flux at the electrode surface(gcm-2s-1)
R_e	Reynolds number
S	scan rate
Sh	Sherwood number
Sc	Schmidt number
v_e	stoichiometric coefficient of the electrons consumed in electrochemical reaction
v	dynamic viscosity(g cm-1 s-1)
W	linear velocity (cm s-1)
W_i	deposited weight of chemical species i(g)
X	distance through the one-dimensional porous electrode(cm)
Z	number of electrons

Greek symbols

α	dimensional coefficient
ε	effectiveness factor for linear polarization
ε_n	effectiveness factor for Tafel polarization
η	potential of cathode (V)

Author details

A. H. Sulaymon
Environmental Engineering Department, Baghdad University, Iraq

A. H. Abbar
Chemical Engineering Department, College of Engineering, Al Qadessyia University, Iraq

7. References

[1] D. Hoormann, J. Jorissen, H. Putter, *Chem. -Ing. -Tech. 77* (2005)1363–1376.
[2] D. J. Pickett, Electrochemical Reactor Design, Elsevier, New York, (1977)12.
[3] E. Buckingham, Physical rev. 4(2) (1914) 345.
[4] C. L. Mantall, "Electrochemical Engineering"4th Mc-Graw Hill, New York, 1960.
[5] R. H. Perry and C. H. Chilton, Chemical Engineers Handbook, 5thed. McGraw Hill, New York, (1973).
[6] H. Rase, Chemical Reactor Design for Process Plants, Wiley-Interscience, New York, (1977).
[7] D. Pletcher and F. Walsh, Industrial Electrochemistry, Chapman and Hall, (1990).
[8] F. Goodridge and K. Scott, Electrochemical Process Engineering, Plenum Press, New York, (1994).
[9] C. Oloman, J. Electrochem. Soc. 126(11) (1979) 1885.
[10] G. Prentice, Electrochemical Engineering Principles, Prentice Hall, (1991)
[11] C. Kadper, Trans. Electrochem. Soc. 78, (1940) 353, ibid, 82, (1942) 153.
[12] J. Newmanm, J. Electrochem. Soc. 113, (1966) 501
[13] J. J. Miksis, Jr. and J. Newman, ibid, 123, (1976) 1030.
[14] P. Pierni and J. Newman, ibid, 126, (1979) 1348.
[15] R. V. Churchill, "Complex variables and application", 2nd Mchrow hill, NewYork, (1960)
[16] E. T. Copson"An introduction to theory of functions of complex variables" Oxford university press, London (1935)
[17] H. Fletcher Moulton, Proceeding of London mathematical society, vol. (3)(1905), 104.
[18] Milton A. Irene A., "Handbook of mathematical functions "Washington National Bearue of standard (1964) 608.
[19] J. S. Newman, 'Electrochemical system"Prenti-hall, NewJersy(1973)
[20] Prentice and Tobias, J. Electrochem. Soc. 129, (1982) 27
[21] W. R. Parrish, J. S. Newman, J. Electrochem. Soc. 117 (1970) 43.
[22] R. Caban, T. W. Chapman, J. Electrochem. Soc. 123 (1976) 1036.
[23] D. J. Pickett, Electrochemical Reactor Design, Elsevier Science Publication Co, New York, 1979.
[24] R. E. White, M. Bain, M. Raible, J. Electrochem. Soc. 130 (1983)1037.
[25] T. V. Nguyen, C. W. Walton, R. E. White, J. Electrochem. Soc. 133(1986) 81.
[26] J. S. Newman, I&EC, vol. 60, no. 4(1968), p. 12
[27] V. G. Levich, `Physiochemical Hydrodynamics', Prentice-Hall, New York (1962).
[28] F. S. Holland, Chem. Ind. (London) (1978) 453.
[29] Idem, UK patents 1 444 367 (1976) and 1 505 736 (1978);US patent 4 028 199 (1977).
[30] F. C. Walsh, N. A. Gardner and D. R. Gabe, ibid. 12 (1982)299.
[31] MVH Cell, van Aspert bv, Kastanjeweg 68, 5401 JP, Uden, The Netherlands.
[32] J-Cl. Puippe, Oberfach. Surf. 32 (2) (1991) 17; EAST Re-port (1990), 38, E. G. Leuze Verlag. Saulgau, Germany;Galvano-Organo-Traitments de Surface 61 (1992) 259.
[33] D. Hemsley, Prod. Fin. 46 (9) (1993) 5; idem, ibid. 47 (5), (1994), 6; idem, ibid. 47 (7), (1994), 9.

[34] M. Mayr, W. Blatt, B. Busse and H. Heinke, Electrolytic Systems for Applications in Fluoride-Containing, Electrolytes, Fourth International Forum on Electrolysis in the Chemical Industry, (1990)..
[35] N. A. Gardner and F. C. Walsh, in `Electrochemical Cell Design', (edited by R. E. White) Plenum Press, NewYork (1984), 225.
[36] F. C. Walsh and D. R. Gabe, I. Chem. E. Symp. Ser. 116(1990) 219.
[37] C. T. J. LowmE. P. L. Roperts F. C. Walsh, Electrochim. Acta 52 (11), (2007) 3831-3840.
[38] C. Madore, M. Matlosz, D. Landolt, J. Appl. Electrochem. 22 (12) (1992)1155.
[39] C. Madore, A. C. West, M. Matlosz, D. Landolt, Electrochim. Acta 37 (1)(1992) 69.
[40] D. N. Bennion, J. Newman,, J. Appl. Electrochem. 2, (1972) 113–122.
[41] T. Doherty, J. G. Sunderland, E. Roberts, ., D. J Pickett, Electrochim. Acta 41 (4), (1996) 519–526.
[42] M. S., El-Deab, M. M., Saleh, B. E., El-Anoduli, B. G., Ateya, J. Electrochem. Soc. 146 (1), (1999) 208–213.
[43] A. Gaunand, D. Hutin,, F. Coeuret, Electrochim. Acta 22 (1), (1977) 93–97.
[44] M. R. V., Lanza, R., Bertazzoli, . J. Appl. Electrochem. 30 (1), (2000) 61–70.
[45] E. J Podlaha., J. M Fenton., J. Appl. Electrochem. 25, 299–306(1995).
[46] C. Ponce de León, D. Pletcher, Electrochim. Acta 41 (4), (1996) 533–541.
[47] M. M. Saleh, J. Phys. Chem., B 108, 13419–13426(2004).
[48] E. A Soltan, S. A. Nosier, A. Y. Salem, I. A. S. Mansour, G. H Sedahmed, ., Chem. Eng. J. 91, (2003) 33–44.
[49] J. S. Newman, W. Tiedemann, in: C. W. Tobias (Ed.), Advances in Electrochemistry and Electrochemical Engineering. Vol. 11: Electrochemical Engineering, John Wiley & Sons, New York, (1978) 353.
[50] T. Risch, J. Newman, J. Electrochem. Soc. 131 (11), (1984) 208–213.
[51] K. Scott, chemical engineering scence. vol. 37 (5), (1982)792.
[52] M.. fleshmann, R. Jansson, Electrochemic. Acta, vol. 127(8), (1982)1029.
[53] Kou-Chuan&J. Jore, J. Electrochem. Soc. 133 (7), (1986) 1394
[54] P. Fedikow, J. Electrochem. Soc. 128 (4), (1981) 831.
[55] F. A. Posey, J. Electrochem. Soc. 111 (10), (1964)1173.
[56] F. Coeuret, D. Hutin, A. Gaunand, J. Appl. Electrochem. 6 (1976) 17.
[57] M. Paulin, D. Hutin, F. Coeuret, J. Electrochem. Soc. 124 (2) (1977)180.
[58] F. Coeuret, D. Hutin, A. Gavnand, Electrochem. Acta. 22 (1977) 93.
[59] A. H. Abbar, A. H. Sulaymon, M. G. Jalhoom, Electrochimica Acta. 53, (2007) 1671–1679.
[60] N. Gupta and C. W. Oloman, Journal of Applied Electrochemistry 36, (2006)1133–1141.
[61] L. Szpyrkowicz and M. Radaell, Journal of Applied Electrochemistry 36, (2006) 1151–1156.
[62] J. L. Nava, M. T. Oropeza, C. Ponce de León J. González-García, A. J. Frías-Ferre, Hydrometallurgy 91, (2008)98–103.
[63] R. Alkire, N. Patrik, J. Electrochem. Soc. 121(1), (1974)95.
[64] R. Alkire, N. Patrik, J. Electrochem. Soc. 124 (8), (1977)1220.
[65] M. Daewon P., D. Chung and JU Jehbeck, Wat. Res. Vol. 35(1), (2001)57-68.
[66] A. H. Sulaymon, A. O. Sharif, T. K. Al-Shalchi, J Chem Technol Biotechnol 86, (2011) 651-657.

Section 3

Environmental Electrolysis

Chapter 10

Ultrasound in Electrochemical Degradation of Pollutants

Gustavo Stoppa Garbellini

Additional information is available at the end of the chapter

http://dx.doi.org/10.5772/47755

1. Introduction

The increase of industrial activities and intensive use of chemical substances such as petroleum oil, polycyclic aromatic hydrocarbons, BTEX (benzene, toluene, ethylbenzene and xylenes), chlorinated hydrocarbons as polychlorinated biphenyls, trichloroethylene and perchloroethylene, pesticides, dyes, dioxines and heavy metals have been contributing to environmental pollution with dramatic consequences in atmosphere, waters and soils (Martínez-Huitle & Ferro, 2006; Megharaj et al., 2011). Electrochemical technologies have been extensively used for degradation of toxic compounds since these technologies present some advantages, among them: versatility, environmental compatibility and potential cost effectiveness (Martínez-Huitle & Ferro, 2006; Chen, 2004; Ghernaout et al., 2011; Panizza & Cerisola, 2009). However, a loss in the efficiency of such degradation processes is observed due to the adsorption and/or insolubilization of the oxidation and/or reduction products on the electrodes surfaces (Garbellini et al., 2010; Lima Leite et al., 2002).

In this sense, power ultrasound has been employed to overcome such electrode fouling problem (passivation) due to the ultrasound ability for cleaning the electrode surface, called sonoelectrochemistry (Compton et al., 1997). The production of ultrasound is a physical phenomenon based on the process of creating, growing and imploding cavities of steam and gases, known as cavitation. During the compression step, the pressure is positive, while the expansion results in vacuum called negative pressure formed in a compression-expansion cycle that generates cavities (Mason, 1990; Martines et al., 2000). In chemistry, ultrasound has been used in organic synthesis, polymerization, sonolysis, preparation of catalysts and sonoelectrosynthesis (Mason, 1990; Martines et al., 2000).

The introduction of ultrasound into electrochemical cells/reactors has a marked effect upon the mass transport and surface activation characteristics of an electrochemical system (Compton et al., 1996a). Mass transport is greatly increased via acoustic streaming and micro jetting (Banks & Compton, 2003a), resulting from cavitational collapse close to the electrode surface (Compton et al., 1997; Compton et al., 1996a). Ultrasound can be combined with electroanalytical determination, for example, of the pesticides and metabolites (Garbellini et al., 2009; Garbellini et al., 2007), metals (Banks & Compton, 2004; Hardcastle & Compton, 2001), nitrite (Oliveira et al., 2007), etc., in complex samples exploring the great effect in mass transport and the cleaning of the electrode surface.

The combination of an ultrasonic field with an electrochemical oxidation and/or reduction can result in a powerful method for pollutant degradation. Power ultrasound can improve the electrochemical degradation of pollutants by physical and chemical mechanisms. The propagation of acoustic waves in a liquid medium induces cavitation and even their violent collapse at high acoustic pressure (Mason & Lorimer, 2002). Among the physical effects of these collapses are the high rates of micromixing, the cleaning of the electrodes surfaces by dissolving or pitting the inhibiting layers. These effects result mainly in an enhancement of the solid-liquid mass transfer between the electrodes and the solution (Compton et al., 1997). The chemical effects are also a consequence of the violent collapses. The "hot spot" theory predicts temperatures of many thousands of Kelvin and pressures of hundreds of atmospheres inside the bubble during the final compression. Under such drastic conditions, oxidizing species are generated by the homolytic cleavage of molecules (gases and solvent). In aqueous media and in the presence of oxygen, radicals such as $HO^{\bullet}$, $HO_2^{\bullet}$ and $O^{\bullet}$ are produced (Dai et al., 2006). Therefore the generation of $\bullet OH$ radicals is the key to the efficient decomposition of organic materials (Martínez-Huitle & Ferro, 2006).

In view of these interesting aspects, the use of ultrasound is a technologically advanced application of oxidation in the treatment of effluents, for accelerating the destruction of contaminants in the liquid phase (Adewuyi & Appaw, 2002; Appaw & Adewuyi, 2002; Lu & Weavers, 2002; Hua & Hoffmann, 1997). There are three major regions which should be considered in relation to the sonochemical processes in aqueous media. The first region is the inside of cavitation bubbles collapse in the extreme thermodynamic conditions which are due to the high temperatures and pressures (Mason & Lorimer, 2002; Flint & Suslick, 1991). In this region, fast pyrolysis of volatile solutes occurs. Water molecules undergo thermal decomposition to produce H atoms and $\bullet OH$ radicals, which are strong and non-selective oxidants of the organic pollutants present in effluents (Makino et al., 1983, Hart & Henglein, 1985). The second region is the interfacial boundary between the liquid and gas phases in which the temperature is lower than inside the bubbles, but still high, to cause thermal decomposition of organic solutes. Possibly, the reactive radicals are mainly located in this region. The third region is the bulk of the solution (usually at room temperature) in which several reactions between radicals and organic solutes can occur (Henglein, 1987). In the liquid phase, a constant concentration of reactive radical species is maintained by continuous irradiation of the ultrasound. Despite the generation of $\bullet OH$ radicals being the key of the efficient decomposition of organic materials, the recombination of $\bullet OH$ to produce H_2O_2 both in the gas phase of the bubbles or in the liquid phase of the solution, is the main process that limits the amount of reactive radicals accessible to organic molecules. This limitation results in a loss of the overall efficiency, since the H_2O_2 sonochemically generated cannot react with the desired organic molecule and thereby decompose it (Abdelsalam & Birkin, 2002).

In this chapter, theoretical aspects of power ultrasound including the effects on mass transport and on the cleaning of the electrodes surfaces, experimental considerations about the ultrasound application to electrochemical experiments and the use of power ultrasound in electrochemical degradation of pollutants including direct sonoelectrochemistry, sonoelectrochemistry with Fenton reactions or ozonation and ultrasound in association with photoelectrocatalysis using different electrode materials, such as boron doped diamond (BDD), lead, platinum and glassy carbon will be presented and discussed.

2. Experimental considerations

Parameters as the frequency (Hz), the intensity (W) and the ultrasonic power (W cm^{-2}), determined by calorimetry (Mason et al., 1992), and the sources of ultrasound should be considered when the power ultrasound is applied to an electrochemical system. Ultrasonic power is the most important parameter since there is a direct relation with the current value of the redox process. On the other hand, the frequency, generally from 20 to 800 kHz, does not strongly affect the current intensity of a redox process. Ultrasonic sources of high frequency produce a great quantity of hydroxyl radicals in aqueous media in relation to those with low frequency using the same ultrasonic power. High frequency generally favours the chemical mechanisms involving radicals (cleavage of substrate ligations or by water sonolysis). On the other hand, low frequencies are more efficient for mechanical effects, as gases elimination and the cleaning of the electrodes surfaces.

Concerning the published papers in the literature which explore the introducing of power ultrasound into an electrochemical cell, the two major sources are ultrasonic baths and ultrasonic immersion horn probes. The ultrasonic bath (Figure 1A) consists of a number of fixed frequency (20-100 kHz) transducers below the physical exterior of the bath unit and it has the capacity to clean surfaces and to help the dissolution of substances (Compton et al., 1997). The bath is filled with distilled water and the conventional electrochemical cell is placed into a fixed position (Walton et al., 1995). In this arrangement, the cell is electrically separated and the sound waves penetrate a glass wall before entering the electrochemical reactor. This type of source has been used in polymerization reactions (Yildiz, 2002), metal electrodepositions (Kobayashi et al., 2000; Agullo et al., 1999) and in studies showing the effect of ultrasound in the sonoelectrochemical response of some compounds (Lorimer et al., 1996; Walton et al., 2000).

On the other hand, studies with the ultrasonic horn transducer as a tip shape (Figure 1B) are very frequent. This titanium alloy tip (Ti-6Al-4V) is properly insert in the electrochemical cell. The instrument which produce ultrasound converts 50/60 Hz at a high frequency of electric energy, which is transmitted to a piezoelectric transducer, transforming in mechanical vibrations. These vibrations are intensified by the probes, creating pressure waves on liquids. This action produces millions of microscopic waves (cavities) that expand during the negative pressure and implode violentally during the positive pressure. As the bubbles implode, millions of shock waves are produced, generating extreme pressures and temperatures in the implosion sites, with a huge energy liberation. In this way, in front of the horn tip, the formation of a cavitation bubbles cloud can be observed at at sufficiently high intensities (Compton et al., 1997) (Figure 2). Alkire and Perusich (1983) and Compton' works (West, 2002; Saterlay et al., 1999; Saterlay et al., 2001; Villagrán, 2005) present diverse configurations for the determination of different analytes, as example, when the working electrode and the tip are placed in the same or in different compartments in the same electrochemical cell. Concerning the degradation of pollutants by association of power ultrasound to electrochemical tecnhiques, many configurations of sonoelectrochemical cells and/or reactors are available and presented in the published papers, being difficult in this book chapter to describe all systems.

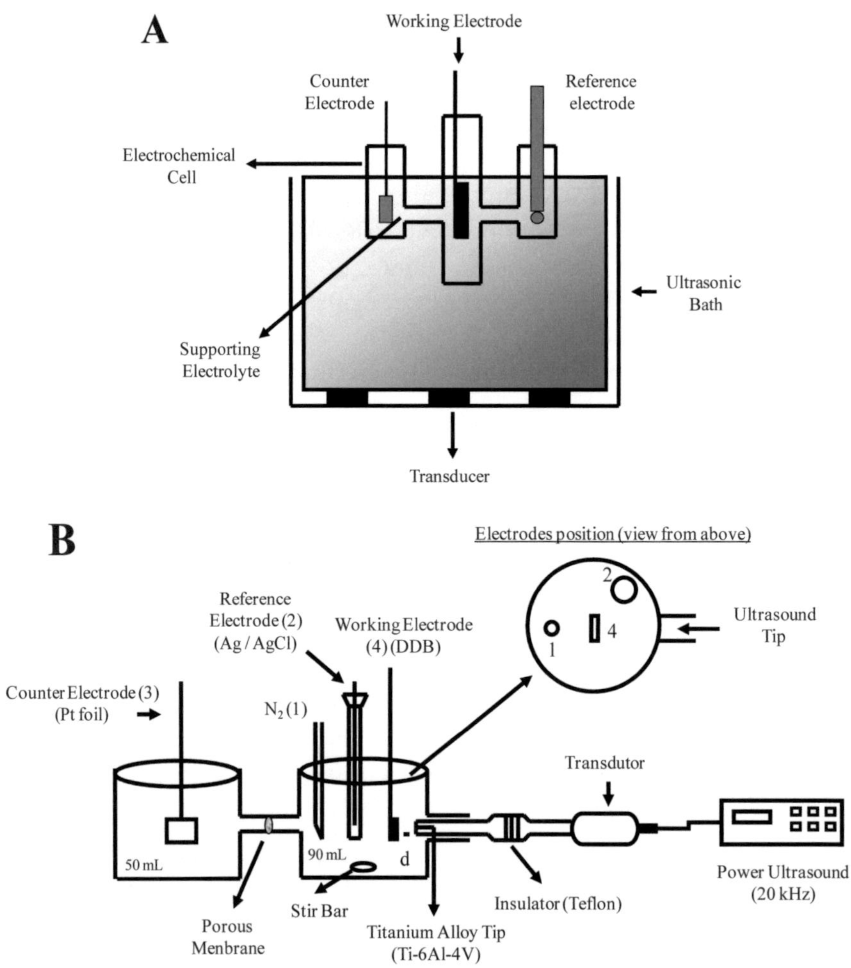

Figure 1. (A) Ultrasonic bath containing an electrochemical cell and (B) Sonoelectrochemical cell whose ultrasound tip is placed in a distance *d* from the electrode surface (BDD electrode in this case) (Garbellini et al., 2010).

The ultrasonic horn has a number of advantages (Compton et al., 1997) over an ultrasonic bath: (i) the horn can supply higher ultrasound intensities (10-1000 W cm^{-2}). This radiation directly applied to the electrochemical system can be controled by the amplitude of the tip vibration; (ii) the distance from the tip to the electrode can be used as a parameter for controlling the ultrasound radiation in the electrochemical system; (iii) the tip of the horn can be used as an electrode and (iv) the geometry and the dimensions of the electrochemical cell has a little effect on the electrode process. However, this transducer has limitations (Compton et al., 1997) as (i) the erosion of the titanium alloy tip that liberates titanium particles in the chemical system of interest causing contamination; (ii) difficulty to maintain the temperature constant during the electrochemical experiment due to the heating effect resulting from the solution agitation and (iii) the need for bipotentiostatic control of the titanium horn tip (Marken & Compton, 1996) due to the direct contact of the tip with the electrochemical system, i.e, the titanium alloy tip present a potential in relation to the reference electrode. This potentiostatic control can be eliminated by insulating the transducer from the probe with a Teflon® disk, as demonstrated in Figure 1B.

Figure 2. Ultrasound tip in front of BDD electrode (distance = 5 mm).

Additionally, the experiments with the system of Figure 1B, and also with other configurantions, can be conduced using alternative geometric arrangements of the tip in relation to the working electrode, (a) face on geometry (Compton et al., 1994; Compton et al., 1995a; Compton et al. 1995b; Lee et al., 1995): the ultrasound tip is placed in front of the working electrode; (b) side on geometry (Eklund et al., 1996): the working electrode is placed perpendicularly to the ultrasond tip and (c) the ultrasound tip can be used with working electrode (sonotrode) (Compton et al., 1996b).

3. Effect of ultrasound on mass transport

The transport of fluids in an ultrasonic field occurs by acoustic flows, movement induced by field and cavitation (Cooper & Coury Jr, 1998). For acoustic flows, the power ultrasound or certain energy intensity generates a significant flow when the ultrasonic energy is absorbed by the media. This effect is the result of the acoustic energy conversion into kinetic energy in the bulk solution. The movement induced by the field occurs when the ultrasound pass through a fluid creating a periodic displacement of fluid particles. If an electrode surface is placed sufficiently close to the ultrasonic transducer, the periodic movement contributes to the signal current through alternate increase or decrease of the transport of species to the electrode surface. This process is analogous to what happens in the rotating disk electrode. As commented, the acoustic cavitation involves the nucleation, growth and collapse of vapor cavities in a liquid subjected to ultrasound. The violent collapse of these bubbles in the electrolyte can produce shock waves that contribute to the mass transport (Cooper & Coury Jr, 1998). If the bubble collapses near the electrode surface, it generates a jet which improves the fluid transport (Figure 3).

In summary, the improvement of mass transport provided by ultrasound is due to two transient processes. The bubble collapses at or near the electrode-solution interface with the formation of high speed flows directed to the electrode surface, besides the movement of the bubble in or near the electrode diffusion layer. It is interesting to note that the acoustic flows are largely responsible for the convective flow to the solid-liquid interface. As example, cyclic voltammograms of the electrochemical process of $K_4[Fe(CN)_6].3H_2O$ were obtained on BDD electrode in 0.1 mol L^{-1} KCl in the absence and presence of ultrasound (Figure 4).

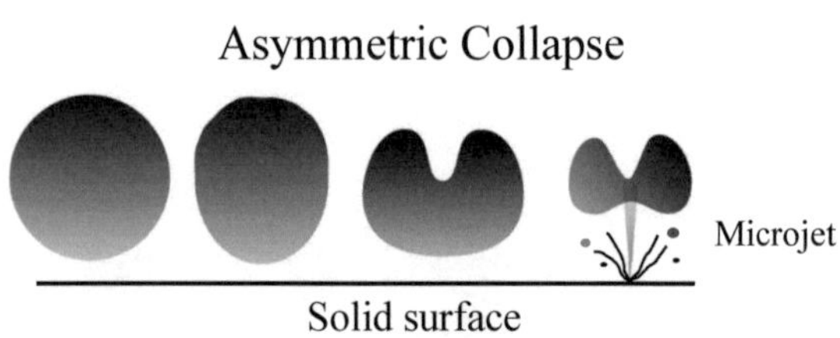

Figure 3. Occurrence of micro-jets when the bubble collapses near the surface (adapted from Banks & Compton, 2003a).

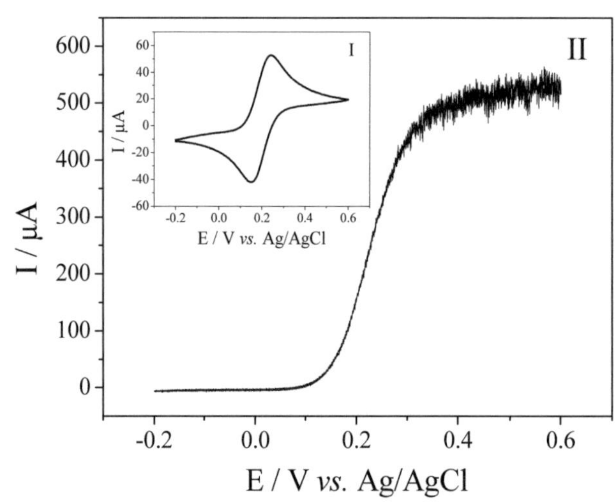

Figure 4. Cyclic voltammograms (50 mV s^{-1}) for 1.0×10^{-3} mol L^{-1} $K_4[Fe(CN)_6]$ obtained in 0.1 mol L^{-1} KCl in the absence (I) and presence of ultrasound, direct and reverse scans (II).

The voltammogram of Figure 4 (I) is characteristic of a reversible system (par redox). Figure 4 (II) shows a measure under the same conditions in the presence of ultrasound directed to the electrode surface of BDD electrode with a distance from the ultrasound probe relative to the electrode of 10 mm. The current intensity of the sonovoltammogram, in which diffusion and convection participate of the mass transport, is significantly greater when compared to the current in voltammogram obtained without the use of ultrasound, in which only the diffusion contributes to the transport of the material. In addition, the profile of response in the absence of ultrasound is modified from a typical cyclic voltammogram to a voltammogram with a sigmoidal shape corresponding to the steady state, and the magnitude of this current is controlled by convective transport of the electroactive species to the electrode surface. It is also important to note that sonovoltammogram from 0.3 V is obtained originally with "noise", which represents cavitation processes occurring in electrode-solution interface and turbulent microflows provided by ultrasound.

Thus, the increase in mass transport to the electrode surface provided by the ultrasound can be measured by assuming a reduction of the diffusion layer thickness (δ). The diffusion layer model enables a description of the mass transport to the interface of the electrode considering a laminar sublayer near the surface and a concentration gradient approximately linear through a thin layer adjacent the electrode. The Equation 1 (where n is the number of electrons transferred, F the Faraday constant, D the diffusion coefficient, A the electrode area and C is concentration in the bulk solution) describes the transport in an electrode subjected to ultrasound based on the model of electrode "accessibly" uniform (δ constant value over the electrode surface), being the electrode larger than the size of δ (Compton et al., 1997; Banks et al., 2003b).

$$I_{lim} = \frac{nFDAC_{bulk}}{\delta} \tag{1}$$

In sonoelectrochemistry, the analysis of the currents based on the Nernst diffusion layer model gives additional criteria regarding the mass transport if electroactive species with different D are compared. The δ value is a function of D, δ(D) = Dx, which exponent can be considered an indicator of the mass transport process nature (Holt et al., 2001). In addition, δ depends on the applied power ultrasound and electrode radius (Compton et al., 1996a). The δ value is considerably smaller for microelectrodes in relation to the electrodes with conventional dimensions, although, in both cases, δ in the presence of ultrasound is significantly lower than in the absence of ultrasound (Compton et al., 1995a). It is noteworthy that electrodes with dimensions on the order of micrometres (between 0.8 and 50.0 μm), known as ultramicroelectrodes, have advantageous properties compared to conventional electrodes (greater than 50.0 μm). Its dimensions are similar to those of the diffusion layer, resulting in a high-speed mass transport, due to the spherical shape of the diffusion layer, which facilitates the study of fast electrochemical reactions under steady state conditions achieved in a shorter time than with conventional electrodes (Correia et al., 1995).

Sonoelectrochemical experiments in which the ultrasound probe is placed in front of the electrode have show that this one behaves as an hydrodinamic electrode. In such cases, the processes of diffusion and convection occur simultaneously, contributing to the transport of electroactive material to the electrode surface (Compton et al., 1997; Banks et al., 2003b). An example of the mass transport increase can be illustrated in the electrolyses of an organochlorine pesticide, pentachlorophenol (PCP) on BDD electrode (Figure 5).

A comparison between electrolyses of the PCP at 3.0 V in the absence and presence of ultrasound on the BDD electrode showed an increase in mass transport promoted by ultrasound, since sonoelectrolyses (curve 2) showed current intensities 1.5 times higher than that obtained by silent electrolyses (in the absence of ultrasound) (curve 1), as shown in Figure 5. In both electrolyses, there was no decay of current due to the occurrence of water oxidation in this high value of applied potential.

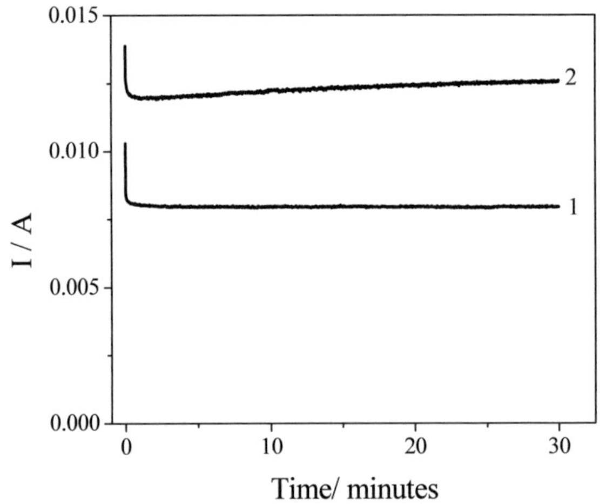

Figure 5. Electrolyses of the PCP at 3.0 V vs. Ag/AgCl in the absence (1) and presence (2) of ultrasound for 30 minutes. Conditions: [PCP] = 5.0×10^{-5} mol L^{-1} in BR buffer 0.1 mol L^{-1} at pH 5.5 and ultrasound tip-BDD electrode distance (d) = 7 mm, (Garbellini et al., 2010).

4. Cavitation effects on electrodes surfaces

Acoustic cavitation is the centre for many cleaning operations in laboratories and for in situ electrode depassivation. The ultrasound, when applied directly to an electrode surface, can provide a severe degradation of the surface by erosion of the electrode material (Compton et al., 1994), as well as it induces the activation and enhancement in the performance due to the electrode cleaning (Compton et al., 1994; Zhang & Coury Jr, 1993).

The beneficial effect of electrode cleaning is widely used for analysis of metals or organic compounds in complex matrices. The organic compounds and/or samples constituents have usually strong interactions with the surface of solid electrodes (adsorption process) thus hindering their use in analytical determinations. Moreover, numerous problems associated with the direct determination of these compounds in real and complex samples are encountered. For example, the presence of fats, proteins and carbohydrates in food matrices, lead to electrode fouling due to adsorption of these species and it can also reduce the sensibility of the detection methods. In this sense, power ultrasound has been employed to overcome such electrode fouling problem (passivation) due to the ultrasound ability for cleaning the surface electrode (Saterlay et al., 2001), allowing sensitive electroanalysis of some analytes to be carried out in a range of hostile media including eggs (Davis & Compton, 2000), blood (Kruusma, et al. 2004a), wine (Akkermans et al., 1998), among others (Hardcastle & Compton, 2001; Kruusma, 2004b). Recently, the direct determination of methylparathion in potato and corn extracts and its degradation product 4-nitrophenol in lemon and orange juices was performed by ultrasound-assisted square wave voltammetry using diamond electrodes (Garbellini et al., 2009). The sonovoltammetric results for both analytes in pure water and in complex food samples showed greater sensitivity and precision and much lower limits of detection and quantification than the silent measurements.

Evaluations of the changes in various types of electrode surfaces after ultrasound application should be performed by ex situ and in situ methods available for the investigation of surface properties, such as, atomic force microscopy (Compton et al. 1997). Studies performed by Compton´s group (Compton et al., 1994), showed that platinum and aluminium electrodes were degraded after some minutes of ultrasound exposure. In materials like glassy carbon (Zhang & Coury Jr, 1993; Marken et al., 1996), no damage was detected after 30 minutes of ultrasound applied. On the other hand, this time of ultrasound application on gold electrodes (Marken et al., 1996) cause a significant rugosity. Nevertheless, these materials can be used for sonoelectrochemical experiments, since the damages depend on, for example, the intensity of the power ultrasound and the distance between the electrode and ultrasound tip. A viable alternative for these materials is the diamond, due to their extreme hardness and corrosion resistance. BDD electrode, since over violent conditions, has not presented indications of erosion or surface damage (Compton et al., 1998; Goeting et al., 1999).

As discussed, the cavitation process produces active radicals, e.g, hydroxyl, strong oxidizing agents which undoubtedly contribute to increase the degradation levels of pollutants. In the next section of the chapter, some works will be presented.

5. Ultrasound in electrochemical degradation of pollutants

Residues of toxic organic compounds in environment certainly cause problems to human health, degradation of natural resources, leading to biological and ecological imbalances, including the contamination of groundwater and aquifers, damage to microorganisms in aquatic and soil and reduce the productivity of marine plants and corals. Therefore, the development of methodologies for the degradation of these compounds and the metabolites is increasingly necessary. Concerning the technologies for this purpose, in this topic, power ultrasound in electrochemical degradation of pollutants will be discussed including direct sonoelectrochemistry, sonoelectrochemistry with Fenton reactions and with ozonation and ultrasound in association with photoelectrocatalysis using different electrode materials.

5.1. Direct sonoelectrochemical degradation

As aforementioned discussed, the association of power ultrasound with electrochemical techniques can be a powerful tool for degradation of pollutants, since the ultrasound is responsible for the increase of mass transport, activation of electrodes surfaces and generation of active hydroxyl radicals due to the water sonolysis. Some works considering the use of sonoelectrochemistry for pollutants degradation are discussed in this section and the type of pollutant and applied techniques are collected in Table 1.

Pollutant	Methodology	Reference
Sandolan Yellow dye	Sonoelectrolyses in galvanostatic mode	Lorimer et al., 2000
2,4-dihydroxybenzoic acid	Sonoelectrolyses in potentiostatic and galvanostatic modes	Lima-Leite et al., 2002
N,N-dimethyl-p-nitrosoaniline	Sonoelectrolyses in galvanostatic mode	Holt et al., 2003
Perchloroethylene	Sonoelectrolyses in galvanostatic mode	Sáez et al., 2010
P-substituted phenols	Sonoelectrolyses in galvanostatic mode	Zhu et al., 2010
Pentachlorophenol	Sonoelectrolyses in potentiostatic mode	Garbellini et al., 2010
Trichloroacetic acid	Sonoelectrolyses in galvanostatic mode	Esclapez et al., 2010
Reactive blue 19 dye	Sonoelectrochemistry	Siddique et al., 2011
Diuron	Sonoelectrolyses in galvanostatic mode	Bringas et al., 2011

Table 1. Some published papers concerning the direct sonoelectrochemical degradation of pollutants.

Lorimer et al. (2000) applied procedures as sonolysis, electrolysis and sonoelectrolysis to degraded solutions of the acidic dye, Sandolan Yellow, using platinum electrodes and sodium chloride as supporting electrolyte. The process involves the liberation of chlorine at the anode and hydroxide ion at the cathode, resulting in the in situ generation of the hypochlorite ion (equations 2-4), which is a powerful oxidant. The electro-oxidation of Sandolan Yellow process is significantly improved by the use of ultrasound in conjunction with electrolysis, and an optimum acoustic ultrasonic power was observed when using a source at a frequency of 20 kHz.

$$\text{Anode: } 2Cl^- \rightarrow Cl_2 + 2e \tag{2}$$

$$\text{Cathode: } 2e + 2H_2O \rightarrow 2OH^- + H_2 \tag{3}$$

$$\text{Overall reaction: } 2OH^- + Cl_2 \rightarrow Cl^- + OCl^- + H_2O \tag{4}$$

Lima Leite et al. (2002) investigated the electroxidation of the 2,4-dihydroxybenzoic acid (2,4-DHBA) by potentiostatic and galvanostatic electrolyses on platinum electrode with the ultrasound at two frequencies (20 e 500 kHz). Potentiostatic sonoelectrooxidation of 2,4-DHBA was carried out at three different potentials (1.2; 1.5 and 2.0 V), according to the rate of dioxygen production. Under 20 kHz ultrasound irradiation alone, 2,4-DHBA is almost unaffected. Degradation rate increases with the applied electrode potential. Improvement in 2,4-DHBA electrodegradation by low-frequency ultrasound proved to be more marked than in the case of high-frequency ultrasound, certainly due to mechanical effects. Variation in the electrolysis current during the process shows that 2,4-DHBA electrooxidation under low-frequency ultrasound irradiation can be carried out at a greater current density than in the case of electrooxidation alone. Current drops are also reduced in the presence of an acoustic field. As previously mentioned, it can be explained by the elimination of the passivating polymer film formed at the electrode surface: the electrode active surface area increases and the intensity, which is directly proportional to it, increases too. The improvement in the mass transfer and the more rapid desorption of the reaction products are also responsible for the lower current drops.

After the study of potentiostatic sonoelectrolysis parameters, the galvanostatic operation was investigated since it is most widely used on an industrial scale. Key parameters such as current density, initial pollutant concentration and ultrasonic power were studied. For a 300 mg l^{-1} initial concentration and a 300 A m^{-2} current density, the TOC decrease was 47% after passing an electricity amount of 1.5 A h at low frequency and only 32% after passing 3.5 A h at high-frequency sonoelectrooxidation or electrooxidation. At low frequency, 2,4-DHBA degradation is accelerated and final TOC is lower: cavitation phenomena ensure the cleaning of the electrode surface thus increasing the active electrode surface. Observed by-products of sonoelectrodegradation are the same as for electrooxidation alone, including the following: 2,3,4- and 2,4,5-trihydroxybenzoic acids (THBA), maleic acid, glyoxylic acid and oxalic acid. Fewer intermediate aromatic compounds are formed at low-frequency irradiation. Moreover, the faradaic yield increases under low-frequency sonication, showing a more efficient use of electrochemical energy. Nevertheless, the overall energy consumption remains high (>200 kW kg^{-1}).

Additionally, power ultrasound in association with electrolyses for degradation of phenols and phthalic acid (Zhao et al., 2008; Zhao et al., 2009), N,N-dimethyl-p-nitrosoaniline (Holt et al., 2003) and reactive dye Procion Blue (Foord et al., 2001) have also been reported. A preliminary study of the 20 kHz sonoelectrochemical degradation of perchloroethylene in 0.05 mol L^{-1} aqueous sodium sulfate has been carried out by Sáez et al. (2010) using controlled current density degradation sonoelectrolyses in batch mode. PCE sonoelectrolysis experiments were performed at different ultrasound intensities of 1.84, 3.39, 5.09, 6.36 and 7.64 W cm^{-2}, with T=20°C, 362 μM, and the value of the working current density fixed at 3.5 mA cm^{-2}. An analysis for all ultrasound intensities, all of these volatile compounds: PCE, trichloroethylene (TCE) and dichloroethylene (DCE), chloride and chlorate anions were totally degraded in the first 2.5 h of the process, in contrast to the electrochemical route, for which, steady state remaining concentrations of TCE and DCE (higher than 10% of the initial concentration of PCE for each one) were routinely detected even after 5 h of treatment. The main contribution of the ultrasound field presence during the ECT (electrochemical treatment) is not only the total degradation of the main volatile compounds (PCE, TCE and DCE), but also the decrease of the effective reaction time. The energetic consumption with sonoelectrochemical treatment is lower than that presented by sonochemical treatment, due to the fact that the treatment time is significantly reduced.

The effects of low-frequency (40 kHz) ultrasound on the electrochemical oxidation of p-substituted phenols (p-nitrophenol, p-hydroxybenzaldehyde, phenol, p-cresol, and p-methoxyphenol) at BDD and PbO_2 anodes were evaluated by Zhu et al. (2010). The oxidation was performed at constant current density (20mA cm^{-2}) and room temperature (25°C) using sodium sulphate as supporting electrolyte. At the BDD anode, the % increase values were in the range 73–83% for p-substituted phenol disappearance and in the range 60–70% for chemical oxygen demand removal. However, at the PbO_2 anode, the corresponding % increase values were in the range 50–70% for disappearance of p-substituted phenols and only 5–25% for chemical oxygen removal, much lower values than obtained at the BDD anode. The hydroxyl radicals were mainly free at the BDD electrodes with a larger reaction zone, but adsorbed at the PbO_2 electrodes with a smaller reaction zone. Therefore, the enhancement due to ultrasound was greater at the BDD anode than at the PbO_2 anode.

In our work (Garbellini et al., 2010), the beneficial effects of the ultrasound (fixed frequency of 20 kHz in a maximum power of 14 W) were evaluated in association with potentiostatic electrolyses for the degradation of pentachlorophenol at 3.0 V vs. Ag/AgCl, using a BDD electrode during 270 minutes using a Britton-Robinson (BR) 0.1 mol L^{-1} buffer solution as electrolyte. The sonoelectrochemical cell used for this work is presented in Figure 1B. Different decay levels of the PCP spectrum bands in 220, 251 and 321 nm, respectively, were observed after application of ultrasound without electrochemical process (18.1, 17.7 and 19.8 %), silent electrolyses (29.3, 71.6 and 70.8 %), pulsed sonoelectrolysis (31.0, 75.1 and 76.3%) and sonoelectrolyses (39.2, 80.0 and 82.6 %). Specifically, the pulsed sonoelectrolysis involved a combination of electrolysis in the presence (30 min) and absence (5 min) of ultrasound. This process was carried out purposely without cleaning/reactivation of the diamond electrode surface. The effects of the electrode surface activation provided by ultrasound and simultaneous generation of hydroxyl radicals by both radiation and polarized BDD surface contributed to higher levels of degradation of the pesticide in comparison to the values obtained for the silent electrolyses.

The sonoelectrochemical treatment of aqueous solutions of trichloroacetic acid (TCAA) has been scaled up from the voltammetric analysis to pre-pilot stage, as it has been reported by Esclapez et al. (2010). All the bulk electrolyses in batch mode were carried out using a home-made galvanostat (120 mA) using a titanium disk as working electrode. Sonoelectrolyses at batch scale (carried out with a horn-transducer 24 kHz positioned at about 3 cm from the electrode surface) achieved little improvement in the degradation. However, when a specifically designed sonoelectrochemical reactor (not optimized) was used during the scale-up, the presence of ultrasound field provided better results (fractional conversion 97%, degradation efficiency 26% and current efficiency 8%) at lower ultrasonic intensities and volumetric flow.

Recently, Siddique et al. (2011) reported the decomposition of un-hydrolyzed and hydrolyzed forms of reactive blue (RB) 19 dye by ultrasound assisted electrochemical process using lead oxide as working electrode. The experiments were conducted at various pH values in the range of 3–9 and various ultrasonic frequencies (20–80 kHz) using 50 mg L^{-1} dye concentration for 120 min. The results showed that almost complete 90% color removal and a maximum of 56% TOC removal for 50 mg L^{-1} dye concentration of un-hydrolyzed RB 19 dye was achieved at an ultrasonic frequency of 80 kHz, pH of 8 after 120 min. In case of hydrolyzed dye, the TOC reduction observed was 81%. The sonoelectrolysis for dye decomposition and decolorization proved to be more effective and the energy consumption reduced to half as compared with the electrolysis/sonochemical decomposition.

Macounova et al. (1998) and Bringas et al. (2011) investigated the effects of power ultrasound in the electrochemical oxidation of the herbicide diuron using the glassy carbon and BDD electrodes, respectively. In Macounova's work, the ultrasound was just used to avoid a total blockage of the electrode surface by a passivating film. A mechanism involving dimmers formation was reported; however, no information about degradation rates of herbicide was mentioned. Bringas's work reports the degradation and mineralization of diuron at a BDD anode enhanced by low frequency (20 kHz) ultrasound. Under the operation conditions, 60 mA cm^{-2}, pH = 12, T = 10 °C and 8 h of experimental running, results demonstrated improvements on the mineralization kinetics of diuron closely to 43% when ultrasound was coupled to the electrochemical treatment. In addition, the results showed that alkaline pH favours the mineralization rate obtaining reductions of the total organic carbon higher than the 92% after 6 h of degradation at a constant current density of 60 mA cm^{-2} using sodium sulphate as supporting electrolyte.

5.2. Sonoelectrochemical degradation in association with different methods

The application of the sonoelectrochemical processes associated to other methods for the oxidation of organic compounds has been developed as a powerful tool to treatment of effluents. This way, the Fenton reaction (equation 5) (Brillas et al., 2009) is well known in the degradation of organic material by extra generation of hydroxyl radicals.

$$Fe^{2+} + H_2O_2 \rightarrow Fe^{3+} + OH^- + {}^{\bullet}OH \quad (5)$$

Hydrogen peroxide generated through cavitation action (ultrasound) or electrochemical reduction of molecular oxygen is not highly active towards the destruction of an organic species. Fenton's reagent system can be applied to circumvent this problem what enabled the maximum amount of free radicals (specifically hydroxyl radicals) to be generated. This was achieved by the addition of Fe^{2+} to the solution that is known to catalyse the destruction of organic material through the generation of extra hydroxyl radicals, according to a Fenton's type mechanism (Brillas et al., 2009). Some works are discussed here and the kind of pollutant and techniques are collected in Table 2.

Pollutant	Methodology	Reference
Meldola blue dye	Sonoelectro-Fenton	Abdelsalam *et al.*, 2002
2,4-dichloro-phenoxyacetic acid and 2,4-dichlorophenol	Sonoelectro-Fenton	Yasman *et al.*, 2004
1,3-dinitrobenzene and 2,4-dinitrotoluene	Sonoelectrochemistry with ozonation	Abramov *et al.*, 2006
Methyl orange dye	Sonophotoelectrocatalysis	Zhang et al., 2008
2,4-dichloro-phenoxyacetic acid, 4,6-dinitro-o-cresol and azobenzene dye	Sonoelectro-Fenton	Oturan *et al.*, 2008
Azure B dye	Sonoelectro-Fenton	Martinez *et al.*, 2012

Table 2. Some published papers concerning the sonoelectrochemical degradation in association with different methods of pollutants.

Abdelsalam et al. (2002) reported the degradation of an organic dye molecule (specifically meldola blue, MDB) under the influence of power ultrasound in combination with electrochemically-generated hydrogen peroxide. A novel flow system was employed to measure the degradation as a function of time while minimizing the disturbance to the acoustics of the sonoelectrochemical reactor employed. The sonoelectrochemical reactor contained a 100 cm^3 solution consisting of 0.1 mmol dm^{-3} MDB, 50 mmol dm^{-3} Na_2SO_4, 10 mmol dm^{-3} H_2SO_4 , 0.5 mmol dm^{-3} $FeSO_4$ sonicated at 124 kHz. A constant potential of -0.7 V vs. Ag was applied to the reticulated vitreous carbon electrode (RVC). The degradation of a model dye species has been shown to be significantly enhanced by both the presence of Fe^{2+}, electrochemically generated hydrogen peroxide and the presence of ultrasound. Prolonged exposure of the solution to the degradation conditions resulted in a 60.7% reduction in the chemical oxygen demand after 100 min. The constant rate for the complete destruction of MDB, determined by chemical oxygen demand, was found to be significantly slower at $(10.2\pm2.6) \times 10^{-3}$ min^{-1}.

Yasman et al. (2004) presented a method for detoxification of hydrophilic chloroorganic pollutants, 2,4-dichlorophenoxyacetic acid (2,4-D) and its derivative 2,4-dichlorophenol (2,4-DCP), in effluent water using a combination of ultrasound waves, electrochemistry and Fenton's reagent. Both cathode and anode were made of nickel foil and the support electrolyte was Na_2SO_4 (0.5 g L^{-1}). Sonoelectroxidation was carried out in the galvanostatic mode at current intensities not exceeding 100 mA and the sonication was achieved at low frequencies (20 kHz). The application of the sonoelectrochemical Fenton process (SEF) using a current density of 100 mA resulted in a practically completed degradation of the 1.2 mmol L^{-1} 2,4-D solution in 600 s, also an considerable oxidation of 2,4-DCP. The efficiency of the SEF process was much higher than the other conventional degradation methods and the required times for the complete degradation were considerably shorter.

Oturan et al. (2008) also reported the use of sonoelectro-Fenton process for the degradation of organic pollutants in aqueous medium, such as herbicides 4,6-dinitro-o-cresol (DNOC) and 2,4-dichlorophenoxyacetic acid (2,4-D) and the azobenzene (AB) dye. A cylindrical Pt mesh used as anode was centered in the electrolytic cell and surrounded by the carbon-felt cathode. Aqueous solutions of 250 ml were prepared for each individual pollutant, containing 1 mM 2,4-D, 0.5 mM DNOC or 0.025 mM AB, with 0.05 M Na_2SO_4 as background electrolyte and 0.1 mM Fe^{3+} as catalyst at pH 3.0 adjusted with H_2SO_4 to be comparatively treated by EF process under galvanostatic conditions at 200 mA. The same experimental conditions mentioned above were used to study sono-EF process, but coupling with ultrasound irradiation at low (28 kHz) or high (460 kHz) frequencies provided by a ceramic piezoelectric transducer. The SEF process allowed the obtainment of a higher degradation rate than that provided by the two techniques separately for 2,4-D and DNOC. Similar results have been obtained at low and high frequency, what suggests that the main contribution to the oxidation process arises from Fenton's reaction, not from the effects of sonication on organics. In contrast, the output power greatly influences the sono-EF process performance, being 20W the optimum power because higher values hamper the dissolved O_2 concentration and, consequently, affects the cathodic H_2O_2 electrogeneration required for Fenton's reaction. On the other hand, readily oxidizable compounds such as AB, which undergoes the easy cleavage of azo bond by ultrasounds, are so quickly destroyed by EF process alone that ultrasound irradiation is unable to improve the treatment.

Martinez et al. (2012) recently presented the degradation of azure B dye ($C_{15}H_{16}ClN_3S$; AB) by Fenton, sonolysis and sono-electroFenton processes employing ultrasound at 23 kHz and the electrogeneration of H_2O_2 at the reticulated vitreous carbon electrode. The best oxidative degradation of AB by Fenton reaction was obtained at pH between 2.6 and 3.0 using 0.8×10^{-3} mol L^{-1} Fe^{2+} and 2.4×10^{-3} mol L^{-1} H_2O_2. The oxidative degradation of AB followed apparent first order kinetics, where the rate constants decreased in the following order: sono-electroFenton > Fenton > sonolysis. The rate constant for AB degradation by sono-electroFenton is 10-fold that of sonolysis and 2-fold the one obtained by Fenton under silent conditions. The chemical oxygen demand was abated 68% and 85% by Fenton and sono-electroFenton, respectively, achieving AB concentration removal over 90% with both processes. This way, the sono-EF process offers optimistic perspectives regarding the development of more efficient sonoelectrochemical treatment processes for organic compounds.

On the other hand, there are other methods associated to sonoelectrochemical processes, such as those in the presence of ozone and in conjuction with photoelectrocatalysis.

The molecular ozone in aqueous solutions is one of the most active oxidizing agents. Interaction of ozone with an electron donor, D, or by reduction at a cathode, leads to the formation of the O^{-3} ion (Abramov et al., 2006):

$$O_3 + e^- (D) \rightarrow O_3^- (+D+) \qquad (6)$$

The O_3^- ion is a stronger oxidant than molecular ozone, and in acid media it rapidly undergoes reaction with the formation of the $O^{-\bullet}$ anion–radical and oxygen:

$$O_3^- \rightarrow O^{-\bullet} + {}^1O_2 \quad (7)$$

Alternatively, it can interact with water, giving oxygen, hydroxyl ion and radical •OH:

$$O_3^- + H_2O \rightarrow HO\bullet + {}^1O_2 + HO^- \quad (8)$$

Thus the cathodic reduction of ozone yields highly active oxidizing agents. Abramov et al. (2006) reported a method for the destruction of 1,3-dinitrobenzene and 2,4-dinitrotoluene in aqueous solutions using sonoelectrochemistry treatment with ozone addition. A titanium ultrasonic horn radiator was used as a cathode (sonotrode) and the intensity used in all sonochemical experiments was 5 W cm^{-2}. 100 cm^3 of 1% acetic acid solution containing DNB, and DNT at concentrations of 100 mg dm^{-3} were introduced into the cathode compartment. The compounds have shown to be stable to reaction with ozone, even under ultrasonic activation. The use of ultrasound enhances the rate of electrochemical reduction but the overall rate of reaction is still slow. However, the simultaneous application of ultrasound and ozonation to the electrochemical reaction allows virtually complete destruction of the compounds in short times. The effect is attributed to the ultrasonic enhancement of the electrochemical process giving intermediates that are susceptible to ozone oxidation.

Among methods that can be associated with power ultrasound in order to destruct organic compounds is the photocatalysis using TiO_2 as photocatalyst. However, the low photocatalytic efficiency limited the application of this technology in degradation treatments. Ultrasonic-assisted photocatalysis can enhance the photocatalytic efficiency (Bejarano-Perez et al., 2007). Zhang et al. (2008) used hybrid processes involving both electro-assisted and ultrasonic assisted ways to enhance the photocatalytic efficiency, sonophotoelectrocatalysis (SPEC), to investigate the degradation of azo dye methyl orange (MeO) in aqueous solution using TiO_2 nanotubes. Different power of ultrasound (from 60 to 150 W) and different dye concentrations (from 1×10^{-5} to 2×10^{-4} mol L^{-1}) for a total period of 60 minutes were applied in the experiments. The results showed that the hybrid processes could efficiently enhance the degradation efficiency of MeO, and followed pseudo-first-order kinetics. Thus the constants rate of decolorization of MeO were 0.0732 min^{-1} for SPEC process; 0.0523 min^{-1} for photoelectrocatalysis process, 0.0073 min^{-1} for sonophotocatalysis process and 0.0035 min^{-1} for photocatalysis process. The constants rate values indicated that there was synergistic effect in the ultrasonic, electro-assisted and photocatalytic processes.

6. Conclusion

Ultrasound combined with electrochemical process and Fenton reactions is a promising method for the degradation of some toxic organic compounds and treatment of wastewater.

The beneficial effects of the power ultrasound use, e.g, cleaning/activation of the electrodes surfaces, enhancement of the mass transport and generation of hydroxyl radicals by water sonolysis contributed to higher levels of pollutants degradation in comparison to the values obtained for the silent electrolyses.

The main drawback of ultrasound is its energetic cost. However, the ultrasonic intensity series has shown that it is not necessary to use very high ultrasound power. Therefore, working at low power ultrasound and with different strategies (pulsed ultrasound) should be planned in order to overcome this drawback and provide economically viable treatments. This way, an optimized sonoelectrochemical reactor design is mandatory. Specifically for the case of pulsed sonoelectrolysis, the absence of reactivation and/or cleaning of the electrode surface could allow future automation of the process for industrial applications for long periods of time.

Author details

Gustavo Stoppa Garbellini
São Paulo State University/Institute of Chemistry/Department of Analytical Chemistry, Brazil

Acknowledgement

The author thanks CNPq (Proc. 142930/2005-9) and Fapesp (Proc. 2009/08161-8) of Brazil, for the scholarships and financial support to this work. In particular, the author would like to thank Dr. Luis Alberto Avaca for teaching electrochemistry and the use of power ultrasound associated to electrochemical techniques.

7. References

Abdelsalam, M. E. & Birkin, P. R. (2002). A Study Investigating the Sonoelectrochemical Degradation of an Organic Compound Employing Fenton's Reagent. *Physical Chemistry Chemical Physics*, Vol. 4, No. 21, pp. 5340-5345, ISSN 1463-9076

Abramov, V. O.; Abramov, O. V.; Gekhman, A. E.; Kuznetsov, V. M. & Price, G. J. (2006). Ultrasonic Intensification of Ozone and Electrochemical Destruction of 1,3-Dinitrobenzene and 2,4-Dinitrotoluene. *Ultrasonics Sonochemistry*, Vol. 13, No. 4, pp. 303–307, ISSN 1350-4177

Adewuyi, Y. G. & Appaw, C. (2002). Sonochemical Oxidation of Carbon Disulfide in Aqueous Solutions: Reaction Kinetics and Pathways. *Industrial & Engineering Chemistry Research*, Vol. 41, No. 20, pp. 4957-4964, ISSN 0888-5885

Agullo, E.; Gonzalez-Garcia, J.; Exposito, E.; Montiel, V & Aldaz, A. (1999). Influence of an Ultrasonic Field on Lead Electrodeposition on Copper Using a Fluoroboric Bath. *New Journal of Chemistry*, Vol. 23, No. 1, pp. 95-101, 1144-0546.

Akkermans, R. P.; Ball, J. C.; Rebbitt, J. O.; Marken F. & Compton, R. G. (1998). Sono-Electroanalysis: Application to the Detection of Lead in Wine. *Electrochimica Acta*, Vol. 43, No. 23, pp. 3443-3449, ISSN 0013-4686

Alkire, R. C. & Perusich, S. (1983). The Effect of Focused Ultrasound on the Electrochemical Passivity of Iron in Sulfuric-Acid. *Corrosion Science*, Vol. 23, No. 10, pp. 1121-1132, ISSN 0010-938X

Appaw, C. & Adewuyi, Y. G. (2002). Destruction of Carbon Disulfide in Aqueous Solutions by Sonochemical Oxidation. *Journal of Hazardous Materials,* Vol. 90, No. 3, pp. 237-249, ISSN 304-3894

Banks, C. E. & Compton, R. G. (2003a). Voltammetric Exploration and Applications of Ultrasonic Cavitation. *ChemPhysChem,* Vol. 4, No. 2, pp. 169-178, ISSN 1439-4235

Banks, C. E. & Compton, R. G. (2003b). Ultrasonically Enhanced Voltammetric Analysis and Applications: an Overview. *Electroanalysis,* Vol. 15, No. 5-6, pp. 329-346, ISSN 1040-0397

Banks, C. E. & Compton, R. G. (2004). Ultrasound: Promoting Electroanalysis in Difficult Real World Media. *Analyst,* Vol. 129, No. 8, pp. 678-683, ISSN 0003-2654

Bejarano-Perez, N. J. & Suarez-Herrera, M. F. (2007). Sonophotocatalytic Degradation of Congo Red and Methyl Orange in the Presence of TiO_2 as a Catalyst. *Ultrasonics Sonochemistry,* Vol. 14, No. 5, pp. 589–595, ISSN 1350-4177.

Brillas, E.; Ignasi, S. & Oturan, M. A. (2009). Electro-Fenton Process and Related Electrochemical Technologies Based on Fenton's Reaction Chemistry. *Chemical Reviews,* Vol. 109, No. 12, pp. 6570-6631, ISSN 0009-2665

Bringas, E.; Saiz, J. & Ortiz, I. (2011). Kinetics of Ultrasound-Enhanced Electrochemical Oxidation of Diuron on Boron-Doped Diamond Electrodes. *Chemical Engineering Journal,* Vol. 172, No. 2-3, pp. 1016–1022, ISSN 1385-8947.

Chen, G. H. (2004). Electrochemical Technologies in Wastewater Treatment. *Separation and Purification Technology,* Vol. 38, No. 1, pp.11-41, ISSN 1383-5866

Compton, R. G.; Eklund, J. C.; Page, S. D.; Sanders, G. H. M. & Booth, J. (1994). Voltammetry in the Presence of Ultrasound - Sonovoltammetry and Surface Effects. *Journal of Physical Chemistry,* Vol. 98, No. 47, pp. 12410-12414, ISSN 0022-3654

Compton, R. G.; Eklund, J. C. & Page, S. D. (1995a). Sonovoltammetry - Heterogeneous Electron-Transfer Processes with Coupled Ultrasonically Induced Chemical-Reaction - the Sono-EC Reaction. *Journal of Physical Chemistry,* Vol. 99, No. 12, pp. 4211-4214, ISSN 0022-3654.

Compton, R. G.; Eklund, J. C.; Page, S. D. & Rebbitt, T. O. (1995b). Sonoelectrochemistry - the Oxidation of Bis(Cyclopentadienyl)Molybdenum Dichloride. *Journal of the Chemical Society-Dalton Transactions,* No. 3, p. 389-393, ISSN 1472-7773

Compton, R. G.; Eklund, J. C.; Page, S. D.; Mason, T. J. & Walton, D. J. (1996a). Voltammetry in the Presence of Ultrasound: Mass Transport Effects. *Journal of Applied Electrochemistry,* Vol. 26, No. 8, pp. 775-784, ISSN 0021-891X

Compton, R. G.; Eklund, J. C.; Marken, F. & Waller, D. N. (1996b). Electrode Processes at the Surfaces of Sonotrodes. *Electrochimica Acta,* Vol. 41, No. 2, pp. 315-320, ISSN 0013-4686

Compton, R. G.; Eklund, J. C. & Marken, F. (1997) Sonoelectrochemical Processes: A Review. *Electroanalysis,* Vol. 9, No. 7, pp. 509-522, ISSN 1040-0397

Compton, R. G.; Marken, F.; Goeting, C. H.; Mckeown, R. A. J.; Foord, J. S.; Scarsbrook, G.; Sussmann, R. S. & Whitehead, A. J. (1998). Sonoelectrochemical Production of Hydrogen Peroxide at Polished Boron-Doped Diamond Electrodes. *Chemical Communications,* Vol. 18, No. 18, pp. 1961-1962, ISSN 1359-7345

Cooper, E. L. & Coury Jr, L. A. (1998). Mass Transport in Sonovoltammetry with Evidence of Hydrodynamic Modulation from Ultrasound. *Journal of the Electrochemical Society,* Vol. 145, No. 6, pp. 1994-1999, ISSN 0013-4651

Correia, A. N.; Mascaro, L. H.; Machado, S. A. S.; Mazo, L. H.; Avaca, L. A. (1995) Ultramicroeletrodos. Parte 1: Revisão Teórica e Perspectivas. *Química Nova*, Vol. 18, No. 5, pp. 475-480, ISSN 0100-4042

Dai, Y.; Li, F.; Ge, F.; Zhu, F.; Wu, L. & Yang, X. (2006). Mechanism of the Enhanced Degradation of Pentachlorophenol by Ultrasound in the Presence of Elemental Iron. *Journal of Hazardous Materials*, Vol. 137, No. 3, pp. 1424-1429, ISSN 0304-3894

Davis, J. & Compton, R. G. (2000). Sonoelectrochemically Enhanced Nitrite Detection. *Analytica Chimica Acta*, Vol. 404, No. 2, pp. 241-247, ISSN 0003-2670

Eklund, J. C.; Marken F.; Waller, D. N. & Compton, R. G. (1996). Voltammetry in the Presence of Ultrasound: a Novel Sono-Electrode Geometry. *Electrochimica Acta*, Vol. 41, No. 9, pp. 1541-1547, ISSN 0013-4686

Esclapez, M. D.; Saez, V.; Milán-Yáñez, D.; Tudela, I.; Louisnard, O. & González-García, J. (2010). Sonoelectrochemical Treatment of Water Polluted with Trichloroacetic Acid: From Sonovoltammetry to Pre-Pilot Plant Scale. *Ultrasonics Sonochemistry*, Vol. 17, No. 6, pp. 1010–1020, ISSN 1350-4177

Flint, E. B. & Suslick, K. S. (1991). The Temperature of Cavitation. *Science*, Vol. 253, No. 5026, pp. 1397-1399, ISSN 0036-8075

Foord, J. S.; Holt, K. B.; Compton, R. G.; Marken, F. & Kim, D. H. (2001). Mechanistic Aspects of the Sonoelectrochemical Degradation of the Reactive Dye Procion Blue at Boron-Doped Diamond Electrodes. *Diamond and Related Materials*, Vol. 10, No. 3-7, pp. 662-666, ISSN 0925-9635

Garbellini, G. S.; Salazar-Banda G. R. & Avaca L. A. (2007). Sonovoltammetric Determination of 4-nitrophenol on Diamond Electrodes. *Journal of the Brazilian Chemical Society*, Vol. 18, No, 6, pp. 1095-1099, ISSN 0103-5053

Garbellini, G. S.; Salazar-Banda G. R. & Avaca L. A. (2009). Sonovoltammetric Determination of Toxic Compounds in Vegetables and Fruits Using Diamond Electrodes. *Food Chemistry*, Vol. 116, No. 4, pp. 1029-1035, ISSN 0308-8146

Garbellini, G. S.; Salazar-Banda, G. R. & Avaca, L. A. (2010). Effects of Ultrasound on the Degradation of Pentachlorophenol by Boron-Doped Diamond Electrodes. *Portugaliae Electrochimica Acta*, Vol. 28, No. 6, pp. 405-415, ISSN 1647-1571

Ghernaout, D.; Naceur, M. W. & Aouabed, A. (2011). On the Dependence of Chlorine By-Products Generated Species Formation of the Electrode Material and Applied Charge During Electrochemical Water Treatment. *Desalination*, Vol. 270, No. 1-3, PP. 9-22. ISSN 0011-9164

Goeting, C. H.; Foord, J. S.; Marken, F. & Compton, R. G. (1999). Sonoelectrochemistry at Tungsten-Supported Boron-Doped CVD Diamond Electrodes. *Diamond and Related Materials*, Vol. 8, No. 2-5, pp. 824-829, ISSN 0925-9635

Hardcastle, J. L. & Compton, R. G. (2001). Sonoelectroanalytical Determination of Heavy Metals in Fish Gill Mucous. *Electroanalysis*, Vol. 13, No, 2, pp. 89-93, ISSN 1040-0397

Hart, E. J. & Henglein, A. (1985). Free-Radical and Free Atom Reactions in the Sonolysis of Aqueous Iodide and Formate Solutions. *Journal of Physical Chemistry*, Vol. 89, No. 20, pp. 4342-4347, ISSN 0022-3654

Henglein, A. (1987). Sonochemistry - Historical Developments and Modern Aspects. *Ultrasonics*, Vol. 25, No. 1, pp. 6-16, ISSN 0041-624X

Holt, K. B.; Del Campo, J.; Foord, J. S.; Compton, R. G. & Marken, F. (2001). Sonoelectrochemistry at Platinum and Boron-Doped Diamond Electrodes: Achieving 'Fast Mass Transport' for 'Slow Diffusers'. *Journal of Electroanalytical Chemistry*, Vol. 513, No. 2, p. 94-99, ISSN 0022-0728

Holt, K. B.; Forryan, C.; Compton, R. G.; Foord, J. S. & Marken, F. (2003). Anodic Activity of Boron-Doped Diamond Electrodes in Bleaching Processes: Effects of Ultrasound and Surface States. *New Journal of Chemistry*, Vol. 27, No. 4, pp. 698-703, ISSN 1144-0546

Hua, I. & Hoffmann, M. R. (1997). Optimization of Ultrasonic Irradiation as an Advanced Oxidation Technology. *Environmental Science & Technology*, Vol. 31, No. 8, pp. 2237-2243, ISSN 0013-936X *in Chemistry and Processing*, ISBN 9783527600540, Weinheim, Germany

Kobayashi, K.; Chiba, A. & Minami, N. (2000). Effects of Ultrasound on Both Electrolytic and Electroless Nickel Depositions. *Ultrasonics*, Vol. 38, No. 1-8, pp. 676-681, ISSN 0041-624X

Kruusma, J.; Nei, L.; Hardcastle, J. L.; Compton, R. G.; Lust, E. & Keis, H. (2004a). Sonoelectroanalysis: Anodic Stripping Voltammetric Determination of Cadmium in Whole Human Blood. *Electroanalysis*, Vol. 16, No. 5, pp. 399-403, ISSN 1040-0397

Kruusma, J.; Tomcik, P.; Banks, C. E. & Compton, R. G. (2004b). Sonoelectroanalysis in Acoustically Emulsified Media: Zinc and Cadmium. *Electroanalysis*, Vol. 16, No. 10, pp. 852-859, ISSN 1040-0397

Lee, C. W.; Compton, R. G.; Eklund, J. C. & Waller, D. N. (1995). Mercury-Electroplated Platinum-Electrodes and Microelectrodes for Sonoelectrochemistry. *Ultrasonics Sonochemistry*, Vol. 2, No. 1, pp. 59-62, ISSN 1350-4177

Lima Leite, R. H.; Cognet, P.; Wilhelm, A. M. & Delmas, H. (2002). Anodic Oxidation of 2,4-Dihydroxybenzoic Acid for Wastewater Treatment: Study of Ultrasound Activation. *Chemical Engineering Science*, Vol. 57, No. 5, pp. 767-778, ISSN 0009-2509

Lorimer, J. P.; Pollet, B.; Phull, S. S.; Mason, T. J.; Walton, D. J. & Geissler, U. (1996). The effect of ultrasonic frequency and intensity upon limiting currents at rotating disc and stationary electrodes. *Electrochimica Acta*, Vol. 41, No. 17, pp. 2737-2741, ISSN 0013-4686

Lorimer, J. P.; Mason, T. J.; Plattes, M. & Phull, S. S. (2000). Dye Effluent Decolourisation Using Ultrasonically Assisted Electro-Oxidation. *Ultrasonics Sonochemistry*, Vol. 7, No. 4, pp. 237–242, ISSN 1350-4177

Lu, Y. & Weavers, L. K. (2002). Sonochemical Desorption and Destruction of 4-Chlorobiphenyl from Synthetic Sediments. *Environmental Science & Technology*, Vol. 36, No. 2, pp. 232-237, ISSN 0013-936X

Macounova, K.; Klima, J.; Bernard, C. & Degrand, C. (1998). Ultrasound-Assisted Anodic Oxidation of Diuron. *Journal of Electroanalytical Chemistry*, Vol. 457, No. 1-2, pp. 141-147, ISSN 0022-0728

Makino, K.; Mossoba, M. M. & Riesz, P. (1983). Chemical Effects of Ultrasound on Aqueous-Solutions - Formation of Hydroxyl Radicals and Hydrogen-Atoms. *Journal of Physical Chemistry*, Vol. 87, No. 8, pp. 1369-1377, ISSN 0022-3654

Marken, F. & Compton, R. G. (1996). Electrochemistry in the Presence of Ultrasound: the Need for Bipotentiostatic Control in Sonovoltammetric Experiments. *Ultrasonics Sonochemistry*, Vol. 3, No. 2, S131-S134, ISSN 1350-4177

Marken, F.; Kumbhat, S.; Sanders, G. H. W. & Compton, R. G. (1996). Voltammetry in the Presence of Ultrasound: Surface and Solution Processes in the Sonovoltammetric Reduction of Nitrobenzene at Glassy Carbon and Gold Electrodes. *Journal of the Electroanalytical Chemistry*, Vol. 414, No. 2, pp. 95-105, ISSN 0022-0728

Martines, M. A. U.; Davolos, M. R. & Jafellici-Junior, M. (2000). O efeito do ultra-som em reações químicas. *Química Nova*, Vol. 23, No. 2, pp. 251-256, ISSN 0100-4042

Martinez, S. S. & Uribe, E. V. (2012). Enhanced Sonochemical Degradation of Azure B Dye by the Electrofenton Process. *Ultrasonics Sonochemistry*, Vol. 19, No. 1, pp. 174–178, ISSN 1350-4177

Martínez-Huitle, C. A. & Ferro, S. (2006). Electrochemical Oxidation of Organic Pollutants for the Wastewater Treatment: Direct and Indirect Processes. *Chemical Society Reviews*, Vol. 35, No. 12, pp. 1324-1340, ISSN 0306-0012

Mason, T. J. (1990). *Sonochemistry: the Uses of Ultrasound in Chemistry*. The Royal Society of Chemistry, ISBN 0-85186-293-4, Cambridge, United Kingdom.

Mason, T. J.; Lorimer, J. P. & Bates, D. M. (1992). Quantifying Sonochemistry - Casting Some Light on a Black Art. *Ultrasonics*, Vol. 30, No. 1, pp. 40-42, ISSN 0041-624X

Mason, T. J. & Lorimer, J. P. (2002). *Applied Sonochemistry: The Uses of Power Ultrasound in Chemistry and Processing*, ISBN 9783527600540, Weinheim, Germany

Megharaj, M.; Ramakrishnan, B.; Ramakrishnan, B.; Venkateswarlu, K.; Sethunathan, N.; Naidu, R. (2011). Bioremediation Approaches for Organic Pollutants: A Critical Perspective. *Environmental International*, Vol. 37, No. 8, pp. 11-41, ISSN 0160-4120

Oliveira R. T. S.; Garbellini, G. S.; Salazar-Banda, Giancarlo R. & Avaca, L. A. (2007). The Use of Ultrasound for the Analytical Determination of Nitrite on Diamond Electrodes by Square Wave Voltammetry. *Analytical Letters*, Vol. 40, No. 14, pp. 2673-2682, ISSN 0003-2719

Oturan, M. A.; Sires, I.; Oturan, N.; Pérocheau, S.; Laborde, J. & Trévin, S. (2008). Sonoelectro-Fenton Process: A Novel Hybrid Technique for the Destruction of Organic Pollutants in Water. *Journal of Electroanalytical Chemistry*, Vol. 624, No. 1-2, pp. 329–332, ISSN 1572-6657

Panizza, M. & Cerisola, G. (2009). Direct and Mediated Anodic Oxidation of Organic Pollutants. *Chemical* Reviews, Vol. 109, No. 12, pp. 6541-6569, ISSN 0009-2665

Saez, V.; Esclapez, M. D.; Tudela, I.; Bonete, P.; Louisnard, O. & González-García, J. (2010). 20 KHz Sonoelectrochemical Degradation of Perchloroethylene in Sodium Sulfate Aqueous Media: Influence of the Operational Variables in Batch Mode. *Journal of Hazardous Materials*, Vol. 183, No. 1-3, pp. 648–654, ISSN 0304-3894

Saterlay, A. J.; Agra-Gutierrez, C.; Taylor, M. P.; Marken, F. & Compton, R. G. (1999). Sono-Cathodic Stripping Voltammetry of Lead at a Polished Boron-Doped Diamond Electrode: Application to the Determination of Lead in River Sediment. *Electroanalysis*, Vol. 11, No. 15, pp. 1083-1088, ISSN 1040-0397

Saterlay, A. J.; Foord, J. S. & Compton, R. G. (2001). An Ultrasonically Facilitated Boron-Doped Diamond Voltammetric Sensor for Analysis of the Priority Pollutant 4-Chlorophenol. *Electroanalysis*, Vol. 13, No. 13, pp. 1065-1070, ISSN 1040-0397

Siddique, M.; Farooq, R.; Khan, Z. M.; Khan, Z. & Shaukat, S. F. (2011). Enhanced Decomposition of Reactive Blue 19 Dye in Ultrasound Assisted Electrochemical Reactor. *Ultrasonics Sonochemistry*, Vol. 18, No. 1, pp. 190–196, ISSN 1350-4177

Villagrán, C.; Banks, C. E.; Pitner, W. R.; Hardacre, C. & Compton, R. G. (2005). Electroreduction of N-Methylphthalimide in Room Temperature Ionic Liquids Under Insonated and Silent Conditions. *Ultrasonics Sonochemistry*, Vol. 12, No. 6, pp. 423-428, ISSN 1350-4177

Walton, D. J.; Phull, S. S.; Chyla, A.; Lorimer, J. P.; Burke, L. D.; Murphy, M.; Compton R. G.; Eklund, J. C. & Page, S. D. (1995). Sonovoltammetry at Platinum Electrodes: Surface Phenomena and Mass Transport Processes. *Journal of Applied Electrochemistry*, Vol. 25, No. 12, pp. 1083-1090, ISSN 0021-891X

Walton, D. J.; Phull, S. S.; Geissler, U.; Chyla, A.; Durham, A.; Ryley, S.; Mason, T. J. & Lorimer, J. P. (2000). Sonoelectrochemistry - Cyclohexanoate Electrooxidation at 38 KHz and 850 KHz Insonation Frequencies Compared. *Electrochemistry Communications*, Vol. 2, No. 6, pp. 431-435, ISSN 1388-2481

West, C. E. Harcastle, J. L. & Compton, R. G. (2002). Sonoelectroanalytical Determination of Lead in Saliva. *Electroanalysis*, Vol. 14, No. 21, pp. 1470-1478, ISSN 1040-0397

Yasman, Y.; Bulatov, V.; Gridin, V. V.; Agur, S.; Galil, N.; Armon, R. & Schechter, I. (2004). A New Sono-Electrochemical Method for Enhanced Detoxification of Hydrophilic Chloroorganic Pollutants in Water. *Ultrasonics Sonochemistry*, Vol. 11, No. 6, pp. 365–372, ISSN 1350-4177

Yildiz, G.; Catalgil-Giz, H. & Giz, A. (2002). Effect of Ultrasound on Electrochemically Initiated Acrylamide Polymerization. *Journal of Applied Polymer Science*, Vol. 84, No. 1, pp. 83-89, ISSN 0021-8995

Zhang, H. & Coury Jr, L. A. (1993). Effects of High-Intensity Ultrasound on Glassy-Carbon Electrodes. *Analytical Chemistry*, Vol. 65, No. 11, pp. 1552-1558, ISSN 0003-2700

Zhang, Z.; Yuan, Y.; Liang, L.; Fang, Y.; Cheng, Y.; Ding, H.; Shi, G. & Jin, L. (2008). Sonophotoelectrocatalytic Degradation of Azo Dye on TiO_2 Nanotube Electrode. *Ultrasonics Sonochemistry*, Vol. 15, No. 4, pp. 370–375, ISSN 1350-4177.

Zhao, G. H.; Shen, S. H.; Li, M. F.; Wu, M. F.; Cao, T. C. & Li, D. M. (2008). The Mechanism and Kinetics of Ultrasound-Enhanced Electrochemical Oxidation of Phenol on Boron-Doped Diamond and Pt Electrodes. *Chemosphere*, Vol. 73, No. 9, pp. 1407-1413, ISSN 0045-6535

Zhao, G. H.; Gao, J. X.; Shen, S. H.; Liu, M. C.; Li, D. M.; Wu, M. F. & Lei, Y. Z. (2009). Ultrasound Enhanced Electrochemical Oxidation of Phenol and Phthalic Acid on Boron-Doped Diamond Electrode. *Journal of Hazardous Materials*, Vol. 172, No. 2-3, pp. 1076-1081, ISSN 0304-3894

Zhu, X.; Ni, J.; Li, H.; Jiang, Y.; Xing, X. & Borthwick, A. G. L. (2010). Effects of Ultrasound on Electrochemical Oxidation Mechanisms of p-Substituted Phenols at BDD and PbO_2 Anodes. *Electrochimica Acta*, Vol. 55, No. 20, pp. 5569–5575, ISSN 0013-4686

Chapter 11

Electrocoagulation for Treatment of Industrial Effluents and Hydrogen Production

Ehsan Ali and Zahira Yaakob

Additional information is available at the end of the chapter

http://dx.doi.org/10.5772/48633

1. Introduction

World has entered into a new era where sustainability is the main factor to encounter the challenges of depletion of our reserves and environmental upsets. Wastewater is not only one of the main causes of irreversible damages to the environmental balances but also contributing to the depletion of fresh water reserves at this planet, generating threats to the next generation. A lot of industrial processes are conducted at the expense of plenty of fresh water which is exhausted as a wastewater, and need to be treated properly to reduce or eradicate the pollutants and achieve the purity level for its reutilization in the industrial process to promote sustainability. A number of wastewater treatment methods are prevailing associated with subsequent advantages and disadvantages. Most commonly wastewater treatments involve biological treatment[1], chemical treatment [2] and Electrocoagulation [3]. Biological and chemical treatments of wastewater are usually associated with the production of green house gases and activated sludge along with some other limitations regarding required area and removal of residual chemicals respectively. On the other hand, Electrocoagulation is an extremely effective wastewater treatment system, removing pollutants and producing hydrogen gas simultaneously as revenue to compensate the operational cost[3]. Electrocoagulation has been documented positively to treat the wastewater from steam cleaners, pressure washers, textile manufacturing, metal platers, meat and poultry processors, commercial laundry, mining operations, municipal sewage system plants and palm oil industrial effluents.

Around the world, 45 million metric tons of palm oil has been produced in 2009 [5]. Approximately 0.65 tons of raw palm oil mill effluent (POME) is produced for every ton of processed fresh fruit bunches (FFB). A large quantity of water is necessary to process the palm fruit for oil production [6]. Furthermore, POME contributes 83% of the industrial organic pollution load in Malaysia (Vigneswaran *et al*, 1999).The POME is rich in organic carbon with a chemical oxygen demand (COD) higher than 40 g/L and nitrogen content around 0.2 to 0.5 g/L as ammonia nitrogen and total nitrogen. POME can also be described as a colloidal suspension of 95–96% water, 0.6–0.7% oil and 4–5% total solids including 2–4% suspended solids [7]. Conventionally, POME is usually treated with open lagoon technology by subjecting it to anaerobic treatment in open pond system to reduce the COD & BOD, this pretreatment method is associated with the risks of production of green house gases i.e. methane as a pollutant to the environment [8]. Usually the existing conventional methods for the pretreatment of POME are expensive or taking long retention time and require a vast pond area.

Parameter	Concentration (mg/L)	Element	Concentration (mg/L)
Oil and grease	4000–6000	Potassium	2,270
Biochemical oxygen demand	25,000	Magnesium	615
Chemical oxygen demand	50,000	Calcium	439
Total solid	40,500	Phosphorus	180
Suspended solids	18,000	Iron	46.5
Total volatile solids	34,000	Boron	7.6
Total Nitrogen	750	Zinc	2.3
Ammonicals nitrogen	35	Manganese	2.0
		Copper	0.89

Table 1. Characteristics of palm oil mill effluent [4]

This chapter emphasizes on the use of Electrocoagulation technique as a tool to promote the trends of sustainability in the existing industrialized world. Electrocoagulation technology was used successfully to pre-treat the Palm oil mill effluent (POME) as an electrolyte for the removal of polluting factors as a result of coagulation and precipitation of suspended solids followed by sedimentation under gravity. Aluminium and iron electrodes were used as sacrificing anodes to be used up in electrolytic oxidation for the production of $Al(OH)_3XH_2O$ and $Fe(OH)_3XH_2O$ respectively in different batch experiments. This study was also partially focused to compare the effectiveness of Aluminium (Al) and Iron (Fe) as electrodes to reduce the polluting nature of Palm Oil Mill Effluent (POME) and simultaneous hydrogen production during Electrocoagulation (EC). The metal (anode) based coagulants were found enough efficient to reduce the chemical oxygen demand (COD) and turbidity of POME. The remarkable pollutants removal was also associated with the hydrogen production as revenue to contribute the operational cost of wastewater treatment. Hydrogen production was also found helpful to remove the lighter suspended solids towards surface. The electrical inputs and findings were subjected to determine the Energy Efficiencies of POME treatment in comparison with water to highlight the associated advantages with EC of POME. This chapter is encompassing a detailed study of the related topics in general linked with experimental findings. Experimental findings have also been discussed in depth with reference to the published articles by other researchers. Concepts and mechanisms of coagulation and Electrocoagulation have been elaborated covering the maximum applications and gains in the industrial sector in context with the literature. Chemical composition of the wastewater and associated risks to the environment and health has been included for the better understanding of the readers. A précised approach was used to make the methodology reproducible and effective by supporting it with diagrammatic representation of the experimental set up. Process description is made conceivable and discussed in context to the general information in the literature. A separate discussion is made to understand the advantageous hydrogen production in addition to the removal of contaminants from the wastewater. Mathematical derivations and graphic representations are frequently used to represent the Energy Efficiency of the Electrocoagulation of the wastewater in comparison with the tap water at different pH. This chapter is presenting a real image of conceptual Electrocoagulation in the light of experimental verification in relation to previous studies.

Hydrogen is considered as an energy carrier like electricity and produces no green house gas or carbon dioxide when burnt in the presence of oxygen in related appliances including fuel cell or combustion engines. Hydrogen can be produced from different feedstock using a variety of techniques. Hydrogen is currently produced in large quantities from natural gas. Although, it is the cheapest way at present to produce hydrogen but the presence of carbon in methane is contributing to increase the global warming. A challenging problem in establishing H_2 as a source of energy for the future is to establish the procedures to produce hydrogen in abundance without creating any environmental threats. This chapter will emphasize on the treatment of wastewater and simultaneous hydrogen production using Electrocoagulation.

2. Technology description

2.1. Mechanism of coagulation and electrocoagulation

Industrial wastewater is in possession of impurities including colloidal particles and dissolved organic substances. The finely dispersed colloids or suspended solids are usually repelled by their outer layer of negative electrical charges and maintain the colloidal nature until treated by flocculants/coagulants for their removal. The process of flocculation and coagulation can be defined as ***"the ionic bridging between the finely divided particles to make flocs followed by their grouping into larger aggregates to be settled under gravity"***. The terms; flocculation and coagulation can separately be restricted to the preparation of flocs and grouping of flocs into aggregates respectively. The mechanism involved is the neutralization of the charges on the suspended solids or compression of the double layer of charges on the suspended solids. Overdose of coagulants may reverse the charge at the outer layer of the colloidal particles to re-stabilize them in a reverse mode. The wastewater treatment and down streaming of industrial fluids can be performed by using a number of flocculating/coagulating agents based on chemical salts and organic polymers.

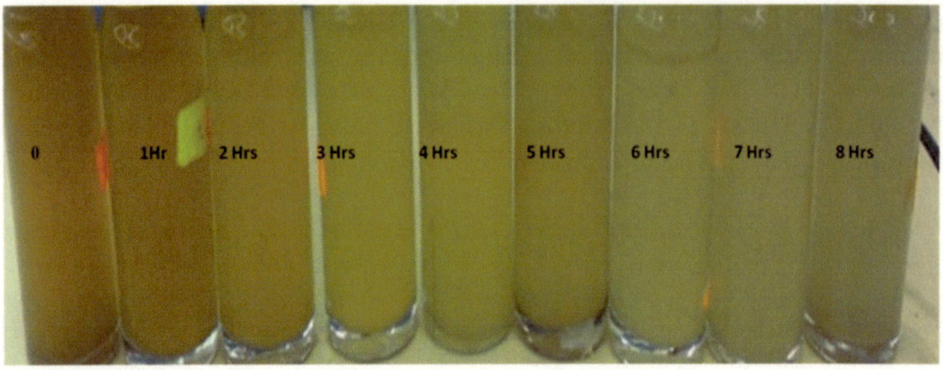

Figure 1. Gradual decrease in COD & turbidity during Electrocoagulation

A wide variety of chemicals and organic compounds have been recognized as efficient agents to remove the suspended solids from the wastewater. Wastewater is a very general term and can be designated to any water after being utilized by the human activities. A range of industrial processes are involved to exhaust a variety of effluents with different nature of pollutants. The treatment by the chemicals as well as organic molecules depends on the nature of pollutants and pH conditions. Because of the different nature of pollutants, no specific strategy can be recognized as versatile treatment to all types of wastewaters. Organic polymers are considerably preferred as coagulating/flocculating agents because of their biodegradable nature as compared to the chemicals causing to produce activated sludge. Coagulation is in routine practice for the treatment of drinking water[9], wastewater and industrial effluents [10].

Treatment of water using electricity was first proposed in UK in 1889 [11]. The application of electrolysis in mineral beneficiation was patented by Elmore in 1904 [11]. Electrocoagulation, precipitation of ions (heavy metals) and colloids (organic and inorganic) using electricity has been known as an ideal technology to upgrade water quality for a long time and successfully applied to a wide range of pollutants in even wider range of reactor designs [12-14]. Electrocoagulation is the technique to create conglomerates of the suspended, dissolved or emulsified particles in aqueous medium using electrical current causing production of metal ions at the expense of sacrificing electrodes and hydroxyl ions as a result of water splitting. Metal hydroxides are produced as a result of EC and acts as coagulant/flocculant for the suspended solids to convert them into flocs of enough density to be sediment under gravity. The electrical current provides the electromotive force to drive the chemical reactions to produce metal hydroxides.

Following reactions are carried out at different electrodes:

Anode:

$$Al - 3e \rightarrow Al^{3+}$$

Alkaline condition:

$$Al^{3+} + 3OH^{-} \rightarrow Al(OH)_3$$

Acidic condition:

$$Al^{3+} + 3H_2O \rightarrow Al(OH)_3 + 3H^+$$

$$2H_2O - 4e \rightarrow O_2 + 4H^+$$

Cathode:

$$2H_2O + 2e \rightarrow H_2 + 2OH^-$$

Dissociation of water by EC generate hydroxide ions which are known as one of the most reactive aqueous radical specie and this radical has the ability to oxidize organic compounds because of its high affinity value of 136 kcal [15]. The generated hydroxides or polyhydroxides have strong attractions towards dispersed particles as well as counter ions to cause coagulation. The gases evolved at the electrodes are also helpful to remove the suspended solids in upward direction [16].

A number of electrochemical reactions are involved within the electrocoagulation reactor. Reduction of metal anodes is responsible to produce hydroxide complexes causing flocculation of suspended solids into stable agglomerates. Production of oxygen and hydrogen as a result of electrolytic dissociation of water molecules cause emulsified oil droplets to be freed from water molecules making a separate layer on the surface. The same mechanism is involved in case of dyes, inks and other type of emulsions. In the presence of chlorine, metal ions can make chlorides which are also helpful in flocculation/coagulation of the wastewater. The production of oxygen in the electrocoagulation chamber can oxidize or bleach the chemicals like dyes.

System components and functions

An Electrocoagulation reactor consists of anode and cathode like a battery cell, metal plates of specific dimensions are used as electrodes and supplied with adequate direct current using power supply. The metal plates known as sacrificial electrodes are usually connected in parallel connection with a specified inter electrode distance (1.5-3.5cm) and supplied electric current is distributed on all the electrodes depending on the resistance of the individual electrodes. Distance of the electrodes has a direct relationship with the consumption of electricity. An electrode is an electrical conductor used to make contact with a nonmetallic part of a circuit. In case of EC, electrodes are known to be sacrificed for the release of metal ions at the anode, and cathode is responsible to produce hydroxyl ions. Metallic electrodes sacrificed to produce ions in the water which ultimately neutralized the charges of suspended particles leading to coagulation. The released ions remove suspended solids by precipitation or flotation. Water molecules are usually in bonding with colloidal particles, oils, or other contaminants in the wastewater leading to stable suspension, EC caused ionization of the water molecules adhering the contaminants to convert them into insoluble moieties to be sediment under gravity or float depending on density.

Experimental data to be presented in this chapter was generated by using the reactor with essential components as below:

Electrocoagulation cell was operated using rectifier, power supply with working range of electric current and voltage 0-60 amp and 0-15 volts respectively, ampere meter with digital working range 0-20 ampere and voltmeter with digital working range of 0-300 volt DC. Electrocoagulation was performed at different voltage (2, 3 and 4 volts). A reactor containing volume 20 liters of POME or water was used to conduct EC experiments (Fig. 1). The twelve aluminum plates were connected to a low voltage power supply. Six alternate plates were connected to the positive pole and the other six were connected to negative pole of the battery, thus acting as anode and cathode respectively. The weight of the plates was measured before and after the electrocoagulation. Aluminium plates were cut from commercial grade sheet (95%-99%) of 3 mm thickness as anode and cathode. POME samples were collected from Sri Ulu Langat Palm Oil Mill with COD, turbidity and pH around 50,000 mg/L, 2800 NTU and 4 respectively. Water samples were collected from usual tap water in the laboratory, the pH of tap water was 6 to 8.5. The pH of the water was adjusted to pH 4 by using 1N HCl.

2.2. Applications for different wastes

Coagulation and precipitation of contaminants can be induced by electrocoagulation technology and addition of coagulation-inducing chemicals. As a result of EC, the liberated hydrogen also took part to remove the lighter suspended solids in upward direction. Electrocoagulation has been employed in treating wastewaters from textile, catering, petroleum, tar sand, and oil shale. It is also used to treat the carpet wastewater, municipal sewage, chemical fiber wastewater, oil–water emulsion, oily wastewater clay suspension, nitrite, and dye stuff from wastewater.

Treatment of wastewater by EC has been practiced from pulp and paper industries[17], vegetable oil industries[18], textile industries [19-20], mining and metal-processing industries[21-22]. In addition, EC has been applied to treat water containing food waste, oily wastes, waste dyes, domestic wastewater etc. Copper reduction, coagulation and separation were also found by a direct current electrolytic process followed by sedimentation of the flocs by using EC [11]. This chapter is encompassing the details of Electrocoagulation of industrial effluent for the pretreatment and hydrogen production as an advantage. It has been explained that how the hydrogen production from industrial effluent may contribute to the cost effectiveness of the treatment process by producing extra revenue.

2.3. Advantages of technology

Electrocoagulation requires simple equipment and small area as compared to the conventional pond system which causes increase in the green house gases. Electrocoagulation is an alternative wastewater treatment that dissolves metal anode using electricity and provide active cations required for coagulation without increasing the salinity of the water [23]. Electrocoagulation has the capability to remove a large number of pollutants under a variety of conditions ranging from: suspended solids, heavy metals, petroleum products, colour from dye-containing solution, aquatic humus and defluoridation of water [23]. Electrocoagulation is usually recognized by ease of operation and reduced production of sludge [24]. Aluminium and iron are suitable electrode materials for the treatment using electrocoagulation [25]. The removal efficiency of electrocoagulation using Aluminium electrodes was higher than that of using Iron electrodes [26]. Electrocoagulation process consists of two stages: (i) electro generation of the metal cations and their physical action on the pollutant, (ii) formation of the flocs, flocculation and settling upon addition of flocculating agents and under low stirring [27].

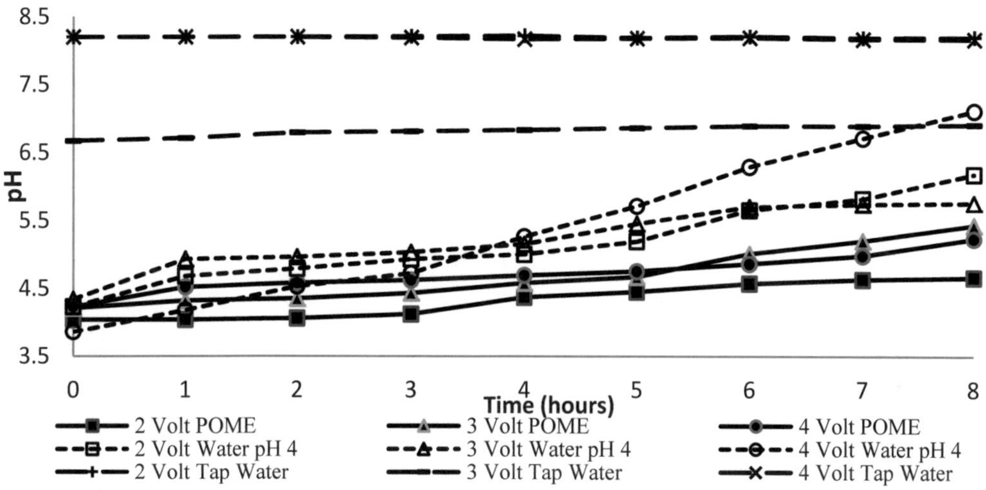

Figure 2. Electrocoagulation of Palm Oil Mill effluent as Wastewater Treatment and Hydrogen Production using Electrode Aluminium

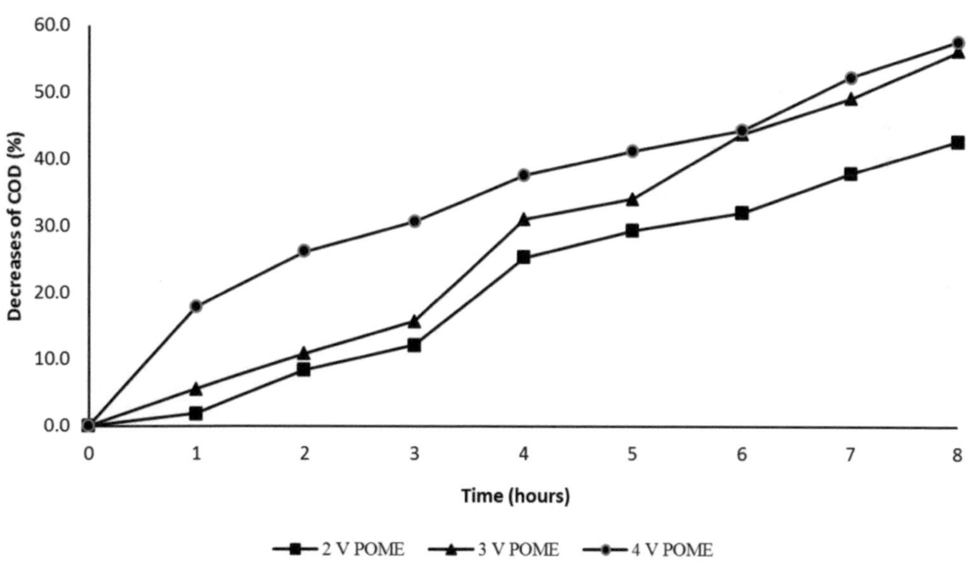

Figure 3. Electrocoagulation of Palm Oil Mill effluent as Wastewater Treatment and Hydrogen Production using Electrode Aluminium

3. Experimental procedures

Palm oil mill effluent (POME) was used as an electrolyte without any additive or pretreatment to perform electrocoagulation (EC) using electricity (Direct current) ranging from 2-4 volts in the presence of aluminum electrodes in a reactor volume of 20 liters. Investigations were made on the removal of pollutant like chemical oxygen demand (COD) and turbidity as a result of electrocoagulation of palm oil mill effluent (POME), and production of hydrogen gas as an advantageous step to meet the energy challenges. The results show that EC was responsible to reduce the COD and turbidity of POME 57% and 62% respectively in addition to the 42% hydrogen production during electrocoagulation. Hydrogen production was also helpful to remove the lighter solids towards surface. The anode reaction was responsible to produce $Al(OH)_3XH_2O$ at aluminium electrode (anode) which is a very reactive agent for flocculation/coagulation of suspended solids. The production of hydrogen gas from POME during electrocoagulation was also compared with hydrogen gas production from tap water at pH 4 and tap water without pH adjustment under the same conditions to highlight the advantageous aspects hydrogen production and wastewater treatment simultaneously. The main advantage of this study was to produce hydrogen gas while treating POME with EC to reduce COD and turbidity effectively. A number of experiments were designed and findings are discussed in different sections.

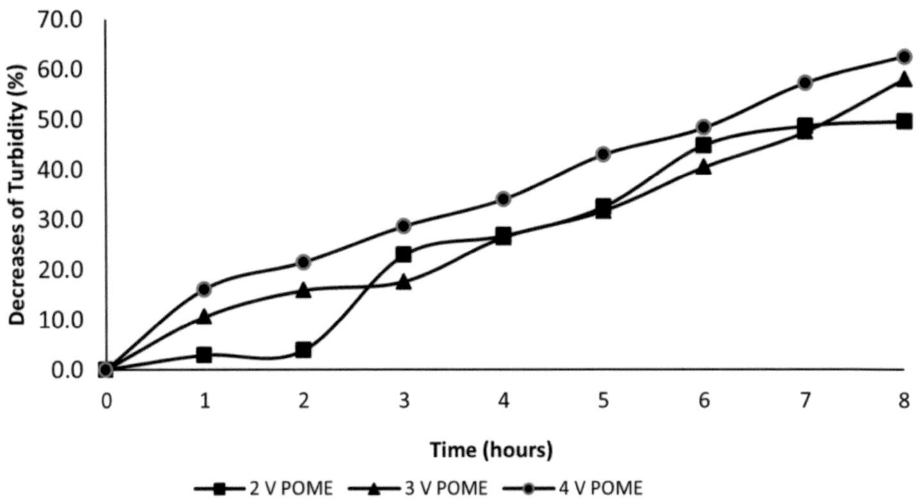

Figure 4. Electrocoagulation of Palm Oil Mill effluent as Wastewater Treatment and Hydrogen Production using Electrode Aluminium

3.1. Methodology

Materials and equipments

Electrocoagulation cell was operated using rectifier, power supply with working range of electric current and voltage 0-60 amp and 0-15 volts respectively, ampere meter with digital working range 0-20 ampere and voltmeter with digital working range of 0-300 volt DC. Electrocoagulation was performed at different voltage (2, 3 and 4 volts). A reactor containing volume 20 liters of POME or water was used to conduct EC experiments (Fig. 1). The twelve aluminum plates were connected to a low voltage power supply. Six alternate plates were connected to the positive pole and the other six were connected to negative pole of the battery, thus acting as anode and cathode respectively. The weight of the plates was measured before and after the electrocoagulation. Aluminium plates were cut from commercial grade sheet (95%-99%) of 3 mm thickness as anode and cathode. POME samples were collected from Sri Ulu Langat Palm Oil Mill with COD, turbidity and pH around 50,000 mg/L, 2800 NTU and 4 respectively. Water samples were collected from usual tap water in the laboratory, the pH of tap water was 6 to 8.5. The pH of the water was adjusted to pH 4 by using 1N HCl.

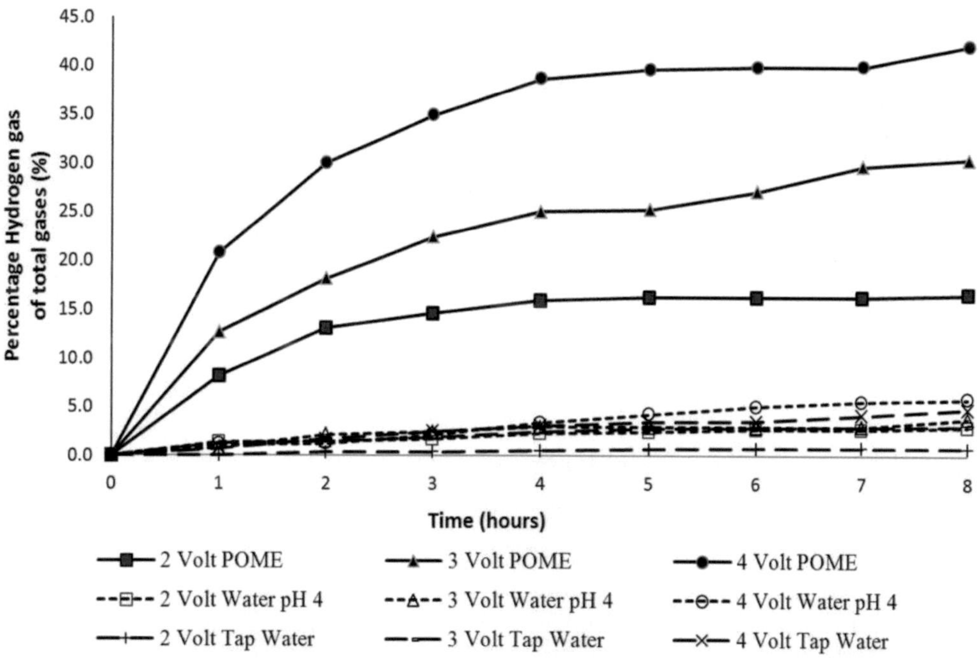

Figure 5. Electrocoagulation of Palm Oil Mill effluent as Wastewater Treatment and Hydrogen Production using Electrode Aluminium

3.2. Removal of pollutants from industrial effluent

Cell operation

A comparative study was conducted by using the POME, tap water at pH 4 (pH was adjusted with acid) and tap water without pH adjustment as electrolyte during different run, each electrolyte was analyzed for pH, COD and turbidity before and after the run. The pH of tap water was found in the range of pH 6.5-8.5 during different runs depending on the source of supply. The experiments were conducted in batch system. Electrodes were put in the reactor as multiple channels; with inter-electrode distance of 3 cm [28]. To perform EC, direct current (DC) was used throughout the experiment which was being converted from alternating current by using power supply and rectifier. POME, tap water at pH 4 and tap water (without pH adjustment) samples were analyzed before and after electrolysis for pH, COD and turbidity using standard techniques and Equipments. Hydrogen concentration was also analyzed using gas chromatography. The electrode surface was mechanically rubbed with 400 grade abrasive paper to remove the rusting or deposits before each run. The experiments were carried out at different voltage values: 2, 3 and 4 volt, and the current were measured during each run. Standard deviations were calculated and plotted to facilitate the reproducibility of the data regarding measurements.

3.3. Hydrogen production

Hydrogen production

A closed container was used to conduct the electrocoagulation; the container was connected to the peristaltic pump to collect the total gas (Fig. 1). The gas was collected at the rate of 900 ml/minute at room temperature in the gas bags equipped with one way valves. The composition of the total gas was analyzed using gas chromatography (SRI 8610C, USA), equipped with a helium ionization detector (15 m length). The temperatures of the oven, injector and detector were 50, 100 and 200 °C respectively.

Cumulative hydrogen gas (Fig. 9) was calculated using the following equation:

$$V_n = \left(QxX_{H_2}\right) + V_{n-1} \tag{1}$$

Where V_n is the volume of hydrogen gas at n hours; Q is the flow rate of total gas; X_{H2} is the concentration of hydrogen gas in total gas; V_{n-1} is the volume of hydrogen gas in total gas.

The electrical energy supplied to the system was calculated using the following equation

$$E_e = VIt \tag{2}$$

Where E_e is the electrical energy supplied by the DC power supply (J); V is the DC voltage applied; I is the current (A) and t (hour) is the duration of the DC voltage applied to the system.

The amount of produced hydrogen gas was calculated using the following equation:

$$PV_{H2} = \left(\frac{m}{M}\right) RT \tag{3}$$

Where P denotes pressure in atm; V_{H2} denotes volume of the cumulative hydrogen calculated from equation (1); m denotes the mass of the cumulative hydrogen (g); M is the molar mass of hydrogen (2 g /mol); R is the gas constant (0.082 L atm. mol^{-1} K^{-1}), T is denoting the room temperature (298 K).

The energy contents of the hydrogen gas were calculated using the equation

$$EH_2 = m\left(122\frac{kJ}{g}\right) \tag{4}$$

where m denotes the mass of the cumulative hydrogen produced within a specified time period.

Energy efficiency of the system was calculated by using the following equation

$$EEf = \frac{EH_2}{E_e} \tag{5}$$

Energy efficiencies were determined by using electrolytes like water at pH 4, tap water, and POME at the expense of electrical inputs of 2, 3 and 4 volts.

4. Results and discussion

Electrocoagulation of POME was performed to reduce its polluting nature as well as hydrogen gas production. It was observed that the POME before electrocoagulation process was brown in color and after electrocoagulation became whitish in color. A remarkable reduction in the turbidity of POME can be visualized after electrocoagulation (Fig.2).

Dynamic response of pH during electrocoagulation

Kılıc M.G. and C. Hosten (2010) has mentioned that the optimum effectiveness of EC can be achieved at pH 9. Chen and Hung (2007) have described pH as an important factor in EC and variation in pH is usually caused by the solubility of metal hydroxides. They further reported that the pH of the effluent after electrocoagulation would increase for acidic influent, however pH would decrease for alkaline effluent [12]. Hydroxides, which are produced as a result of dissociation of water are known as one of the most reactive aqueous radical specie and this radical has the ability to oxidize almost all of the organic compounds because of its high affinity value of 136 kcal [15]. Figure 3 has shown the dynamic response of pH of POME at pH 4, water at pH 4 and tap water (pH 6.6 to 8.2) using electricity at 2, 3 and 4 volt inputs. The pH of the POME under the influence of 2 volts was found near about constant, however a slight increase in pH was found using 3 and 4 volts of electricity with POME. Agustin et al (2008) has performed the EC of de-oiled POME in the presence of additional sodium chloride as electrolyte aid and reported the increase in pH value from 4.3 to 7.63. In our case, the study was performed by using raw POME as it was obtained from palm oil mill and no additional salts were added to enhance the conductivity. It was assumed that the formation of aluminium hydroxide at aluminium electrodes was leading to a simultaneous coagulation of the suspended solids followed by effective sedimentation under gravity. In case of EC of water (pH 4), the 2 volt input was able to increase the pH (4.23 to 6.18) as compared to the 3 volts input (4.34 to 5.76). However the use of 4 volts input was responsible to increase the pH value up to more than 6.5. Tap water at pH (6.6 to 8.2) was also investigated but there was no remarkable change in the pH of tap water after EC. EC of water at pH 4 and water at pH 6.5-8.5 was conducted to compare the efficiency of hydrogen production at different pH of water and ultimately to compare with POME at pH 4.0. It was observed that the EC of POME at pH 4.0 is presenting better results as compared to the water at different pH. Electrolysis/electrocoagulation is closely associated with the variation of the pH and its effects on the experimental solutions. Different aluminium species are formed at the aluminium cathode (electrode) by the combination of the electro-dissolved Al^{3+} ions with hydroxyl ions to affect the pH [29]. The influence of the pH while studying the EC has also been reported by some other researcher's e.g. Kobya et al (2003), has reported the pH increase from 3 to 11 while conducting the EC using aluminium electrode with textile wastewater. A like trend was also achieved by some other researchers [30]. Hence, it has been concluded that the effect of pH is an important parameter influencing the performance of the EC process.

Electrocoagulation for the removal of COD and turbidity from POME

Reduction in the chemical oxygen demand (COD) is a key factor in waste water treatment. EC was performed to investigate the effects of electrochemical treatment of POME. The electrical inputs of 2, 3 and 4 volts were used to proceed the EC of POME to remove the COD and turbidity as well as hydrogen production. As a result of electrocoagulation, a gradual reduction in the color intensity and turbidity of the POME can be visualized with respect to time (Fig. 3). EC is responsible for the electrolytic dissociation of water to produce reactive specie $(OH)^-$ which facilitate the process of flocculation/coagulation of the potential pollutants in the POME. The reactivity of the $(OH)^-$ ions and zeta potential has been described by Wang et al, Li, et al (2003). According to figure 4, a higher reduction in COD of POME was observed while proceeding electrocoagulation at 4 volts rather than at 2 and 3 volts. Electrocoagulation was efficiently responsible to decrease the COD to 57.66% at 4 volts in 8 hours, on the other hand COD reduction at 2 and 3 volts were 42.8% and 56.16% respectively under same conditions. The combination of the Al^{3+} ions and highly reactive specie $(OH)^-$ is effectively known as flocculating/coagulating agent to remove the suspended solids from the waste water [12]. However the total reduction of the COD and turbidity was also contributed by the upward flow of the hydrogen gas during electrocoagulation. Agustin et al (2008) have reported the removal of 30% COD as a result of EC of POME in six hours of operation time but our study has shown a greater reduction in the COD of POME as compared to their study [31].

In this study, neither any additive was used to enhance the electrolytic efficiency of the electrolyte nor was the POME subjected to the extraction of oil or pH adjustment. This study was designed to treat the POME at the industrial level as an effective and primary treatment without any extra treatment or addition of chemicals. Agustin et al (2008) have reported a 100% reduction in turbidity and only 30% reduction in COD after electrocoagulation of POME in the presence of sodium chloride, the high residual COD value in the transparent fluid might be attributed to the presence of some soluble salts due to sodium chloride. The efficiency of EC is also depending on the nature of effluent and processing time. Ugurlu et al (2008) has reported 75% removal of COD with paper mill effluent treatment but the initial COD of this effluent was 86 times lower than initial COD of POME [31]. O.T. Can has also reported 50% COD removal by conducting EC of textile wastewater with 10 minute operating time [32].

Turbidity and COD have straight relation as caused by the presence of suspended solids. Removal of the COD might automatically reduce the turbidity to the lower level accordingly. According to Fig. 5, the maximum removal of turbidity achieved at electrical inputs of 4 volt in 8 hours operating time was 62%. The operating time and value of electrical input have a direct influence on the removal of COD and turbidity. Agustin et al (2008) has reported a transparent fluid after six hours EC operating time but the experimental solution was still in possession of 70% residual COD. In our study removal of 57.66% of COD and 62% turbidity was not able to create the transparency; apparently a remarkable decrease in the color intensity was observed (Fig. 2). It was also observed during the experiments that long operating times, high voltage values, bubbling of hydrogen gas at cathode were supporting factors to remove the turbidity and COD from the electrolyte (POME), as it was previously reported by Kilic et al. [33]. Kilic and Hosten has also reported the removal of 90% turbidity while conducting EC of aqueous suspensions of kaolinite powders with concentration of 0.2 g/L using electrical input of 40 volts, 20 minute operating time and NaCl as additive [34].

Hydrogen production using electrocoagulation

Production of hydrogen by the electrolytic dissociation of water is a usual practice but the production cost is considerably high [35-36]. It was assumed that any advantage accompanying with the electrolysis of water to produce hydrogen may compensate the actual operational cost partially. This study was launched with a specific objective to produce hydrogen gas as well as the pretreatment of palm oil mill effluent simultaneously to maintain the cost effectiveness of the process. The EC was designed to make pretreatment of POME in a closed container specially equipped with a gas collection system Fig. 1. Tap water at pH 4, tap water at pH 6.5-8.5 and POME were used as an electrolyte to conduct the EC with electrical inputs of 2, 3 and 4 volts separately. The pH of the POME was not adjusted but the pH of tap water was adjusted nearer to POME (pH 4) to compare the hydrogen production under the same conditions. Tap water pH (6.5 -8.2) was found varied at different times but not subjected to any adjustment of pH (Fig.7). The tap water was used to compare the efficiency of hydrogen production from water at different pH and ultimately to compare this hydrogen production efficiency from POME while simultaneous removal of pollutants. The above mentioned EC experiments have generated the data which is clearly representing a difference in hydrogen production from water and POME. The gas was collected at the rate of 900 ml/minute at room temperature in the plastic gas bags. The rate of total gas production was 54 L/h. The overall hydrogen production from tap water at pH 4 and pH 6.5-8.5 was found below than 5% (v/v) of the total gas. In case of POME as an electrolyte, the maximum hydrogen production was estimated as 15% (v/v), 30% (v/v) and 42% (v/v) at different electrical inputs of 2, 3 and 4 volts respectively. Phalakornkule et al have reported 0.521 x 10^3 m^3 (6.252 x 10^6 liter/hour) hydrogen gas production using EC in five minutes from waste water containing dyes [37]. Take et al have investigated hydrogen production by using methanol-water solution as an electrolyte keeping cathode and anode separate from each other by a membrane and reported that the hydrogen in cathode exhaust gas was 95.5-97.2 mol% [38]. In our study, the maximum hydrogen gas produced was about

22.68 liters/hour and an efficient reduction of COD and turbidity of POME by as much as 57% and 62% was achieved respectively.

To determine energy efficiency

Energy efficiency (EEf) can be defined as the output obtained in the form of hydrogen gas on the expense of electrical energy provided to the reactor for a certain time period. The energy efficiency was calculated as described in material and equipments using equations (1-5). The EEf showed some variations while using water at pH 4, water at pH (6.5-8.5) and POME during electrolysis/electrocoagulation. The highest energy efficiencies were determined while using water at pH 4.0, water at pH (6.5-8.5) and POME with electrical inputs of 2 volts (Fig. 8). Energy efficiency for the treatment of POME as well as hydrogen production was also determined and plotted separately (Fig. 9). Although, energy efficiency was not so high while using POME as an electrolyte but can be further improved by standardizing the conditions regarding inter-electrode distance, nature of electrodes and proper dilution of POME. EEf was found increasing with the passage of time while using water at pH 4 and water at pH (6.5-8.5), however the EEf of POME was not subjected to any remarkable increment with respect to time (hours). Kargi et al (2001) have reported the hydrogen production by using electrolysis of anaerobic sludge with EEf of 74%, but they have used the serum bottles containing 1L sludge [39]. The low EEf values in our study might be due to the large volume of the electrolyte and can be improved further by standardizing the conditions regarding inter-electrode distance, nature of electrodes and proper dilution of POME.

Palm oil mill effluent can be treated by using environment friendly electrocoagulation, and hydrogen gas can be obtained as revenue to compensate the treatment cost of POME. EC of POME can be performed by using small area as compared to the conventional aerobic/anaerobic pond system. Hydrogen gas was also found helpful to remove the suspended solids towards surface. This study is presenting an approach towards environment friendly treatment of POME and hydrogen production as an alternative energy.

Author details

Ehsan Ali* and Zahira Yaakob
Department of Chemical and Process Engineering. Faculty of Engineering and Built Environment. University Kebangsaan Malaysia. Selangor Darul Ehsan Bangi 43600 Malaysia

5. References

Eddy, M.a., *Inc Wastewater Engineering, Treatment and Reuse.* 4th Edition, 2003. McGraw-Hill, New York, (2003) 545-644.

Norulaini, N.A.N., et al., *Chemical coagulation of settleable solid-free palm oil mill effluent (POME) for organic load reduction.* J. Ind. Technol., 2001. 10: p. 55-72.

* Corresponding Author

Nasution, M.A., et al., *Electrocoagulation of Palm Oil Mill Effluent as Wastewater Treatment and Hydrogen Production Using Electrode Aluminum.* ournal of Environmental Quality, 2011. J 2011 40: 4: 1332-1339 doi:10.2134/jeq2011.0002.

Sumathi, S., S.P. Chai, and A.R. Mohamed, *Utilization of oil palm as a source of renewable energy in Malaysia.* Renewable and Sustainable Energy Reviews, 2008. 12 p. 2404–2421.

Stichnothe, H. and F. Schuchardt, *Comparison of different treatment options for palm oil production waste on a life cycle basis.* Int J Life Cycle Assess, 2010.

Ahmad, A.L., S. Ismail, and S. Bhatia, *Water recycling from palm oil mill effluent (POME) using membrane technology.* Desalination 2003. 157 p. 87-95.

Ahmad, A.L., M.F. Chong, and S. Bhatia, *A comparative study on the membrane based palm oil mill effluent (POME) treatment plant.* Journal of Hazardous Materials 2009. 171: p. 166–174.

Wulfert, K., et al., *Treatment of POME in Anaerobic Fixed Bed Digesters*, in *International Oil Palm Conference*. 2002.

Zainal-Abideen, M., et al., *Optimizing the coagulation process in a drinking water treatment plant -- comparison between traditional and statistical experimental design jar tests.* Water Sci Technol, 2012. 65(3): p. 496-503.

Dovletoglou, O., C. Philippopoulos, and H. Grigoropoulou, *Coagulation for treatment of paint industry wastewater.* J Environ Sci Health A Tox Hazard Subst Environ Eng, 2002. 37(7): p. 1361-77.

Chen, G., *Electrochemical technologies in wastewater treatment.* Separation and Purification Technology, 2004. 38: p. 11-41.

Chen, G. and Y.-T. Hung, *Electrochemical Wastewater Treatment Processes*, in *Handbook of Environmental Engineering, Volume 5: Advanced Physicochemical Treatment Technologies*, L.K. Wang, Y.-T. Hung, and N.K. Shammas, Editors. 2007, The Humana Press Inc.,: Totowa, NJ. p. 57-105.

Kobya, M., O.T. Can, and M. Bayramoglu, *Treatment of textile wastewaters by electrocoagulation using iron and aluminum electrodes.* Journal of Hazardous Materials 2003. B100: p. 163–178.

Ayhan, I., Sengil, and M. ozacar, *Treatment of dairy wastewaters by electrocoagulation using mild steel electrodes.* Journal of Hazardous Materials, 2006. B137: p. 1197–1205.

Li, Y., et al., *Aniline degradation by electrocatalytic oxidation.* Chemosphere, 2003. 53: p. 1229–1234.

Liu, H., X. Zhao, and J. Qu, *Electrocoagulation in Water Treatment*, in *Electrochemistry for the Environment*, C. Comninellis and G. Chen, Editors. 2010, Springer Science+Business Media: New York. p. 245-262.

Vepsalainen, M., et al., *Precipitation of dissolved sulphide in pulp and paper mill wastewater by electrocoagulation.* Environ Technol, 2011. 32(11-12): p. 1393-400.

Nasution, M.A., et al., *Electrocoagulation of Palm Oil Mill Effluent as Wastewater Treatment and Hydrogen Production Using Electrode Aluminum.* J. Environ. Qual., 2011. 40(4): p. 1332-1339.

Kobya, M., M. Bayramoglu, and M. Eyvaz, *Techno-economical evaluation of electrocoagulation for the textile wastewater using different electrode connections.* J Hazard Mater, 2007. 148(1-2): p. 311-8.

Aouni, A., et al., *Treatment of textile wastewater by a hybrid electrocoagulation/nanofiltration process.* J Hazard Mater, 2009. 168(2-3): p. 868-74.

Kumarasinghe, D., L. Pettigrew, and L.D. Nghiem, *Removal of heavy metals from mining impacted water by an electrocoagulation-ultrafiltration hybrid process.* Desalination and Water Treatment, 2009. 11(1-3): p. 66-72.

Parga, J.R., et al., *Cyanide Detoxification of Mining Wastewaters with TiO(2) Nanoparticles and Its Recovery by Electrocoagulation.* Chemical Engineering & Technology, 2009. 32(12): p. 1901-1908.

Holt, P., G. Barton, and C. Mitchell, *Electrocoagulation as Wastewater Treatment*, in *The Third Annual Australian Environmental Engineering Research Event*. 1999: Castlemaine, Victoria

Emamjomeh, M.M. and M. Sivakumar, *Review of pollutants removed by electrocoagulation and electrocoagulation/flotation processes.* Journal of Environmental Management, 2009. 90(5): p. 1663-1679.

Zongo, I., et al., *Removal of hexavalent chromium from industrial wastewater by electrocoagulation: A comprehensive comparison of aluminium and iron electrodes.* Separation and Purification Technology, 2009. 66(Elsevier): p. 159–166.

Wang, C.-T., W.-L. Chou, and Y.-M. Kuo, *Removal of COD from laundry wastewater by electrocoagulation/electroflotation.* Journal of Hazardous Materials, 2009. 164(1): p. 81-86.

Khemis, M., et al., *Treatment of industrial liquidwastes by electrocoagulation: Experimental investigations and an overall interpretation model.* Chemical Engineering Science, 2006. 61: p. 3602-3609.

Cerqueira, A., C. Russo, and M.R.C. Marques, *Electroflocculation for textile wastewater treatment* Braz. J. Chem. Eng. , 2009. 26(4).

Pornsawad, A. and N. Pisutpaisal, *Treatment of palm oil mill effluent using electrocoagulation technique with aluminium plates as elctrodes*, in *35th Congress on Science and Technology of Thailand*. 2009: Chonburi, Thailand.

Othman, F., et al., *Enhancing Suspended Solids Removal From Waswater using Fe Electrodes.* Malaysian Journal of Civil Engineering, 2006. 18(2): p. 139-148.

Agustin, M.B., W.P. Sengpracha, and W. Phutdhawong, *Electrocoagulation of Palm Oil Mill Effluent.* International Journal of Environmental Research and Public Health, 2008. 5(3): p. 177-180.

Can, O.T., et al., *Treatment of the textile wastewater by combined electrocoagulation.* Chemosphere, 2006. 62: p. 181–187.

Kılıc, M.G., C. Hosten, and S. Demirci, *A parametric comparative study of electrocoagulation and coagulation using ultrafine quartz suspensions.* Journal of Hazardous Materials, 2009. 171: p. 247–252.

Kılıc, M.G. and C. Hosten, *A comparative study of electrocoagulation and coagulation of aqueous suspensions of kaolinite powders.* Journal of Hazardous Materials, 2010. 176: p. 735-740.

Barbir, F., *PEM electrolysis for production of hydrogen from renewable energy sources.* Solar Energy, 2005. 78: p. 661–669.

Stojic, D.L., et al., *Hydrogen generation from water electrolysis—possibilities of energy saving.* Journal of Power Sources, 2003. 118: p. 315–319.

Phalakornkule, C., P. Sukkasem, and C. Mutchimsattha, *Hydrogen recovery from the electrocoagulation treatment of dye-containing wastewater.* International Journal of Hydrogen Energy 2010. 35: p. 10394-10943.

Take, T., K. Tsurutani, and M. Umeda, *Hydrogen production by methanol–water solution electrolysis.* Journal of Power Sources, 2007. 164: p. 9–16.

Kargi, F., E.C. Catalkaya, and S. Uzuncar, *Hydrogen gas production from waste anaerobic sludge by electrohydrolysis: Effects of applied DC voltage.* international journal o f hydrogen energy 2011.36: p. 2049 -2056

Chapter 12

Electrolysis for Ozone Water Production

Fumio Okada and Kazunari Naya

Additional information is available at the end of the chapter

http://dx.doi.org/10.5772/51945

1. Introduction

OH radicals produced by ozone decomposition in water are the second strongest oxidizers after fluorine. Therefore, ozone water is used in a wide range of applications, such as the sterilization of medical instruments, the oxidation and detoxification of wastewater, the deodorization and decoloration of well water, the cleaning of electronic components, and food treatment [1, 2]. For example, the possible applications of ozone water to the control of microbiological safety and the preservation of food quality are well summarized in a review paper [3]. · OH radicals can also kill various bacteria, such as O157, MRSA, *Pseudomonas aeruginosa*, *Bacillus subtilis*, *Bacillus anthracis*, *Bacillus cereus* and *Legionella pneumophila* [4].

There are two major processes for ozone water production as shown in Fig. 1. One is a gas phase ozone production followed by mixing ozone with water [Fig. 1(a)]. In this process ozone gas is produced from oxygen gas using electric discharge, and then mixed with water using hollow fibers or mechanical mixers. Although this process can supply ozone water at a low electric cost, it requires a high-voltage power supply, an oxygen cylinder, and a mixing process. Therefore, the system is large and cumbersome in operation, and the facility cost is high. The other process is direct water electrolysis using a polymer electrolyte membrane (PEM) [Fig. 1(b)]. In this process, water is introduced to the anode side of the electrolysis cell, electrolytically decomposed, converted to ozone, and mixed with the ozone gas formed in the compact cell. Such apparatuses can enable ubiquitous and low-voltage operation as home electronics.

The ozone formation reactions and standard electrode potentials in the direct water electrolysis are as follows:

Anode:

$$2H_2O \rightarrow O_2 + 4H^+ + 4e^- \quad E^0 = 1.229\ V \tag{1}$$

$$3H_2O \rightarrow O_3 + 6H^+ + 6e^- \quad E^0 = 1.511\ V \tag{2}$$

$$O_2 + H_2O \rightarrow O_3 + 2H^+ + 2e^- \qquad E^0 = 2.075\ V \tag{3}$$

Cathode:

$$2H^+ + 2e^- \rightarrow H_2 \qquad E^0 = 0\ V. \tag{4}$$

These reactions are fundamental in electrochemistry and have been extensively studied. Ozone gas evolves by a six-electron reaction in Eq. (2) at a voltage higher than 1.511 V, accompanied by oxygen evolution. By increasing voltage to above 2.075 V, the oxidation of O_2 gas to form O_3 is also expected as shown in Eq. (3). Since O_2 evolution occurs at lower potential than O_3 evolution, the production rate and electric power consumption in O_2 evolution are much higher than those in O_3 evolution. Therefore, catalytic activity in O_3 evolution reactions is indispensable for electrolysis cell design.

Current efficiency or charge yield, $\Phi^e_{O_3}$ (%), and power efficiency or specific energy consumption, $w^e_{O_3}$ (kWh/kg-O_3), were calculated using

$$\Phi^e_{O_3} = \frac{5 \times F \times Q}{3 \times M_{O_3} \times I} \times 100\ (\%), \tag{5}$$

$$w^e_{O_3} = \frac{I \times U}{1000 \times Q}\ (\text{kWh/kg-}O_3)\,, \tag{6}$$

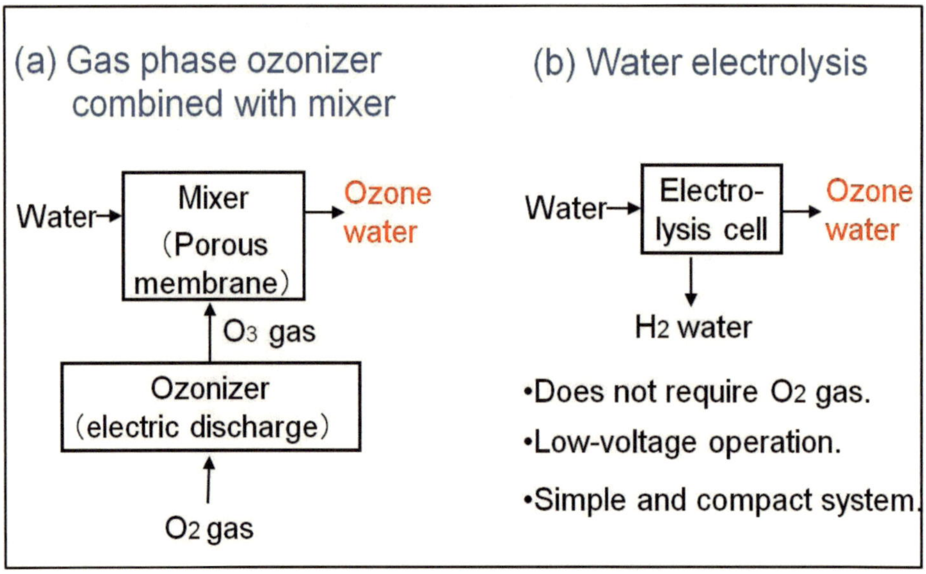

Figure 1. Two types of representative ozone water production systems.

where F is Faraday's constant (9.6485 x 10^4 C/mol), Q is the O_3 production rate (kg/h), M_{O_3} is the molecular weight of O_3 (48), I is the current (A), and U is the potential difference between the anode and the cathode (V). The derivation of Eq. (5) is as follows:

When the current I (A) is applied to the cell, the electron flow rate per hour is written as

$$\frac{I}{F}\times 3600\left(\mathrm{mol/h}\right). \tag{7}$$

O_3 produced per hour, Q (kg/h), can be converted to mol/h using

$$\frac{Q}{M_{O_3}}\times 10^3\left(\mathrm{mol/h}\right). \tag{8}$$

Because six electrons are required to produce one O_3 molecule, 1/6 of the electron flow rate calculated using Eq. (7) becomes the theoretical and maximum O_3 molecule production rate, which corresponds to 100 % efficiency. Therefore, real current efficiency is defined by

$$\Phi_{O_3}^{e}=\frac{\dfrac{Q}{M_{O_3}}\times 10^3}{\dfrac{I}{F}\times 3600\times\dfrac{1}{6}}\times 100\ (\%). \tag{9}$$

Calculating the right-hand side of the equation above, we obtain

$$\Phi_{O_3}^{e}=\frac{5\times F\times Q}{3\times M_{O_3}\times I}\times 100\ (\%).$$

The electrochemically produced O_3 molecules decompose in anode water to form OH radicals. The decomposition reaction is very complicated, and more than 30 elemental reactions are involved in the process. The reaction set proposed by Wittmann, Horvath and Dombi (WHD model) consists of 33 elemental reactions [5]. Although many reactions are taken into consideration, the decomposition behavior of O_3 in neutral pH water cannot be reproduced adequately using the WHD model.

The actual half-life of O_3, *i.e.*, about 2 h in ion-exchanged water and 4 h in pure water, are much longer than the calculated values. To improve the accuracy of the elemental reaction set, Mizuno *et al.* proposed new rate constants in a few elementary reactions [6]. They reduced the rate constants in OH^- formation reactions up to three orders of magnitude to fit their experimental data. We also noticed a discrepancy between the actual and estimated half-lifes, and obtained findings similar to those by Mizuno et al. However, we were not as radical as them. We would not write a paper, because we did not obtain experimental evidence to confirm the small rate constants of OH^- formation reactions.

For readers of this chapter, a simplified image of the·OH radical formation reactions is presented below:

$$O_3 + OH^- \rightarrow HO_2^- + O_2 \quad (10)$$

$$O_3 + HO_2^- \rightarrow O_3^- + HO_2 \quad (11)$$

$$O_3^- + H_2O \rightarrow \cdot OH + O_2 + OH^- \quad (12)$$

Sum of Eqs. (10) to (12):

$$2\,O_3 + H_2O \rightarrow \cdot OH + 2\,O_2 + HO_2 \quad (13)$$

O_3 decomposition is started by the reaction of O_3 with OH^-, forming HO_2^- as in Eq. (10). The initiation of O_3 decomposition is pH-dependent, because OH^- concentration determines the reaction rate in Eq. (10). Therefore, water pH strongly affects the half-life of O_3, *i.e.*, O_3 has a shorter half-life in alkaline water and a longer one in acidic water. HO_2^- reacts with another O_3 molecule to form O_3^- [Eq. (11)]. Finally, O_3^- is converted to a · OH radical by reacting with H_2O [Eq. (12)]. Eq. (13), the summation of Eqs. (10) to (12), indicates that one · OH radical is produced from two O_3 molecules, which means that · OH radical formation is a second-order reaction of O_3. Therefore, four times higher · OH radical concentration will be obtained by doubling the O_3 concentration in water. The bactericidal power of ozone water is also the second order of the O_3 concentration in water.

The development of electrochemical ozone production (EOP) systems using a PEM has shown significant progress. The main interest in studies of EOP systems has been in the preparation of effective anodes aiming at high catalytic activities for O_3 evolution. Many anode substances have been tested in a PEM system. The first demonstration of highly concentrated and efficient EOP using a PEM system was reported by Stucki *et al.* [7], who used a PbO_2 layer as an anode catalyst. They obtained a current efficiency of more than 15 %. Since then, studies on ozone water production using PEM systems have become an active area. Feng *et al.* utilized an Fe-doped PbO_2 catalyst electrodeposited on a Ti tube, and obtained a current efficiency of more than 12 % in 0.52 M K_2PO_4/0.22 M KH_2PO_4 buffer containing 2.5 mM KF [8]. Santana *et al.* reported that an IrO_2-Nb_2O_5 catalyst deposited on a Ti anode shows a current efficiency of 18 % in 3 M H_2SO_4 + 0.03 M KPF_4 solution, but its lifetime is only about 1 h: the corrosion of the anode catalyst prevents the system from longer operation [9]. Chang *et al.* utilized an Sb-doped SnO_2 anode catalyst, and obtained a current efficiency of more than 15 % in 0.1 M HCl solution [10]. They improved the efficiency up to 36.3 % in 0.1 M H_2SO_4 solution by changing the anode catalyst to Ni- and Sb-codoped SnO_2 [11]. Awad *et al.* showed a current efficiency of 11.7 % using a TaOx/Pt/Ti anode electrode in tap water, and reported that the system could be operated for as long as 65 h [12]. B-doped diamond substrates have been used as an anode, and Arihara *et al.* reported a stable operation with a current efficiency of 29 % in 1 M H_2SO_4 solution [13]. Recently, Kitsuka *et al.* have reported that a high current efficiency of 9% was achieved at a low current density of 8.9 mA cm^{-2} in 0.01 M $HClO_4$ at 15°C using n-type TiO_2 thin films deposited on Pt/TiO_x/Si/Ti substrates [14].

The generation mechanism of · OH radicals in a B-doped diamond electrode system was investigated by Marselli *et al.* by ESR spectroscopy in the presence of the spin trap 5,5-dimethyl-1-pyrroline-*N*-oxide (DMPO), and the formation of · OH radicals during water electrolysis was confirmed [15].

However, several problems remain unsolved for the commercial applications of the PEM system: Some catalysts, such as PbO_2 and Sb, are harmful to the human body and cannot be used for systems where contact of the human body with ozone water is inevitable; the use of an electrolyte, such as HCl or H_2SO_4, is also undesirable for the same reason; diamond substrates are inferior in forming water flow channels that enable good contact between ozone and water in the apparatus; and no catalyst lifetime longer than 500 h has been reported yet. Therefore, improvements in the performance and stability of the PEM system are still necessary.

Pt electrodes have been a classic substance for water electrolysis, and thus we have reinvestigated the potential of a Pt anode used in a conventional PEM system. There is an encyclopedia of studies of O_2 formation by the electrochemical splitting of water using Pt electrodes [16]. However, research on EOP is a rather limitted. The current efficiencies of Pt electrodes were reported in the above-mentioned studies, because researchers measured the efficiencies of Pt electrodes as references for comparison with those obtained using their original anode catalysts.

The current efficiency range of the Pt anode is from 0.7 to 7 %, and strongly depends on the system setup. If a high efficiency and a long lifetime could be achieved using a Pt electrode, we would eventually obtain a reasonably priced and safe EOP system.

In our research, an EOP system using Pt mesh electrodes combined with a conventional PEM, *i.e.*, Nafion 117 membrane, was investigated in terms of current efficiency and lifetime. We found that Pt particles that dissolve from the Pt anode migrate through the membrane, and are dendritically deposited on the cathode surface in contact with the membrane. The deposited Pt particles decompose the membrane, decreasing the current efficiency and lifetime of the EOP system. As a solution to solve this problem, one of the key improvements is achieved in our EOP system: A quartz felt separator is inserted between the membrane and the cathode. Pt particles are captured and isolated in the separator. Therefore, no decomposition occurs in the membrane [17].

Another important improvement is realized by the appropriate selection of a cathode electrolyte (catholyte). When a NaCl catholyte with concentrations higher than 0.085 M is circulated on the cathode side of the cell, the formation rate of Pt oxide on the anode surface is markedly reduced. Cl^- ions that migrated from the catholyte to the anode surface inhibit PtO_2 formation. The anode surface is kept clean during the EOP operation, and the clean Pt surface provides both good current efficiency and long lifetime [18]. As a result, the new EOP system achieved a high current efficiency of more than 25 % and a lifetime longer than 2,000 h. These findings and experimental results are presented in this chapter.

2. Conventional electrolysis system and its performance

2.1. Structure of conventional electrolysis cell

Photographs of electrolysis cells for the EOP system are shown in Fig. 2. A small-capacity cell that can produce ozone water up to a flow rate of 1 L/min is shown in Fig. 2(a), and a large-capacity cell for 20 L/min ozone water production is shown in Fig. 2(b). Both cells are designed and manufactured by FRD, Inc. The electrolysis cells consist of a Nafion 117 membrane sandwiched by Pt #80 mesh electrodes and have Ti and stainless terminal plates.

The cross-sectional cell configuration and the cathode structure of the small cell are shown in Figs. 3(a) and 3(b), respectively. The #80 Pt mesh electrodes are 4 x 2 cm^2 in area, 0.2 mm in thickness, and 0.3 mm in pitch distance. To form a waterway, a Ti #40 mesh is inserted between the Ti plate and the Pt mesh on the anode side, and a SUS304 #100 mesh is inserted between the SUS304 plate and the Pt mesh on the cathode side. The electrodes are held by SUS304 and Ti plates, which act as the cathode and anode terminals, respectively. The use of the mesh and plate made of Ti is important, since SUS304 cannot withstand oxidative corrosion caused by anode electrolysis reactions.

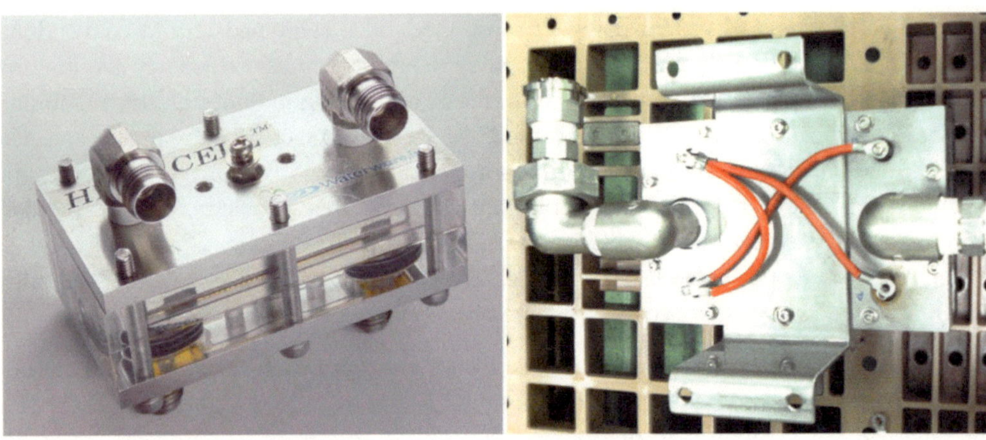

(a) 1 L/min O_3 water production cell (b) 20 L/min O_3 water production cell

Figure 2. O_3 water production cells using Pt electrodes and Nafion membrane. Copyright: FRD, Inc.

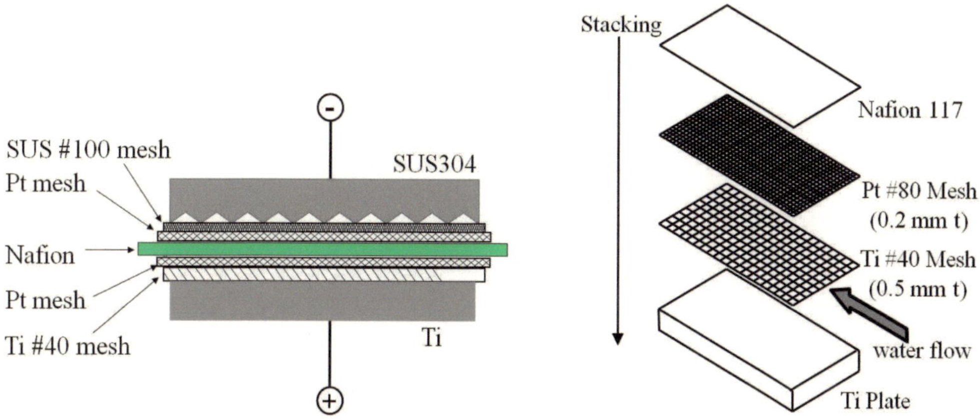

(a) Configuration of 1 L/min cell

(b) Cathode structure of 1 L/min cell

Figure 3. Cross-sectional views of 1 L/min electrolysis cell [17]. Reproduced with permission from The Electrochemical Society.

The flowchart of the EOP system is shown in Fig. 4. Tap water was passed through Na-type ion-exchange resin (Rohm and Haas, Amberlite IR-120B Na) to eliminate Ca^{2+} ions that lead to Ca_2CO_3 deposition on the anode and decrease the efficiency of the system. The ion-exchanged water is introduced into the anode and cathode at flow rates of 300-900 mL/min. Since the solubility and decomposition rate of ozone in water depend on water temperature, the temperature of the supplied water was controlled to 20 ± 1 ºC using a heat exchanger (HB Corporation, TEC1-127B) . However, temperature control is unnecessary for the practical use of the system.

The voltage and current of the electrolysis cell are controlled by a programmable power supply (TDK-Lambda Corporation, ZUP20-40) and monitored by using an A/D converter (Pico Technology, ADC-16). Ozone water and gas phase ozone were sampled in a 1-cm-optical-length cell, and ozone concentration was spectroscopically determined using a compact UV-VIS absorptiometer (Shimazu UV mini-1240). Ozone-containing gas was sampled by replacing water in the optical cell. Although water droplets remained in the cell after the gas sampling, the absorption and scattering caused by the droplets were measured by replacing the captured O_2 and O_3 gas mixture with air, and they were then subtracted from the gas spectra. To convert the absorbance of ozone to ozone concentration, 2950 $mol^{-1}cm^{-1}$ at 258 nm was used as the absorption coefficient of ozone in the gas and liquid phases [6].

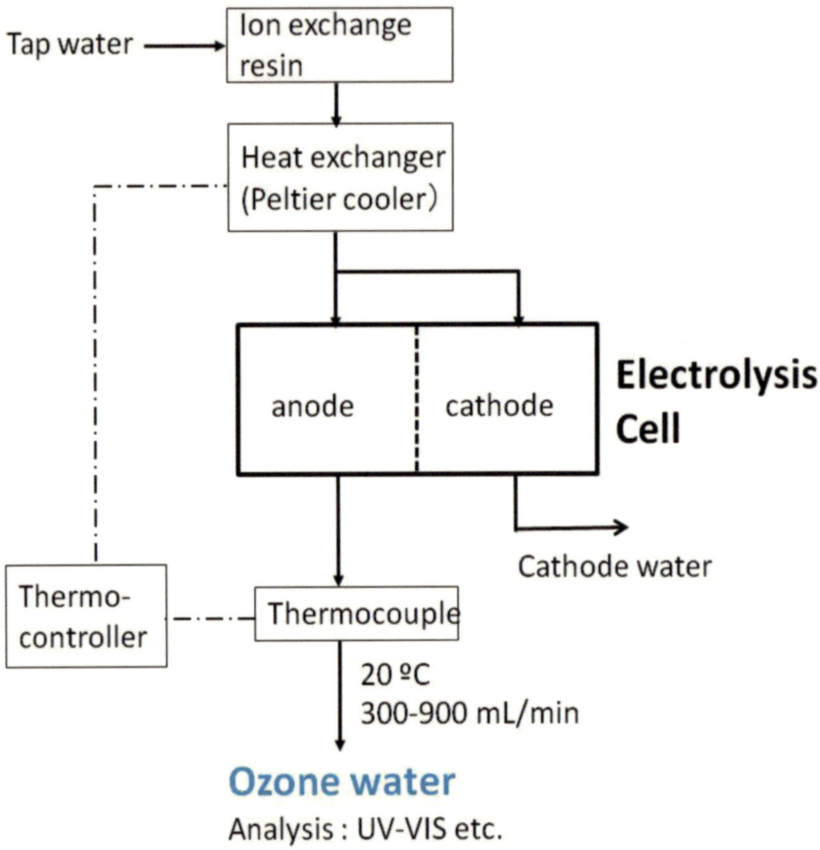

Figure 4. Flowchart of conventional EOP system

2.2. Efficiency of conventional electrolysis cell

The conventional cell can produce highly concentrated ozone water in the initial operation period. The obtained ozone water shows clear absorption by O_3. An example of O_3 decay in the anode water is shown in Fig. 5. In this experiment, O_3 electrochemically produced in the anode water was sampled in the optical cell, set in the UV-VIS absorptiometer, and measured of its decay every 10 min after the sampling time. As is evident from the spectra, ozone decay in the anode water is slow enough to neglect the sampling and scan times of the absorptiometer: The 1 min required for the absorption measurement does not affect the correct determination of ozone concentration. The typical half-life of ozone, 2 h in ion-exchanged water and 4 h in pure water, are in good agreement with those reported by Mizuno *et al.* [6].

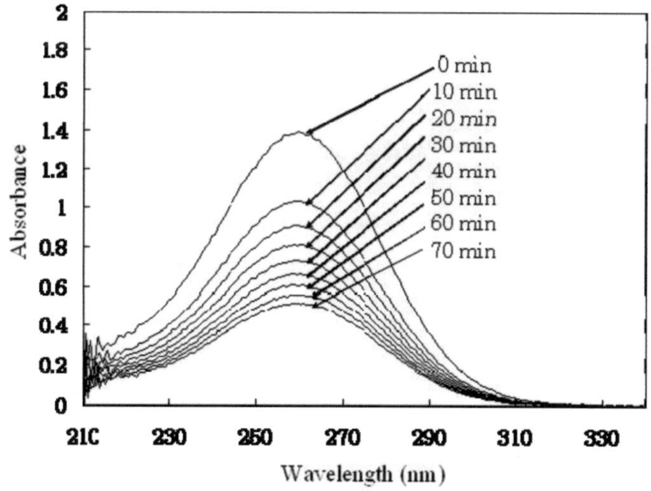

Figure 5. Time dependence of ozone absorption spectra [17]. Reproduced with permission from The Electrochemical Society.

To investigate the current efficiency of O_3 production with the conventional cell configuration and to check its run time dependence, ion-exchanged water was electrochemically decomposed in the cell at 5 A (0.625 A/cm^2) for 100 h. The flow rate of the anode water was 348 mL/min. The pH at the exit of the anode water was 6.0. The current efficiency of O_3 production in anode water is plotted as a function of run time in Fig. 6(a). No gas phase ozone was included in the efficiency calculation; only water-phase ozone was accounted for.

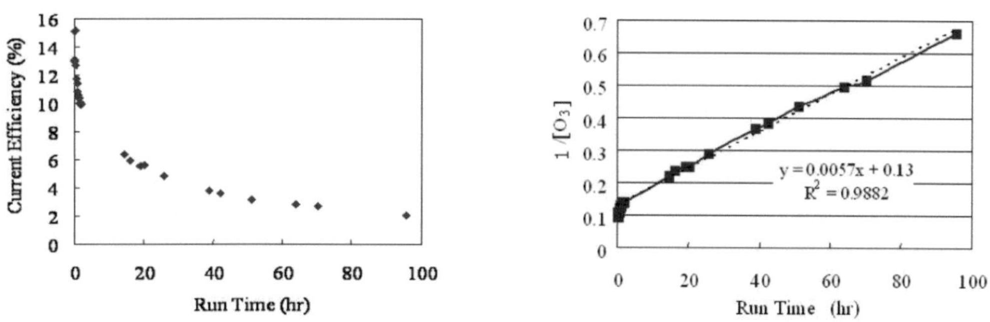

(a) Time dependence of current efficiency (b) Time dependence of reciprocal O_3 concentration

Figure 6. Time dependence of efficiency and reciprocal O_3 concentration obtained using conventional cell [17]. Reproduced with permission from The Electrochemical Society.

As current efficiency decreased, voltage was increased from 6 V to 13 V during the experiment. The initial efficiencies of the system were markedly high: 13 to 15 % were observed in the first 1 h of operation. However, current efficiency decreased with run time to as low as 2 % after a 100 h operation. The initial ozone concentration was about 10 mg/L, but it decreased to 1.5 mg/L at the end of the run. It is clear that the lifetime of a conventional cell is on the order of 10 h. The O_3 concentrations in Fig. 6(a) are converted to the reciprocal O_3 concentration, $1/[O_3]$ (L/mg), and are shown in Fig. 6(b). Such a figure is commonly used to estimate the order of reactions. The relation between $1/[O_3]$ and run time was found to be linear except for the initial 2-3 h of operation. A rapid degradation of the system occurs in the initial operation, and a steady degradation of the system starts after the rapid one. The linear increase in $1/[O_3]$ with run time after 2-3 h suggests the existence of a second-order reaction of the membrane with O_3 or its byproducts.

2.3. Mechanism of Nafion decomposition

The Nafion 117 membrane is a transparent film before the EOP operation. However, this membrane became black after the time dependence experiment. The image of the degraded Nafion 117 membrane is shown in Fig. 7. The cathode side of the membrane has a severely damaged (black) part and a slightly damaged (metallic silver) part.

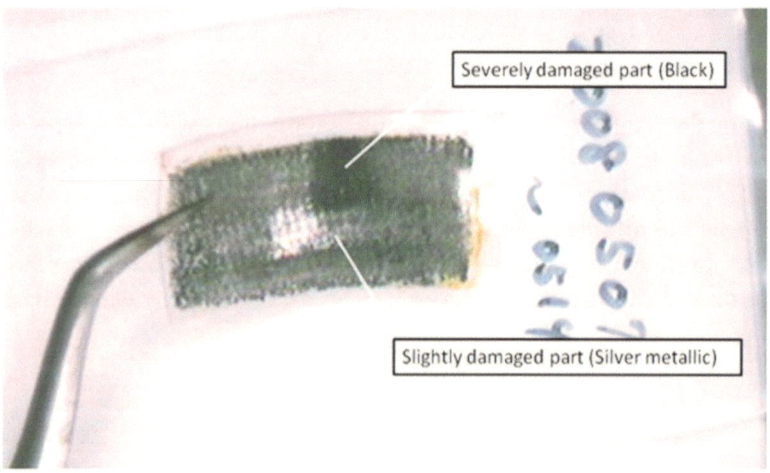

Figure 7. Degraded Nafion 117 membrane after 100-h operation at 0.625 A/cm^2 (view from cathode side).

FESEM images of the degraded Nafion 117 membrane are shown in Fig. 8. The anode side of the Nafion 117 membrane showed no morphological change, as shown in Fig. 8(a). However, holes were observed where the Nafion 117 membrane made contact with the Pt cathode, as shown in Fig. 8(b). The holes seem to be a result of the decomposition of the Nafion 117 membrane, and dendritic Pt nanoparticles were observed in the hole, as shown on the lower left side of Fig. 8(c).

EPMA measurements revealed that the Pt anode surface was covered with flakes of films consisting of Pt oxides, as shown in Fig. 9. The existence of Pt was also confirmed in the black region of the Nafion 117 membrane, but the composition could not be determined by EPMA because of the larger diameter of the probe beam of EPMA than of Pt nanoparticles. O, C, F and S signals originating from the Nafion 117 membrane also masked the real composition of Pt nanoparticles. However, semiquantitative XRF analysis indicated that the

Pt concentration on cathode side of the Nafion 117 membrane is about four times higher than that on the anode side. It is clear that the decomposition of the Nafion 117 membrane and the subsequent formation of holes decrease the efficiency of ozone production in the system.

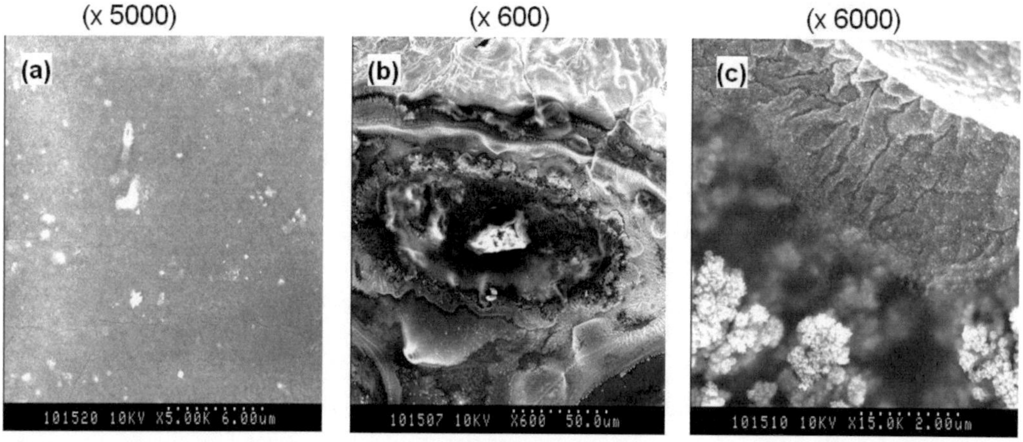

(a) Anode-side surface (b) Cathode-side surface (c) Inside hole of the cathode surface

Figure 8. FESEM images of degraded Nafion 117 membrane after time dependence experiment [17]. Reproduced with permission from The Electrochemical Society.

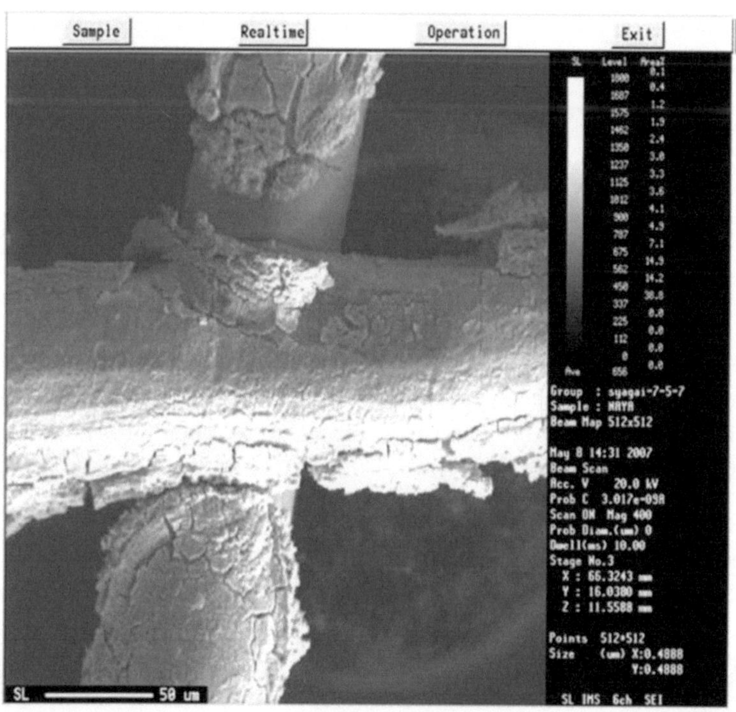

Figure 9. Surface of anode Pt mesh after time dependence experiment.

To determine the distribution of Pt black nanoparticles in the Nafion 117 membrane, four Nafion 117 membranes were set in the cell in piles, and the electrolysis of ion-exchanged water was performed for 50 h. A constant voltage of 12 V was applied, and current decreased from 17 A to 7 A according to the degradation of the cell. The ozone production efficiency calculated from the ozone concentration in the anode water was 10 % in the initial state, which decreased to 2 % at the end of the run. After the experiment, the Nafion 117 membrane on the anode side was light gray and almost transparent, but that on the cathode side was completely black. The sheets between them showed gradual tanning toward the cathode side.

From the above observations, we concluded that the rapid degradation is caused by the oxidation of the Pt anode to form Pt oxides, such as PtO_2. The formation of Pt oxides on the anode has been confirmed by the EPMA measurements. From the phase diagram of Pt in electrochemical equilibrium, PtO_2 and PtO_3 are expected to be stable substances in the electrolysis process where a voltage higher than 2 V is applied [19]. The poorer catalytic activity of Pt oxides than that of Pt probably decreases the current efficiency in this stage. During the oxidation of the Pt anode, ionic species, such as $Pt(OH)x^{+}$, can be formed where the local potential is in the range of 0-1 V, since Pt(OH) is also stable in this lower-potential region [19]. The ionic species could be released from the Pt anode surface, electrophoresed through the membrane, and concentrated on the cathode side of the Nafion 117 membrane. Finally, the ionic species are reduced by hydrogen and may work as a catalyst to decompose the Nafion 117 membrane by a mechanism similar to that reported in a fuel cell system [20]. An image of such a degradation mechanism is shown in Fig. 10.

H_2O_2 formed either in the self-decomposition process of O_3 or in the reduction of dissolved O_2 by H_2 should be a candidate source of $\cdot$ OH radicals via Fenton's reaction on reduced Pt particles that are deposited on the cathode in contact with the Nafion membrane. To check the former possibility, O_3 self-decomposition simulation was performed using the WHD model [5]. A couple of the rate constants were adjusted to reproduce the O_3 decay profiles observed in our experiments (Fig. 5). The H_2O_2 concentrations calculated in the self-decomposition of saturated ozone water are very low: 4.38×10^{-12} mol/L at 1 sec and 4.35×10^{-11} mol/L at 10 sec. Moreover, the concentrations of H_2O_2 at 1 sec and 10 sec linearly increase with increasing initial O_3 concentration, and the second-order dependence of H_2O_2 formation with O_3 concentration cannot be reproduced. Therefore, it is clear that H_2O_2 formed via the self-decomposition of ozone is irrelevant to the degradation mechanism of the system.

The electrochemical generation of H_2O_2 during water electrolysis is reported by Kusakabe *et al.* [21] and Yamanaka *et al.* [22]. In particular, the Nafion 117 membrane was used as the solid polymer electrolyte, and the successful reduction of dissolved O_2 by H_2 to form H_2O_2 was reported in the latter's paper. Although the potentials applied in the latter's paper, 0-0.6 V, are much lower than those in our experiments, the inevitable formation of H_2O_2 can be expected in our system. Moreover, both the O_2 and H_2 concentrations should be proportional to the O_3 concentration in our degraded system, since a lower O_3 production leads to a lower production of both O_2 and H_2. Therefore, the second-order formation of H_2O_2 with ozone concentration can be expected, and the second-order degradation of the system with ozone concentration can be explained by the electrochemical generation of H_2O_2.

Although we tried to check the H_2O_2 concentration in the cathode solution, the H_2O_2 concentration was too low to prove the existence of H_2O_2. The detectable limit of H_2O_2 in our lab is 0.02 mg/L, *i.e.*, 6×10^{-7} mol/L. A detailed elucidation of the degradation mechanisms of the system was not possible in this study. Therefore, we note the constants obtained from the linear part of Fig. 6(b) for future study: the intercept of the y-axis, which corresponds to the initial ozone concentration in the steady-degradation mode, was 7.7 mg/L; the slope, which indicates the rate constant of the second-order reaction of the Nafion 117 membrane with ozone, was 0.0057 L $mg^{-1} h^{-1}$.

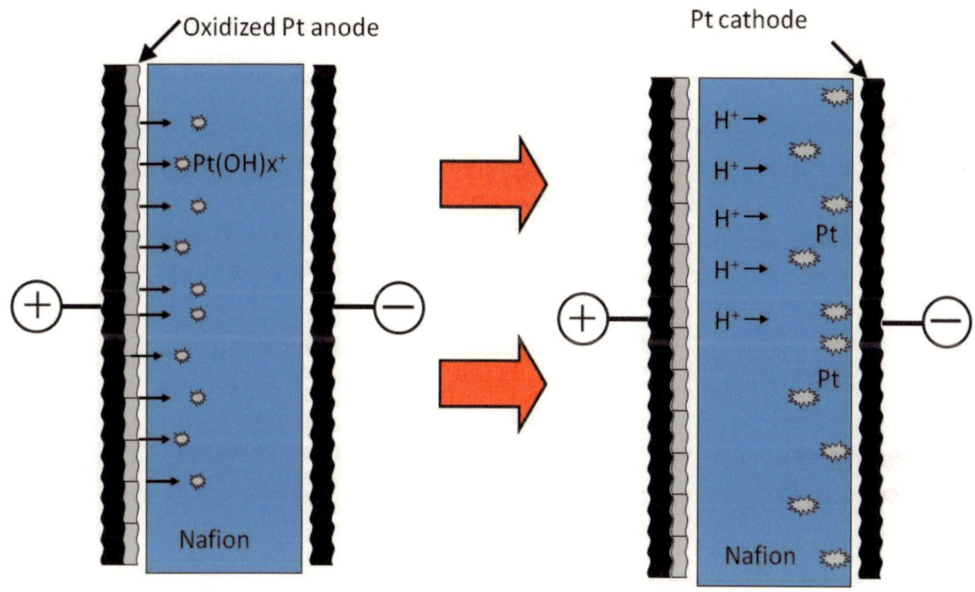

Figure 10. Mechanism of Nafion decomposition in electrolysis cell.

3. New EOP system

With the hypothesis that Pt nanoparticles deposited on the cathode side of the Nafion 117 membrane decompose the membrane, a new EOP cell was designed and tested for EOP operation. We inserted a quartz felt separator that captures Pt particles and prevents the degradation of the Nafion 117 membrane. Fundamental data for understanding the characteristics of the new electrolysis system were acquired during its first 50 h of operation, and the time dependence of current efficiency was measured in the subsequent experiment.

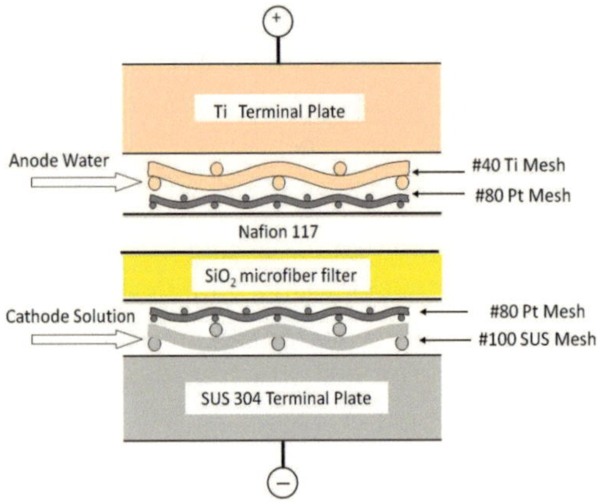

(a) Cross-sectional image of new electrolysis cell

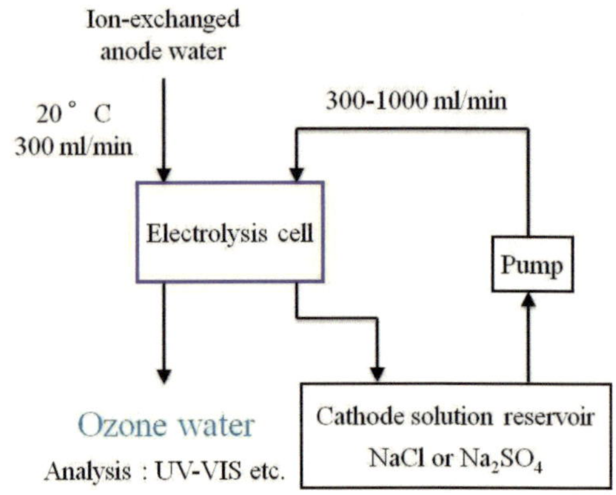

(b) Flowchart of new electrolysis system

Figure 11. Configuration of new electrolysis system

3.1. Insertion of quartz felt separator

The new electrolysis cell consists of a Nafion 117 membrane and a quartz felt separator sandwiched by Pt #80 mesh electrodes. Two sheets of quartz filters (Whatman, QM-A 2-335-05) are stacked in piles, and used as the separator. The thickness of the separator is about 1 mm under ambient condition. The configuration of the new cell is illustrated in Fig. 11(a). The quartz filter is labeled as "SiO_2 microfiber filter" in the figure. To compensate for the low conductivity of the quartz separator, guaranteed reagent-grade NaCl and Na_2SO_4 (Kanto Chemical Co. Inc.) are used as cathode electrolytes (catholytes) without further purification. The purities of NaCl and Na_2SO_4 are more than 99 % and 99.5 %, respectively. These substances are dissolved in pure water, diluted to 0.01-1 M, and circulated through the cathode electrode at flow rates of 300-1000 mL/min by a water pump (Kamihata Fish Industries, Rio 50). The flowchart of the new system is shown in Fig. 11(b). The conductivities of the pure water and ion-exchanged water are 0.1 μS/cm and 240 μS/cm, respectively.

3.2. Characteristics of new EOP system

The O_3 production rates in the anode water and in the gas phase obtained with two catholytes are shown in Fig. 12(a) as functions of current. The current efficiencies calculated from these data in Fig. 12(a) are plotted in Fig. 12(b). In the case of 0.01 M catholytes, water electrolysis did not proceed. When the concentrations of NaCl and Na_2SO_4 were increased above 0.1 M, efficient water electrolysis was observed regardless of the solute and concentration. However, as is shown in Fig. 12(b), the current efficiencies at currents lower than 4 A obtained with Na_2SO_4 solutions were inferior to those obtained with NaCl solutions. Therefore, the 0.5 M NaCl solution was used in the time dependence experiment. The origin of such a discrepancy is further investigated, and explained in section 4. O_3 production rates in the anode water were about two times higher than those in the gas phase regardless of the solute and concentration, and they showed a saturation behavior in the high-current region, while those in the gas phase increased in contrast [Fig. 12(a)].

Current efficiencies of more than 20 % were obtained at currents higher than 6 A. The efficiency is higher than those reported by other researchers who used Pt or PbO_2 electrodes, and close to that obtained using diamond electrodes. The power efficiency at 6 A and 7.74 V using the 0.5 M NaCl catholyte was 220 kWh/kg-O_3.

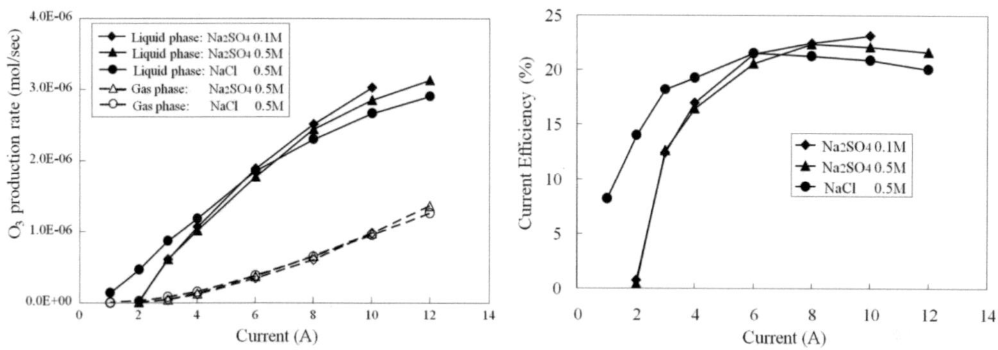

(a) O_3 production rates in liquid and gas phases (b) Current efficiencies in O_3 production

Figure 12. O_3 production rates and current efficiencies of new electrolysis cell [17]. Reproduced with permission from The Electrochemical Society.

Since both current and power efficiency are functions of water flow rate and applied current, high efficiencies can be expected when water flow rate and current increase. An example of the relationship between water flow rate and current efficiency calculated from the ozone concentration in the anode water is shown in Fig. 13. In the experiments, the 0.5 M NaCl catholyte was circulated on the cathode side of the cell. Current efficiency increases with increasing water flow rate and current. At currents higher than 10 A, O_3 concentration starts to saturate in the anode water, and the differences in current efficiency become difficult to detect. If the water flow rate is increased to more than 1000 mL/min, and if O_3 becomes unsaturated in the anode water, the differences in current efficiency between 10 A and 14A would be apparent. A high voltage, such as above 7 V, applied to our system gave sufficient overpotential for O_3 production, and was probably one of the origins of good current efficiency. However, ozone concentration decreases with increasing water flow rate. Therefore, appropriate conditions should be chosen according to the required ozone concentration and water flow rate for practical use.

One of the possible efficiency loss is due to the solution resistance and/or contact resistance in the system. To measure such a loss, the overall applied voltage and potential difference between the Pt anode and cathode side of the Nafion 117 membrane were measured by placing electric wires on them using tiny Kapton tapes. The overall voltage and potential difference between the electrodes were measured at various currents using 0.5 M NaCl catholyte. Representative results are shown in Fig. 14. The potential difference between the anode and cathode side of the membrane was roughly 30 % smaller than the overall voltage applied to the anode and cathode terminals of the cell. Since observable O_3 evolution occurs at a current of 1 A, as shown in Figs. 12 (a) and (b), nearly 4 V of the potential difference is required for EOP in the new system, suggesting that the involvement of the reaction in Eq. (3) is important for effective EOP.

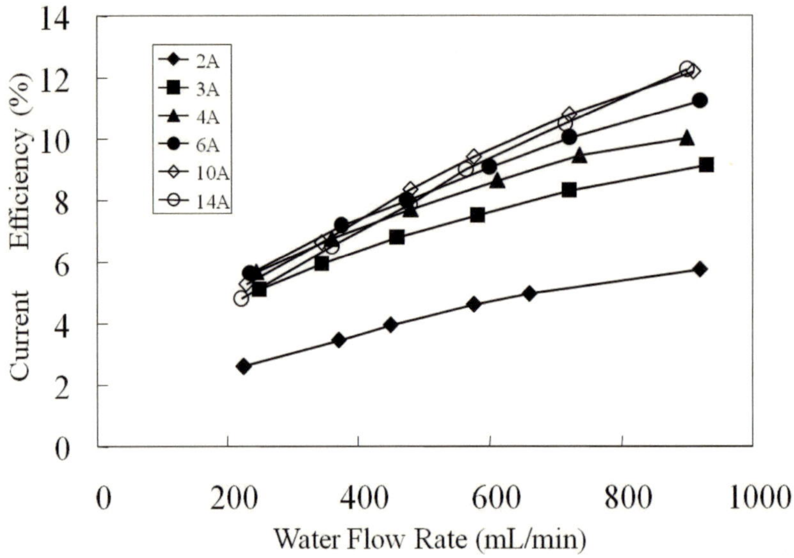

Figure 13. Current efficiency of O_3 formation in anode water as a function of water flow rate and current [17]. Reproduced with permission from The Electrochemical Society.

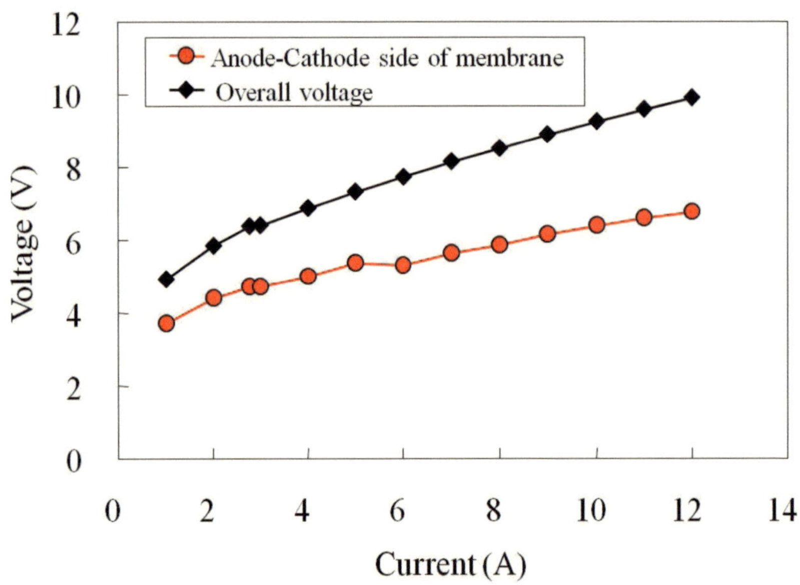

Figure 14. Relationship between applied voltage and potential difference of the electrodes

3.3. Lifetime of new system

After the characterization of the new system for 50 h, the accelerated degradation experiments were carried out using the 0.5 M NaCl catholyte. The flow rates of the anode water and catholyte were both 500 mL/min. A high current of 14 A was applied to accelerate the degradation of the system. The O_3 concentration in the anode water showed little degradation for 150 h of operation, as shown in the upper plots in Fig. 15. O_3 concentrations of 12 to 15 mg/L and a current efficiency of 18 to 20 % were maintained during the operation with a quartz felt separator. In this figure, the data in Fig. 6(a), which were obtained at a mild current of 5 A, are also plotted to compare the O_3 concentrations in the conventional and new cells. The separator works to maintain the high efficiency of the system and supports our hypothesis concerning the degradation mechanism. Provided that the degradation is a second-order process with O_3 concentration and that the O_3 concentration becomes one third at 4A compared with that at 14 A, as shown in Fig. 12(a), the lifetime at 4 A will be nine times longer than that at 14 A. Therefore, a stable operation of more than 1350 h (150 h multiplied by 9) could be expected under low-current operation.

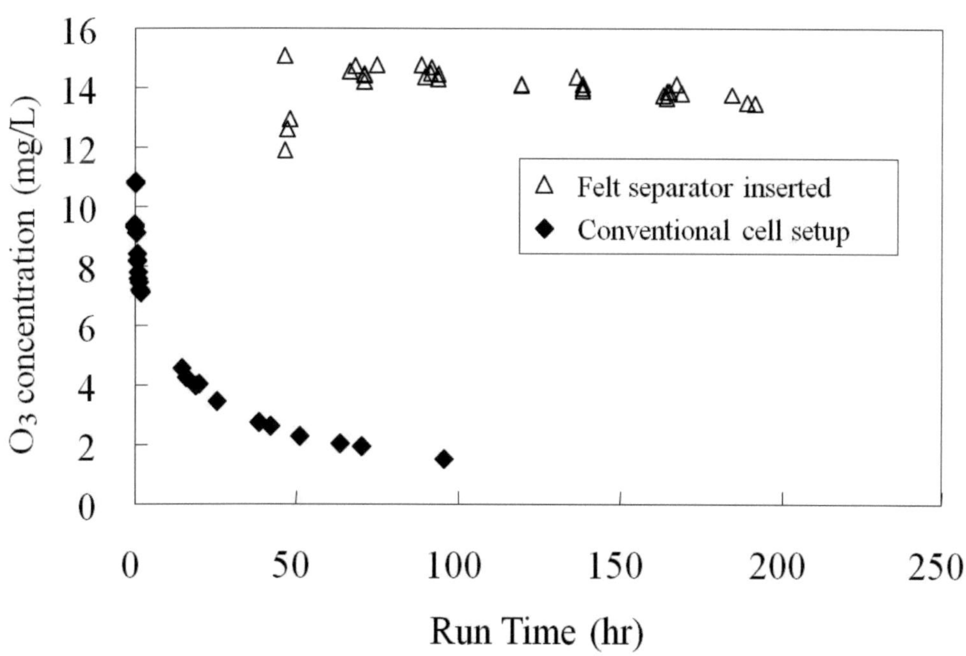

Figure 15. Run time dependence of O_3 concentration obtained in conventional and new cells [17]. Reproduced with permission from The Electrochemical Society.

The concentrations of ionic species in the anode water sampled in the middle of the run time dependence measurement were analyzed by ion chromatography (IC) and atomic absorption spectrometry (AAS), and the results are summarized in Table 1. Ion concentrations at the exit of the anode water were almost the same as those at the exit of the ion-exchange resin, *i.e.*, no significant change in ion concentration was detected except for small changes in Cl^- and Na^+ concentrations: Cl^- concentration slightly increased from 11.3 ppm at the exit of the ion exchange resin to 13.4 ppm at the exit of the anode water; on the other hand, Na^+ concentration decreased from 31.5 ppm at the exit of the ion exchange resin to 24.5 ppm at the exit of the anode water. The pHs of the anode water and cathode water in the middle of the run were 5.8 and 13.2, respectively. These changes were caused by the electrophoretic migration of ionic species.

The migration of the anode water associated with proton transfer through the membrane to the cathode side was also confirmed. The volume change in the circulated cathode solution was measured after operating for 42 h at 10 A. The volume increased from 1 L at the initial stage to 2 L after the operation. Therefore, the adjustment of the volume and pH of the cathode water is required at one or two day intervals during long runs of the system.

Ion	Tap water	Exit of ion-exchange resin	Exit of anode	(tw ppm) Method of analysis
Cl^-	11.5	11.3	13.4	IC
NO_3^-	8.1	7.7	7.8	IC
SO_4^{2-}	20.4	21.0	21.1	IC
Na^+	7.6	31.5	24.5	AAS
K^+	1.9	0.1	0.1	AAS
Mg^{2+}	4.0	0.0	0.0	IC
Ca^{2+}	19.9	0.0	0.0	IC
NH_4^+	0.0	0.2	0.2	IC

Table 1. Ion concentrations in tap water, exit of ion-exchange resin, and exit of anode water (IC: ion chromatography, AAS: atomic absorption spectrometry) [17]. Reproduced with permission from The Electrochemical Society.

4. Selection of catholyte

The key questions arising from the experiments in the above sections are as follows:

a. We noticed that the lifetime of the EOP system using the Na_2SO_4 catholyte is much shorter than that using the NaCl catholyte. How do these catholytes affect the lifetime of the system?
b. How does the NaCl catholyte concentration affect the efficiency and lifetime of the system?
c. The dissolution of the anode Pt electrode will be the limiting factor for the lifetime of the NaCl catholyte system after eliminating the problem of the decomposition of the membrane using the quartz felt separator. What is the dissolution rate of the Pt anode, and how long can the system be operated without changing the anode?
d. The contaminant concentrations in the ozone water produced from ion-exchanged water are lower than the allowable levels for drinking water. If pure water is introduced into the anode instead of ion-exchanged water, what will be the contaminant concentrations? Will the contaminants be at allowable concentrations for the sterilization of medical instruments and for the cleaning of electronic components?

To answer the above important questions, the dependences of catholyte composition and concentration on the efficiency and lifetime of the EOP system were investigated, and the dissolution rate of the Pt anode and the contaminant concentration in ozone water were measured. All the experiments were carried out using pure anode water instead of ion-exchanged water, because the analysis of ions that migrated from the catholyte to the anode Pt surface seemed to be essential to understand the effects of the catholyte on the efficiency and lifetime of the system. The other experimental setup and measurements used are the same as these in the last section (see Fig. 11).

4.1. Effects of catholytes on EOP efficiency

Pure water produced by passing tap water through Na-type ion-exchange resin (Rohm and Haas, Amberlite IR-120B Na) and reverse osmosis (Dow Chemical, FILMTEC TW30-2514) is introduced into the anode at a flow rate of 300 mL/min. The conductivity of the pure water is less than 0.1 μS/cm. The residence time and flow speed of the anode water in the cell are estimated to be 0.085 s and 0.94 m/s, respectively. The O_3 concentrations obtained using 0.5 M NaCl and 0.5 M Na_2SO_4 catholytes as functions of current density are shown in Figs. 16(a) and (b), respectively. In these experiments, current was slowly increased and held for 20 min to stabilize the system at each measurement point. Therefore, 2 h was required to obtain a series of NaCl or Na_2SO_4 data. The holding time should be long enough to observe the degradation of the system, if any.

When the NaCl catholyte was used, the O_3 concentrations in the anode water and gas phase increased with current, as shown in Fig. 16(a). The concentration of ozone water was higher than 20 mg-O_3/L-water (mg/L) at a current higher than 1 A/cm^2 (8.0 V); and it was 24.8 mg/L at 1.25 A/cm^2 (9.1 V). On the other hand, the O_3 concentrations obtained using the Na_2SO_4

catholyte were lower than those obtained using the NaCl catholyte in the entire current range in Fig. 16(b). No O_3 concentration in the anode water higher than 20 mg/L was obtained using the Na_2SO_4 catholyte. Moreover, the O_3 concentrations in the anode water and gas phase decreased at 1.25 A/cm², suggesting that the system was degraded within 2 h of the measurement.

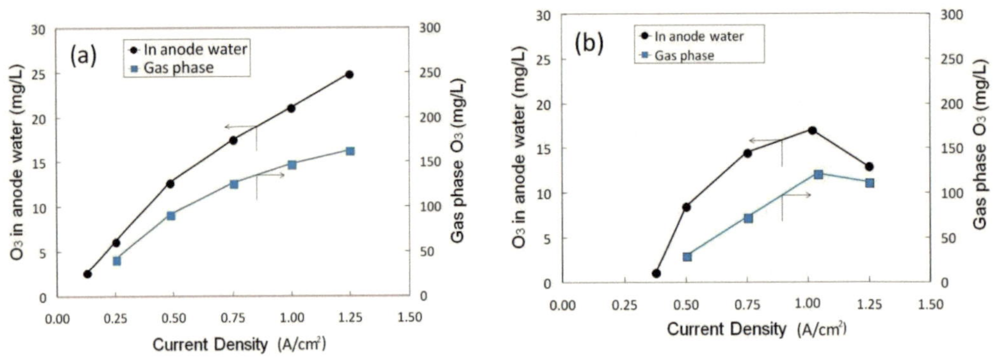

(a) 0.5 M NaCl catholyte (b) 0.5 M Na_2SO_4 catholyte

Figure 16. O_3 production rates in anode water and gas phase

The current efficiencies calculated from the data in Fig. 16 are plotted in Figs. 17. As seen in Fig. 17(b), the current efficiencies obtained using the Na_2SO_4 catholyte were lower than those obtained using the NaCl catholyte [Fig. 17(a)]. Although the electric conductivity of a Na_2SO_4 electrolyte, 8.23 Sm^{-1}, is higher than that of a NaCl electrolyte, 7.06 Sm^{-1}, at 1 M and 20 ºC, no improvement in current efficiency was observed in the experiments. The origin of the differences observed in the EOP efficiencies caused by changing the catholyte will be discussed in section 4.4.

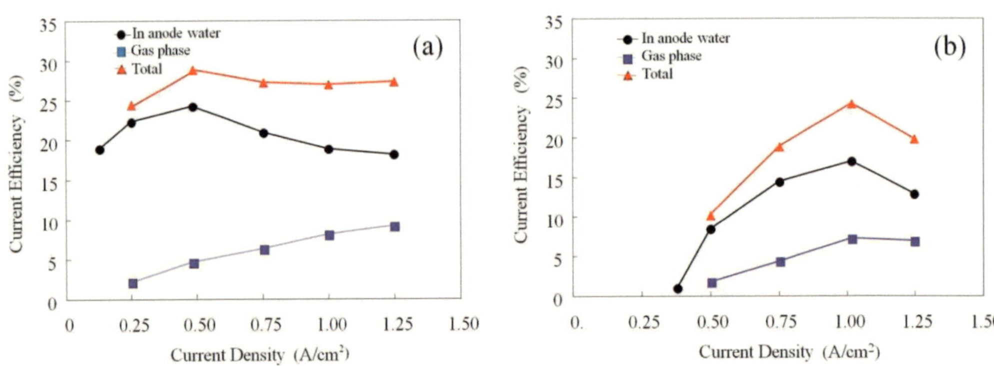

(a) 0.5 M NaCl catholyte (b) 0.5 M Na_2SO_4 catholyte

Figure 17. Current efficiency of O_3 production

The current efficiencies shown in Fig. 17(a) were higher than those reported in the previous data in section 3.2. The current efficiency of 29 % at 0.5 A/cm² (6.6 V) was markedly high, and efficiencies higher than 25 % were obtained in the entire current range. The O_3 production rates in the anode water and gas phase at 0.5 A/cm^2 were measured to be 274 mg/h and 52 mg/h, respectively. From these data, the power efficiency of the system at 0.5 A/cm^2 was determined to be 76 kWh/kg-O_3, which is lower than that (220 kWh/kg-O_3) obtained in a previous experiment and those reported using other anode materials in EOP systems. The theoretical electric power required for a perfect EOP system is estimated to be 5.06 kWh/kg-O_3, supposing 100 % consumption of the current for producing O_3 and the standard voltage of O_3 formation, 1.511 V. Further improvement in power efficiency is possible by introducing other measures, such as O_2 gas recycling into the cathode water [23]. However, the cost of electricity is reasonably low even in the present EOP system. The NaCl catholyte system can produce 1 kg-O_3 at an electricity cost of 7 US dollars assuming that the price of electricity is 10 cents/kWh. 100 kL of 10 mg/L-ozone water will be produced from 1 kg-O_3.

4.2. Effects of catholytes on lifetime

To observe the time dependence of the O_3 concentration and to check the morphological change of the Pt anode after a long operation, accelerated degradation experiments were performed. During the experiments, the flow rates of the anode water and the catholytes were both set to 300 mL/min, and a high current of 1.25 A/cm^2 (10 A) was applied to accelerate the degradation. The concentrations of the NaCl and Na_2SO_4 catholytes were set to 0.5 M. The concentrations of Na^+, Cl^- and SO_4^{2-} in the anode water were measured in the middle of the accelerated-degradation experiments, and are summarized in Table 2.

Ion	Na_2SO_4 0 A	Na_2SO_4 10 A	NaCl 10 A	(wt ppm) Method of analysis
Cl^-	0.06	0.01	7.4	IC
SO_4^{2-}	0.14	1.15	-	
Na^+	0.11	-	0.02	
K^+	0.04	-	-	
Ca^{2+}	0.02	0.01	0.01	
NH_4^+	0.01	0.01	0.01	
Pt	0.023 wt ppb	0.13 wt ppb	0.62 wt ppb	ICP-MS

Table 2. Impurity concentrations in anode water sampled during electrolysis at 0 A and 10 A using the Na_2SO_4 and NaCl catholyte. (Method of analysis: IC, ion chromatography; ICP-MS, inductively coupled plasma mass spectrometry)

The ion concentrations at the exit of the anode water sampled at 0 A/cm^2 using the Na_2SO_4 catholyte should be the same as those in pure water, except for slight increases in SO_4^{2-} and Na^+ concentrations. These ions came from the catholyte by diffusing through the Nafion 117 membrane. SO_4^{2-} concentration increased to 1.15 wt ppm at an electrolysis current of 10 A (1.25 A/cm^2) using the Na_2SO_4 catholyte. Cl^- concentration also increased to 7.4 wt ppm, and Na^+ concentration decreased to 0.02 wt ppm at 10 A (1.25 A/cm^2) using the NaCl catholyte. These changes were caused by the electrophoretic migration of these ionic species.

The results of the degradation experiments are shown in Fig. 18. Note that the O_3 production in the gas phase is not included in the plots. When the 0.5 M NaCl catholyte was used, the O_3 concentration showed only a slight degradation during the operation for 130 h, as is shown in the upper plots in Fig. 18. A high O_3 concentration ranging from 20 to 25 mg/L was maintained during the experiment (the Cl^- concentration in the anode water was 7.4 ppm). The slight decrease in the concentration is probably due to the dissolution of the Pt anode electrode resulting in poor contact between the anode and the Nafion 117 membrane. However, the use of the 0.5 M Na_2SO_4 catholyte caused a rapid decrease in O_3 concentration: The initial concentration of 18 mg/L decreased to 7 mg/L after operating the system for 60 h, as shown in the lower plots in Fig. 18. In this experiment, only 0.01 wt ppm Cl^- was found in the exit of anode water. The degradation data reported in section 3.3 using ion-exchanged water and the 0.5 M NaCl catholyte are also shown in the middle plots in Fig. 18. A stable EOP was observed in our previous experiment (the SO_4^{2-} and Cl^- concentrations in the anode water were 21.1 ppm and 13.4 ppm, respectively) [17].

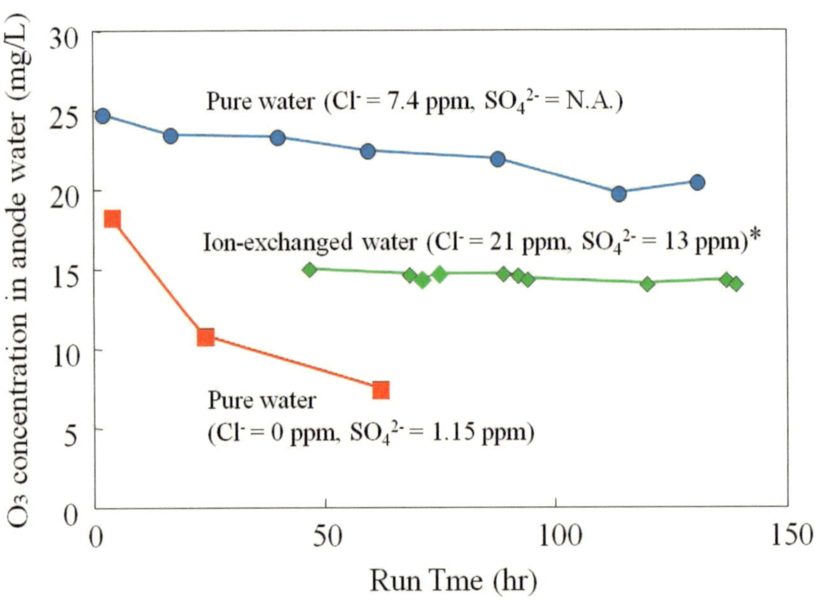

Figure 18. Run time dependence of O_3 concentration in anode water. The data set indicated by "*" is the same as that in Fig. 15.

After the accelerated-degradation experiments, the anode surface in contact with the Nafion 117 membrane became brown when the 0.5 M Na_2SO_4 catholyte was used. In contrast, the anode surface was clean and only a small portion of that in contact with the Nafion 117 membrane became yellow when the 0.5 M NaCl catholyte was used. SEM images of the Pt anode electrodes after the degradation experiments are shown in Fig. 19. As is clear in Fig. 19(c), the surface of the anode electrode in contact with the Nafion 117 membrane was covered with an approximately 10-µm-thick oxide layer, when the system was operated using the 0.5 M Na_2SO_4 electrolyte. The surface of the anode electrode remained clean [Fig. 19(a)], and only a thin layer of oxide deposit was observed in a small area [Fig. 19(b)], when the system was operated using the 0.5 M NaCl catholyte.

The above oxide layers were assigned to amorphous PtO_2 (a-PtO_2) by microscopic Raman spectroscopy, in which the characteristic broad spectrum of a-PtO_2 peaking at 600 cm^{-1} with an FWHM of 200 cm^{-1} was observed. No sharp peak originating from PtO or α-PtO_2 crystals was observed [24]. EDX measurements indicated that 3.5 wt % Cl was present in the a-PtO_2 layer formed using the 0.5 M NaCl catholyte [Fig. 19(b)]. Because our EDX apparatus could not detect oxygen, the rest of the material measured in the oxide layer was 96.5 wt % Pt. No Cl was observed in the oxide layer formed using the 0.5 M Na_2SO_4 catholyte. The high concentration of Cl adsorbed onto the a-PtO_2 layer suggests that Cl^- that migrated on the anode surface inhibited the formation of the a-PtO_2 layer. When chloride is added to a sulfuric medium, its adsorption inhibits the formation of oxide films in the low-voltage range, *i.e.*, at voltages lower than 1.5 V vs RHE [25, 26]. Because the anode voltages applied in our EOP system were much higher than those in the literature, a more efficient adsorption of chloride is expected.

SEM images of the captured Pt particles on the cathode side of the quartz separator are shown in Fig. 20. Small and spinous Pt particles were captured in the separator using the 0.5 M NaCl catholyte, as shown in Fig. 20(a). On the other hand, granular Pt aggregates were observed using the 0.5 M Na_2SO_4 catholyte, as shown in Fig. 20(b). The difference in the shape of the captured Pt particles suggests that the precursors of the particles may be different. However, further analysis should be required to determine the origin of the difference in the shape of the captured Pt particles.

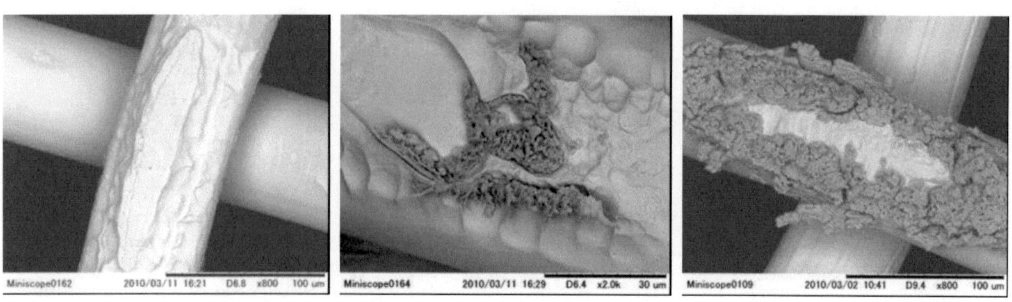

(a) Clean part (0.5 M NaCl catholyte) (b) Yellow part (0.5 M NaCl catholyte) (c) Brown part (0.5 M Na_2SO_4)

Figure 19. SEM images of Pt anodes after accelerated degradation experiments

(a) 0.5 M NaCl catholyte (x 2,000) (b) 0.5 M Na_2SO_4 catholyte (x 3,000)

Figure 20. SEM images of captured Pt particles in quartz felt separator

4.3. Effects of NaCl concentration

The effects of catholyte concentration on the current efficiency and lifetime of EOP were measured by changing the NaCl catholyte concentrations from 0.003 M to 0.085 M. These measurements were performed at a constant current density of 0.625 A/cm^2 (5 A) for 20 h. The results are shown in Fig. 21. The current efficiency of EOP increases with increasing concentration of the NaCl catholytes probably owing to the increase in catholyte conductivity. The decrease in the current efficiency with run time was observed for NaCl concentrations lower than 0.028 M. The efficiency linearly decreases with run time using these dilute catholytes. However, in the case of the 0.085 M catholyte, the degradation speed of the system became lower, and the efficiency was kept higher than 23 % for 20 h. This efficiency is similar to that obtained using the 0.5 M NaCl catholyte [Fig. 17(a)], and the low degradation speed is indicative of the realization of the long and stable operation of the system. Therefore, a NaCl concentration higher than 0.085 M is adequate for the EOP system.

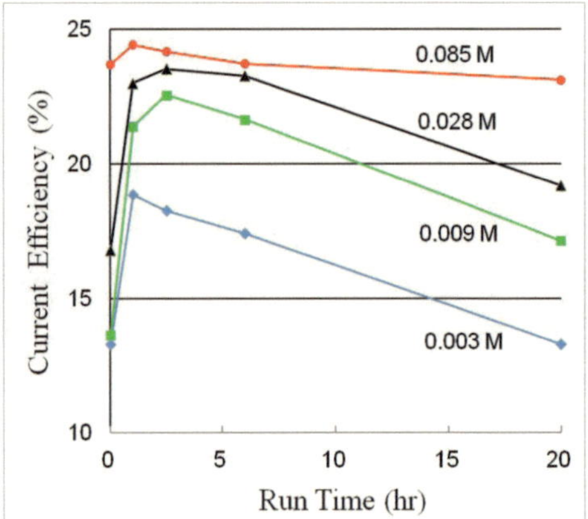

Figure 21. Concentration dependence of catholyte on the efficiency and lifetime measured at 0.625 A/cm^2 (5 A).

4.4. Inhibition mechanism of Pt oxide formation

From these observations, we conclude that Cl^- that migrated from the catholyte to the anode surface inhibited the formation of the a-PtO_2 layer. As a result, the anode surface was kept clean during the accelerated degradation experiments. The poor performance of the system using the Na_2SO_4 catholyte, shown in Figs. 16(b), 17(b) and 18, suggests that the anodes were rapidly oxidized from the beginning of the experiments owing to the absence of Cl^-. Therefore, Cl^- that migrated from the catholyte to the anode surface is indispensable for a long and stable operation of the EOP system, whereas SO_4^{2-} concentration is probably irrelevant for long lifetime as long as sufficient amount of Cl^- is supplied to the anode, as shown in the middle plots in Fig. 18.

The standard electrode potentials for chlorination and oxidation reactions of Pt are as follows [27]:

$$Pt + 4Cl^- \rightarrow \left[PtCl_4\right]^{2-} + 2e^- \qquad E^0 = 0.758\ V \qquad (14)$$

$$\left[PtCl_4\right]^{2-} + 2Cl^- \rightarrow \left[PtCl_6\right]^{4-} + 2e^- \ E^0 = 0.726\ V \qquad (15)$$

$$Pt + H_2O \rightarrow PtO + 2H^+ + 2e^- \qquad E^0 = 0.980\ V \qquad (16)$$

$$PtO + H_2O \rightarrow PtO_2 + 2H^+ + 2e^- \qquad E^0 = 1.045\ V. \qquad (17)$$

Because chlorination reactions take place at lower potentials than oxidation reactions, Pt is chlorinated, ionized, and dissolved in anode water in the presence of Cl^- ions. Thus, the formation of Pt oxide films can be inhibited.

If the efficiency of the system is degraded by the formation of the a-PtO_2 layer on the anode surface using the Na_2SO_4 catholyte, an important question arises: Does the a-PtO_2 layer have a lower catalytic activity than Pt in EOP reactions? The formation mechanisms and resultant structures of oxide layers on a Pt anode have been extensively studied to understand the electrocatalytic properties of such layers [28-34]. Pt oxide layers are considered to have slightly beneficial effects on oxygen evolution reactions. Shibata reported that the oxidation treatment of a Pt anode surface with 1 M H_2SO_4 solution for 28 h enables a stable electrolysis activity of the Pt anode for more than 30 h [28]. Tremiliosi-Filho *et al.* reported that the rate of O_2 evolution increases with the period of oxide film formation at various electrolysis potentials [33]. Gottesfeld *et al.* found that Pt oxide films formed after a long-term polarization of potential between 2.1-2.25 V vs. RHE in 0.5 M H_2SO_4 solution could be identified as film β (it is now assigned to a-PtO_2), and that the films shows good electric conductivity as well as an improvement in the performance of oxygen evolution reaction with the lowering of the oxygen evolution voltage from 2.1 V to 2.0 V at a current of 10 mA/cm² [31]. Therefore, we did not concern ourselves with the negative effect of Pt oxides on EOP reactions.

If Pt oxide layers significantly decrease oxygen overpotential in water electrolysis reactions and lead to higher O_2 evolution, the efficiency of O_2 formation would increase and the negative effect on EOP could be expected. To determine the effects of a-PtO_2 layers on EOP, we are now conducting experiments to measure the electric properties and electrocatalytic activities of a-PtO_2 layers by fabricating such layers on the Pt anode. We have found that a-PtO_2 layers have a much lower catalytic activity than pure Pt in EOP reactions; the results will be reported soon [35].

4.5. Current balance

The current balance of the 0.5 M NaCl catholyte system was measured and checked by comparing the calculated currents for O_3 production in the anode water and gas phase as well as those for O_2 production. In this experiment, the concentration of dissolved O_2 in anode water was measured using an electrochemical O_2 sensor (Hach Ultra, Orbisphere 3600 and 31124). The O_2 sensor was calibrated to match the current balance of the system at high current density operation. The gas phase O_2 production rate was obtained by subtracting O_3 production from the overall anode gas production. To double-check the current balance, we tried to measure H_2 production rate. However, it was not possible to capture H_2 gas in a cylinder, because H_2 formed very tiny bubbles that dispersed in the catholyte. Current balances calculated at seven currents are summarized in Fig. 22. The calculated currents are in good agreement with the measured ones, *i.e*, the 100 % line in Fig. 22, except that in the current range lower than 0.25 A/cm^2, where the low gas production rates could be the source of error. As is clear from Fig. 22, the system has neither an apparent current leakage nor markedly abundant products other than O_3, O_2, and H_2. More than 70 % of the current was consumed in producing O_2. It is also clear from Fig. 22 that the current consumption for O_2 formation in anode water decreases with applied potential, while that for O_3 formation in gas phase and anode water increases. Therefore, it is suggested that the oxidation of O_2 in anode water to O_3 [Eq. (3)] is the dominant process for O_3 formation. However, more detailed kinetic studies on O_2 and O_3 formation reactions, such as the branching ratio of Eqs. (2) and (3), will be required for the further understanding of the system.

4.6. Pt dissolution rate and lifetime of new system

The weight of the Pt anode mesh was about 0.4 g before the experiments. They decreased by 5.4 mg and 7.7 mg, corresponding to dissolution rates of 5.4 µg/Ah and 6.9 µg/Ah, after the accelerated experiments using the Na_2SO_4 and NaCl catholytes, respectively. These values are in agreement with the reported one, 5 µg/Ah, in which the corrosion rate of the Pt electrode with a 1 cm^2 surface area was measured in the H_2SO_4 electrolyte up to 1 M [36]. It is clear from Fig. 19(a) that the anode surface in contact with the Nafion 117 membrane dissolved during the operation, and that the surface area of our Pt anode mesh in contact with the Nafion 117 membrane was also on the order of approximately 1cm^2. Although the accuracy was not sufficient for the precise comparison of Pt concentrations in anode water listed in Table II because of the infinitesimal quantity of Pt determined by ICP-MS, the tendency coincides with the results of the weight measurements: The Pt concentration obtained using the 0.5 M NaCl catholyte is higher than that obtained using the 0.5 M Na_2SO_4 catholyte. From these observations it can be assumed that the Pt anode dissolves in anode water faster when the NaCl catholyte was used than when the Na_2SO_4 electrolyte is used. In the case of electrolysis using the NaCl catholyte, Pt mass balance calculation indicates that 20 % of the dissolved Pt is carried out of the system in anode water, and the rest of the dissolved Pt diffuses and migrates to the cathode side. The lifetime of the system can be estimated in terms of Pt weight loss that leads to poor contact between the anode and the Nafion 117 membrane. Because the Pt anode is pressed against the membrane surface using a Ti #40 mesh and a Ti terminal plate, the contact can be maintained fairly well when the Pt weight loss is small. If a 10 % loss in Pt weight is allowed to maintain good contact, the system can be operated up to a weight loss of 40 mg which corresponds to an operation period of 5800 Ah, *i.e.*, 2900 h of operation at 0.25 A/cm^2, using the NaCl catholyte. 6 mg-O_3/L ozone water will be obtained at a current density of 0.25 A/cm^2, and the system can be operated for 8 h a day for 365 days without changing the anode.

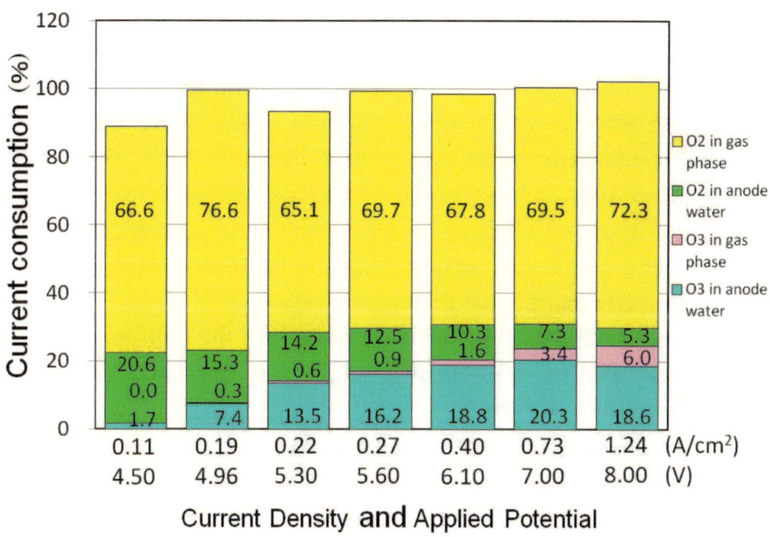

Figure 22. Current consumption balances calculated from production rates of O_3 and O_2 (0.5 M NaCl catholyte was used).

5. Impurities in ozone water

Impurity levels in the exit of the anode water obtained using ion-exchanged water are comparable to those in drinking water, as shown in Table 1. Therefore, ozone water should be safe enough for family use, such as gargles and washing hands. On the other hand, the concentrations of the impurities listed in Table 2, such as Na^+, Ca^{2+}, and NH^{4+} in anode water using the 0.5 M NaCl catholyte, are on the order of 0.01-0.02 wt ppm. They are much lower than those in ion-exchanged water (see Table I). The ozone water obtained in the pure water system consists of highly concentrated O_3 and 7.4 wt ppm Cl^-. Therefore, ozone water can be used in various processes, such as the sterilization of medical instruments. Because Cl^- is undesirable in the cleaning processes of electronic components, Cl^- washout will be required after using the ozone water for such processes.

6. Conclusion

The efficiency of ozone production using a conventional Pt electrode system decreased with run time. The linear increase in $1/[O_3]$ with run time after 2-3 h suggests the existence of a second-order reaction of the Nafion 117 membrane with ozone or its decomposition products. The degradation was caused by Pt particles concentrated on the cathode side surface of the Nafion 117 membrane. The particles catalyze the reaction between the Nafion 117 membrane and ozone to produce holes on the membrane's cathode side surface. To prevent the decomposition of the membrane, a quartz felt separator was inserted between the Nafion 117 membrane and the cathode electrode. As a result, a high current efficiency of more than 20 % and a lifetime longer than 150 h have been achieved in the new electrolysis system. The quartz felt separator succeeded in capturing Pt particles and preventing the membrane from decomposition.

Then the effects of the NaCl and Na_2SO_4 catholytes on the current efficiency and lifetime of the new EOP system were investigated using pure anode water. When a 0.5 M NaCl catholyte was used, a high current efficiency of 29 % and a high power efficiency of 76 kWh/kg-O_3 were achieved at an electrolysis current of 0.5 A/cm^2 (6.6 V). The accelerated degradation experiments indicated that the combined use of pure anode water and the 0.5 M NaCl catholyte kept the Pt anode surface clean and enabled an efficient operation longer than 130 h. On the other hand, the Pt anode surface in contact with the Nafion 117 membrane was covered with a 10-μm-thick a-PtO_2 film and the efficiency decreased from 13 % to 5 % after the accelerated experiments for 60 h using the Na_2SO_4 catholyte. It is suggested that the formation of a-PtO_2 films decreases the electrocatalytic activity of the Pt anode in the O_3 formation reaction. The NaCl concentration dependence measurements indicate that NaCl concentrations higher than 0.085 M will be adequate for the EOP system. Pt dissolution rates of 5.4 μg/Ah and 6.9 μg/Ah were measured using the Na_2SO_4 and NaCl catholytes, respectively. If a 10 % loss in Pt weight is allowed to maintain good contact between the anode and the Nafion 117 membrane, the system using the NaCl catholyte can be operated for 5800 Ah, *i.e.*, 8 h per day for 365 days at 0.25 A/cm^2 without changing the anode. The long lifetime, good current and power efficiencies, very low impurity concentrations, the compactness of equipment and low-voltage operation achieved by the new system will enable a wide expansion of the application of ozone water.

7. Future research

Important breakthroughs remain to be achieved. The authors plan to develop the following advanced EOP technologies: a direct tap water electrolysis system that does not require the use of ion-exchange resin or reverse osmosis; a once-through catholyte flow system that can eliminate the formation and discharge of a strongly alkaline catholyte solution; an effective utilization system of gas phase ozone; a new and effective anode catalyst system. Such kinds of improvements will be achieved soon, because we have ideas and strategies for realizing advanced systems.

To clarify the effectiveness of ozone water for sterilization processes, we are now conducting experiments and developing a simulation model to obtain the orders and rate constants of the reactions of O_3 with bacteria. The results will be reported soon [37].

Author details

Fumio Okada*
Department of Chemical Systems Engineering, University of Tokyo, Tokyo, Japan

Kazunari Naya
Research & Development Office, FRD, Inc., Toda-shi, Japan

Acknowledgement

The authors thank Prof. Y. Yamaguchi, Associate Prof. Y. Tsuji and Assistant Prof. S. Inasawa of the Department of Chemical Systems Engineering, University of Tokyo, for their kind help in the instrumental analysis and for the fruitful discussion. Measurements of the impurity concentrations in water were conducted by the analytical laboratory group in the 5th building of the School of Engineering, University of Tokyo. The content in section 4 was published in Electrochim. Acta 78, pp. 495-501, K. Naya and F. Okada, "Effects of NaCl and Na_2SO_4 cathode electrolytes on electrochemical ozone production" [18], Copyright Elsevier (2012).

8. References

[1] Rice RG and Netzer A, eds., "Handbook of Ozone Technology and Application", Ann Arbor Science (Michigan, 1982).
[2] 2004/EPRI/Global Ozone Handbook: Agriculture and Food Industries; Global Energy Partners, LLC, Lafayette, CA, 2004, 1282-2-04.
[3] Kim JB, Yousef AE, and David S (1999), J. Food Prot. 62: 1071.
[4] Kataoka Y, Miyagawa Y, Koyama S, Naya K, and Fukui K (2004), Proceeding of the 3rd Ann. Meeting of Japanese Society for Functional Water, in Japanese.
[5] Wittmann G, Horvath I, and Dombi A (2002), Ozone Science & Eng. 24: 281.
[6] Mizuno T, Tsuno H, and Yamada H (2007), Ozone Sci. & Eng. 29: 55.
[7] Stucki S, Theis G, Kotz R, Devantay H, and Christen H. J (1985), J. Electrochem. Soc. 132(2): 367.
[8] Feng J, Johnson DC, Lowery S N, and Carey J. J (1994), J. Electrochem. Soc. 141: 2708.

* Corresponding Author

[9] Santana MHP, De Faria LA, and Boodts FC (2004), Electrochim. Acta 49: 1925.
[10] Cheng SA, and Chan KY (2004), Electrochem. and Solid-State Lett. 7(3): 134.
[11] Wang YH, Cheng S, Chan KY, and Li XY(2005), J. Electrochem. Soc. 162: D197.
[12] Awad M. I, Seta S, Kaneda K, Ikematsu M, Okajima T, and Ohsaka T(2006), Electrochem. Commun. 8: 1263.
[13] Arihara K, Terashima C, and Fujishima A (2006), Electrochem. and Solid-State Lett. 9(8): D17.
[14] Kitsuka K, Kaneda K, Ikematsu M, Iseki M, Mushiake K, and Ohsaka T (2010), J. Electrochem Soc. 157: F30.
[15] Marselli B, Garcia-Gomez J, Michaud PA, Rodrigo MA., and Comninellis C (2003), J. Electrochem. Soc. 150(3): D79.
[16] Ed. by Allen J. Bard (1974) "Encyclopedia of Electrochemistry of the Elements, Vol. II, Chapter II-5: Oxygen": M. Dekker, New York.
[17] Okada F and Naya K (2009), J. Electrochem. Soc. 156: E125.
[18] Naya K and Okada F, Electrochim. Acta 78: 495.
[19] Pourbaix M (1966), "Atlas of Electrochemical Equilibria in Aqueous Solutions": Pergamon Press.
[20] Miyake N, Wakizoe M, Honda E, and Ohta T (2005), ECS Transactions 1(8): 249.
[21] Kusakabe K, Kawaguchi K, Maehara S, and Taneda M(2007), J. Chem. Eng. Japan 40(6): 523.
[22] Yamanaka I and Murayama T, Angrew. Chem(2008). Int. Ed. 47: 1900.
[23] Onda K, Ohba T, Kusunoki H, Takezawa S, Sunakawa D, and Araki T (2005), J. Electrochem. Soc. 152: D177.
[24] McBride JR, Graham GW, Peters CR, and Weber WH (1991), J. Appl. Phys. 69: 1596.
[25] Birss VI, Chang M, and Segal J(1993), J. Electroanal. Chem. 355: 181.
[26] Novak DM and Conway BE (1981), J. Chem. Soc. Faraday Trans. I, 77: 2341.
[27] Ed. by the Electrochemical Society of Japan (2000), "Denki-Kagaku Binran 5th Edition": Maruzen, in Japanese.
[28] Shibata S (1963), Bull. Chem. Soc. Japan 36: 525.
[29] Shibata S (1964), Bull. Chem. Soc. Japan 37: 410.
[30] Shibata S (1976), Electrochimica Acta 22: 175.
[31] Gottesfeld S, Yariv M, Laser D, and Srinivasan S (1977), J. Phys. Colloques 38: C5-145.
[32] Gottesfeld S, Maia G, Floriano JB, Tremiliosi-Filho G, Ticianelli EA, and Gonzalez ER (1991), J. Electrochem. Soc. 138: 3219.
[33] Tremiliosi-Filho G, Jerkiewicz G, and Conway BE (1992), Langmuir 8: 658.
[34] Jerkiewicz G, Vatankhah G, Lessard J, Soriaga MP, and Park YS (2004), Electrochimica Acta 49: 1451.
[35] Okada F, Tsuji Y, and Naya K, in preparation.
[36] Ota K, Nishigori S, and Kamiya N (1988), J. Electroanal. Chem. 257: 205.
[37] Kataoka Y, Naya K, and Okada F, in preparation.

Chapter 13

Marine Electrolysis for Building Materials and Environmental Restoration

Thomas J. Goreau

Additional information is available at the end of the chapter

http://dx.doi.org/10.5772/48783

1. Introduction

Within weeks after Alessandro Volta developed the battery in 1800, William Nicholson and Anthony Carlisle applied it to the electrolysis of water, producing hydrogen at the cathode and oxygen at the anode, and thereby showing that water was not an irreducible element, as had been thought, but a chemical compound made up of two elements with very different properties. It was quickly found that adding salts to the water greatly accelerated reaction rates. We now know this is caused by increased electrical conductivity and reduced resistivity, thereby increasing the electrical current flowing for a given applied battery voltage according to Ohms's Law. Humphrey Davy soon applied electrolysis to the practical problem of oxidative corrosion of copper plates used to sheath ships and protect the wood from boring organisms, founding the field of galvanic protection of metals from corrosion, now widely used to protect steel ships, oil rigs, bridges, and subsea pipes from failure.

Seawater electrolysis for galvanic protection can use sacrificial anodes, driven by the voltage potential difference between different metals, or actively impressed currents driven by a battery or a direct current power supply. In the first case the voltage differences are small, usually only tenths of a volt, according to the difference in electromotive potentials of the various metals or alloys used. The metal acting as the cathode is completely protected from rusting and corrosion as long as the electrical current flows. The metal acting as the anode usually dissolves away as the reaction proceeds, and needs to be periodically replaced in order to continue to prevent corrosion of the cathode. Increased currents accelerate reaction rates, which can cause mineral growth or scale, something most uses of cathodic protection wish to avoid. For example, if a boiler is being cathodically protected from rusting, one does not want to precipitate a mineral scale layer on it, because that is less thermally conductive than the metal, and reduces heat transfer and boiler efficiency. Therefore most uses of cathode corrosion protection use the lowest possible voltages and currents needed to prevent rusting, in order to avoid growth of scale.

There is a "natural" analog of cathode protection that is crucial for marine archaeology. A shipwreck invariably contains objects of several different metals, such as various steel alloys, copper, brass, bronze, aluminum, and others. The metal that acts as the strongest anode, according to its electromotive potential, proceeds to dissolve, releasing electrons that flow to the cathode metals, protecting them from oxidation. When the anode has completely dissolved, the next metal in the electromotive series then plays that role until there are no more anodic metals left, and at that point corrosion can take place on the last cathode metal. The process causes growth of limestone scale on the cathode, which protects and conceals it. Metal artifacts preserved in marine shipwrecks have been protected because they acted as cathodes. Despite the popular image of treasure hunters finding shiny golden coins, in fact the treasure is completely encrusted in limestone, appearing as irregular white crystalline lumps with the metal surface completely concealed. The first thing marine archaeologists do with these lumps is to throw them into an acid bath to dissolve away the limestone, in some cases speeding the process up by wiring them up as an anode of a battery, although that risks destroying the artifact if it proceeds too far. Only once the limestone has dissolved can the archaeologist see the metal artifact.

Later applications of aquatic electrolysis included making chlorine and bleach (sodium hypochlorite) from seawater and chloride brines, and purification of metals, but largely under highly controlled conditions in limited volumes, often from fused salts or acid solutions rather than from seawater. Following the First World War, the Nobel Prize winning German Jewish chemist, Fritz Haber, whose work on industrial nitrogen fixation via the Haber-Bosch process is the basis for almost all fertilizer nitrogen production, and hence for our global food supplies, sought to use electrolysis of sea water to extract traces of gold from the ocean to pay back war reparations imposed by the victors. He found that concentrations were too low to be economic, and was then hounded to death by the Nazis.

2. Physical properties of mineral production from sea water

Michael Faraday was the first to precipitate solid minerals by electrolysis of seawater. It was not until 1976 that Wolf Hilbertz recognized that these minerals, under the right conditions, could be a resource rather than a problem to be avoided. Hilbertz, an innovative architect working on self-growing construction materials, experimented with electrolysis of sea water and discovered that by varying the voltage and current applied he could grow different minerals on the cathode, ranging from soft to hard (Hilbertz, 1979). His inspiration was biological: if marine organisms could grow shells and skeletons of precisely controlled architecture from minerals dissolved in seawater, we should be able to figure out how to do so as well. Limestone does not precipitate naturally from seawater, so marine organisms must use their metabolic energy resources in order to create special internal chemical conditions that cause shell growth.

Hilbertz found that under low electrical current conditions he could grow extremely hard calcium carbonate limestone deposits, made up of crystals of the mineral aragonite, the same compound that makes up coral skeletons and the bulk of tropical white sand beaches.

Higher currents caused the growth of the mineral brucite, or magnesium hydroxide, which is soft and tends to easily break off. Through experimentation it proved possible to grow rock-hard limestone coatings of any desired thickness on steel frames of any desired shape or size, at up to 1-2 cm per year, with compressive (load-bearing) strength up to 80 Newtons per square millimeter (MegaPascals), or about three times the strength of concrete made from ordinary Portland Cement.

This material, which Hilbertz first called "Seacrete" or "Seament", is now called "Biorock®" in order to emphasize that this is the only GROWING marine construction material that gets larger and stronger with age, and is self-repairing, like biological materials, but unlike any other marine construction material. This unique property causes any damaged or broken portion to grow back preferentially over growth of undamaged sections.

Figure 1. Biorock materials grown at Ihuru, North Male Atoll, Maldives, around a 6mm diameter steel bar in approximately one year. The darker surface color is a thin film of dried algae that migrates on the surface as it grows outward, leaving the interior bright white. The piece was hacksawed out of a growing structure. There is no corrosion at all on the steel. Photograph by Wolf H. HIlbertz.

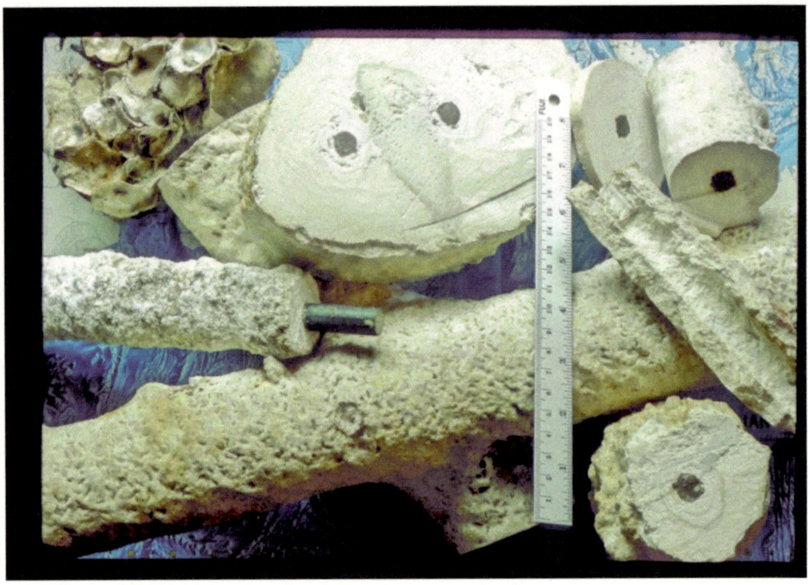

Figure 2. Biorock materials from various locations. The piece at mid left is the one shown in the previous photo. The one at top left, completely overgrown with oysters, is from Louisiana, and all the rest were grown in a two and a half year period at Ihuru Island, North Male Atoll, Maldives. Samples tested from that set of samples in the Materials Testing Laboratory of the University of Graz, Austria, had compressive strength of 60-80 Mega Pascals, around three times the load bearing strength of ordinary Portland Cement concrete. Photograph Wolf W. Hilbertz.

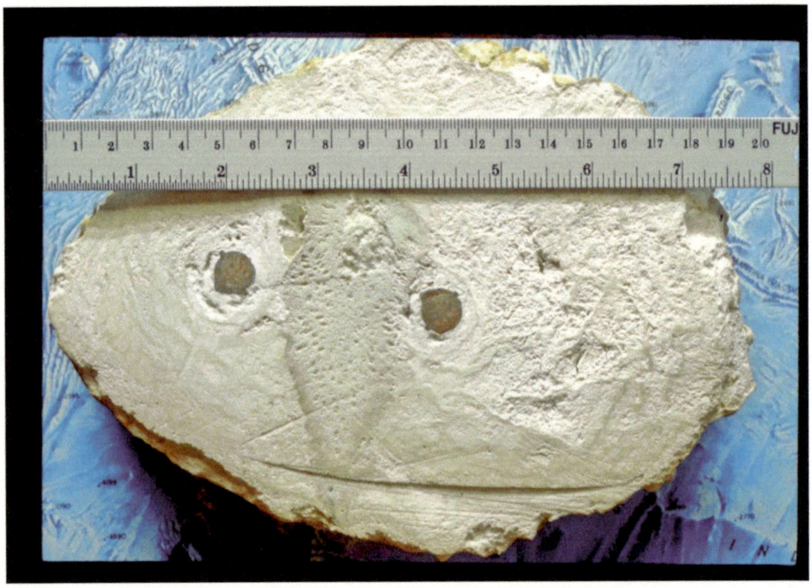

Figure 3. This piece was cut near where two steel bars crossed. We had wedged a coral between the bars after a few months of growth. The coral skeleton is the slightly darker vertically oriented area in the center. After 2.5 years it was completely overgrown and encased by electrochemically produced minerals. Photograph Wolf H. Hilbertz.

22/4/2011

19/4/2011

8/5/2011

27/5/2011

1/6/2011

14/6/2011

15/10/2011

Figure 4. Self repair of Biorock damaged by big boat impact. Time series photographs by Rani Morrow-Wuigk. This structure was installed in June 2000. Note there is no rust on the steel after nearly 11 years in seawater.

The remarkable property of self-healing structures results from the distribution of the electrical field. Initially the electrical gradient between the anode and the cathode results in growth of mineral layers all over the cathode, starting at the closest points, or at sharp extremities that focus electrical field gradients, or at sites where water currents preferentially transport electrons.

Unlike the steel, the minerals are poor electrical conductors, and act as partial insulators. Nevertheless, electrons continue to flow because of the imposed electrical gradient. Although the electrolytic reactions generate hydroxyl ions and alkalinity in the water that are neutralized by mineral deposition taking place at the surface of the metal (see next section), production of hydrogen gas at the cathode surface causes creation of tiny pores and channels from the metal surface to the seawater, out of which hydrogen bubbles emerge (such bubbling provides visible proof that the reaction is working properly). Even Biorock material with three times the load-bearing strength of ordinary concrete has around 20% porosity. While it might be thought that minerals might insulate the cathode and prevent further growth, the imposed electrical gradient ensures that growth continues, in part because electrons flow through the hydrogen escape pores. We observe no long term decrease in the rate of bubbling or the growth of minerals, even in cases where more than 30 cm of hard minerals have grown over the cathode.

When the mineral growth is broken off, whether by severe storm wave damage, boat impacts, or deliberately by pliers, hammers, or hacksaws, and the bare metal is exposed, there is greatly increased growth at that point, until the newly deposited minerals are as thick as adjacent unbroken material. The metal is all at the same voltage potential, but reduced or absent mineral coatings cause the increased electrical current and mass transfer to flow through the water at that point. When the newly grown material is as resistive as the old coating the increased growth rate is self-limiting. In some cases new material is more porous due intense hydrogen bubbling, and the repaired area may grow thicker than adjacent harder and less porous material. We first recognized this focusing of current to freshly exposed surfaces in an experiment using multiple lengths of rebar as cathodes. We would periodically remove one rebar in order to measure the thickness of the material growth, replacing it with a fresh rebar, in an attempt to measure long-term growth rates and changes in chemical composition. The bare steel surface focused the current on the new rebar, which grew at the expense of all the others, stopping their growth. While the experiment did not work as intended, it provided valuable insight into the process.

3. Chemical mechanisms of mineral deposition

A minimum voltage of 1.23 Volts (at standard conditions, plus junction potentials) is needed to initiate electrolysis of water. Water is broken down at the anode to make oxygen gas and hydrogen ion, making the local environment both oxidizing and acidic:

$$2H_2O=O_2+4H^++4e^- \tag{1}$$

Water is broken down at the cathode to make hydrogen gas and hydroxyl ion, making the local environment both alkaline and reducing:

$$4\,H_2O + 4e^- = 2H_2 + 4OH^- \tag{2}$$

The net reaction satisfying charge and pH balance is:

$$6H_2O = 2H_2 + O_2 + 4\,H^+ + 4\,OH^- \tag{3}$$

Above the threshold voltage (the sum of the half cell reactions and the junction potentials), the rate of reaction is proportional to the impressed electrical current, but the voltage determines which reactions can take place.

The hydrogen gas produced at the cathode bubbles out of the water, and could be trapped and used as a valuable side product. This is easy in an enclosed system, but practically impossible in the ocean due to currents and wave surge. The oxygen gas produced at the anode, being more soluble in water than hydrogen, tends to dissolve in the water rather than bubble out (under low to moderate current, bubbling will be greater at high current). Oxygen produced at the anode provides organisms in surrounding areas with this essential element and acts to reduce anoxia and dead zones in the ocean.

Hydrogen ions produced at the anode dissolve in the water until they react with limestone sediments in surrounding areas and are neutralized:

$$H^+ + CaCO_3 = Ca^{++} + HCO_3^- \tag{4}$$

On the other hand the hydroxyl ions produced at the cathode are rapidly consumed by precipitation of limestone directly on the cathode surface:

$$Ca^{++} + HCO_3^- + OH^- = CaCO_3 + H_2O \quad (5)$$

The net reaction is neutral with regard to pH and alkalinity and hence to ocean CO_2 content and acidification (Hilbertz, 1992). Note however that the net effect causes limestone to be deposited in a specific and controlled location on the cathode at the expense of dissolution of limestone in sediment surrounding the anode, so this amounts to moving limestone around to a more useful location and with a controlled shape determined by the size and shape of the cathode.

Limestone deposition does not happen naturally in seawater even though the surface ocean is several times supersaturated with regard to the mineral calcite. Calcite precipitation should take place on thermodynamic grounds, but kinetic factors prevent its nucleation and growth in seawater. This occurs because magnesium ions cover the surface of calcite crystal nuclei, changing the surface free energy to make the seed crystals more soluble and preventing crystal growth (Berner, 1971; Berner et al. 1979). Seawater must be supersaturated several times over to precipitate limestone, and what precipitates is not calcite but the metastable mineral aragonite, whose chemical composition is the same but whose cell lattice is denser, and is more stable at high pressure conditions deep inside the earth than at the surface. Aragonite crystal nuclei do not adsorb a surface layer of soluble magnesium ions and hence can grow in seawater. In fresh water with low magnesium concentrations, calcite will precipitate even though it will not grow in seawater, which has 5 atoms of dissolved magnesium ions for each calcium ion. Organisms that make limestone shells and skeletons, like snails, clams, and corals, must use up metabolic energy to create internal chemical conditions that overcome these nucleation barriers, and also control the form of calcium carbonate produced.

The cathode can be made of any kind of electrically conductive metal or material, which will be completely protected against corrosion by the electrical current, with the sole exception of Aluminum. Aluminum is an amphoteric oxide and is the only common metal that readily dissolves under both alkaline and acidic conditions, so it can't be used as either an anode or a cathode. The anode, being acidic and oxidizing, creates highly corrosive conditions, so most anode metals will dissolve, usually releasing biologically toxic ions into the environment. Either the anode must be replaced as needed, or a special non-corrodible and non-toxic material must be used.

4. Effects of competing electrolytic side reactions

Overcharging the cathode with higher electrical current densities greatly increases hydroxyl ion concentrations, which causes precipitation of the mineral brucite, $Mg(OH)_2$ instead of aragonite. Brucite requires very high pH to precipitate, appears to have little or no kinetic barrier to precipitation, and should grow at a rate proportional to the square of the microsite pH next to the cathode. Brucite, a white mineral similar in appearance to limestone, is structurally weak and flakes off. In seawater of normal pH brucite dissolves, the hydroxyl ions

Figure 5. Brucite crystals grown on Biorock. Scanning electron micrograph by Noreen Buster, US Geological Survey.

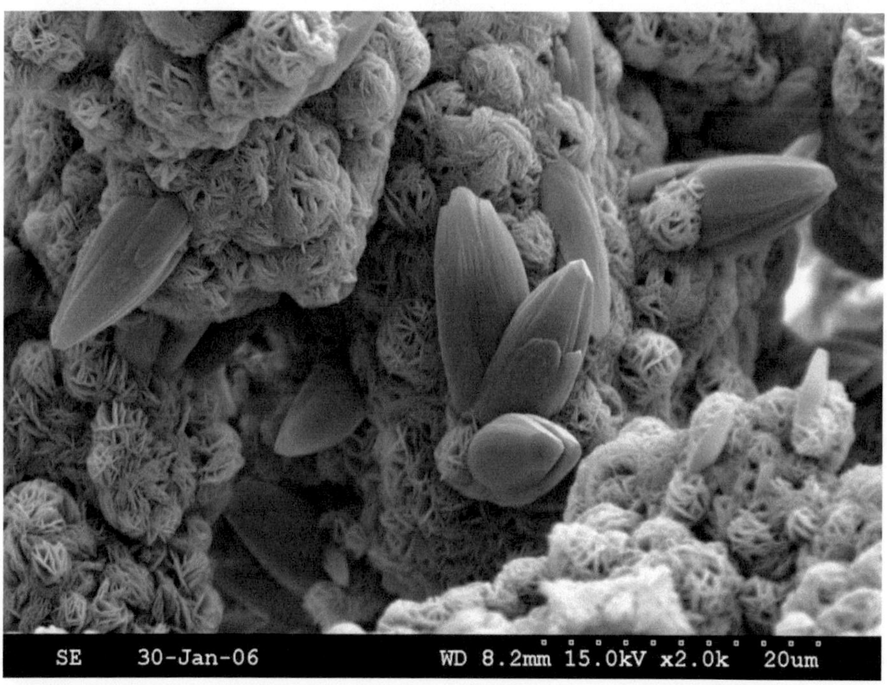

Figure 6. Mixture of Brucite crystals (rosettes) and Aragonite crystals (elongated needles). Scanning electron micrograph by Noreen Buster, US Geological Survey.

raise the pH, and convert bicarbonate ion to carbonate ion, which reacts with calcium ions. Consequently as the material grows and brucite ages it is replaced by aragonite. To optimize strength the Biorock minerals are grown at a low charging rate to produce hard limestone rather than soft brucite. We find experimentally that a growth rate of not more than 1-2 cm/per year provides maximum growth and structural strength, and above that brucite dominates.

These results are strongly affected by temperature, because brucite is a normal mineral whose solubility increases with temperature, while calcium carbonate minerals are extremely unusual in having retrograde solubility, being more soluble in cold water than hot water. As a result materials grow faster and harder in warm tropical waters than in cold boreal waters. In addition the electrical conductivity is directly proportional to the salinity, so growth rates are highest in very salty waters and brines, lower in brackish waters, and very small in pure fresh water. The aragonite chemical composition, as measured by X-ray fluorescence, is indistinguishable from that of coral skeletons, being essentially pure calcium carbonate with about one percent strontium substitution in the aragonite lattice and only trace amounts of magnesium and other metals. However growth of minerals can trap sediment material suspended in the water that lands on the limestone as it grows, affecting the color of the Biorock minerals. They are pure white on remote limestone islands, but grey where there are a lot of clay minerals in the water, and can even be red where there are lots of iron oxide minerals in suspended sediments.

Because at high current densities direct brucite precipitation removes hydroxyl ion without converting bicarbonate to carbonate ion, it also reduces the amount of CO_2 produced by limestone deposition:

$$Ca^{++} + 2HCO_3^{-} = CaCO_3 + H_2O + CO_2 \tag{6}$$

This is an interesting point because limestone deposition is, along with volcanic outgassing, the major source of atmospheric CO_2 on a geological time scale, while dissolution of limestone, along with weathering of aluminosilicate minerals, is the major sink. This is widely misunderstood by those not knowledgeable about the chemistry of the carbon cycle. Almost everyone seems to think that limestone deposition, which is a sink of oceanic bicarbonate, must also be a sink of atmospheric CO_2, when in fact it is a source! This common error is due to the fact that bicarbonate is the major form of inorganic carbon in the ocean, and because the ocean is a pH-buffered chemical system. In effect for each molecule of bicarbonate precipitated as limestone one molecule is released as CO_2 in order to maintain charge and pH balance. Therefore brucite formation at the expense of aragonite has a net effect of reducing the effects of ocean acidification caused by increased CO_2 in the atmosphere.

However, to put this into perspective, about half of all the net limestone burial in the ocean used to take place in coral reefs (Milliman, 1993), at least back when coral reefs were healthy and growing, before global warming, new diseases, and pollution killed most of them. About an order of magnitude more limestone was formed by planktonic organisms, but almost all of that dissolves when their microscopic skeletons fall into deep water, where they dissolve because of the lower temperature, higher pressure, and the higher acidity of deep waters caused by decomposition of organic matter that is formed at the ocean surface by photosynthesis and falls to the deep sea where it is oxidized by decomposing organisms and bacteria. However, the rate at which we are now adding CO_2 to the atmosphere from fossil fuel combustion is about 100 times greater than the natural sources from global limestone burial (Ware et al., 1991), indicating how greatly human pollution has overwhelmed natural sources. Consequently global ocean acidification caused by fossil fuel-caused CO_2 buildup cannot be effectively countered by manipulating limestone deposition, unless fossil fuel CO2 sources are greatly reduced and a mechanism is developed to directly remove CO2 from the atmosphere. If allowed to build up in the atmosphere, fossil fuel CO2 will only be very slowly neutralized over hundreds of millennia to millions of years by dissolution of terrestrial limestone rocks on land and marine limestone sediments.

Above 1.36 volts chloride, the most abundant anion in sea water, is converted to chlorine gas at the anode:

$$2Cl^- = Cl_2 + 2e^- \quad (7)$$

If the voltage could be maintained between 1.23 and 1.36 volts (ignoring junction potentials) then chlorine production can be avoided entirely, but this requires very precise regulation and is made more complicated by junction potentials. In practice, sufficient overvoltage to overcome junction potentials makes some chlorine production unavoidable, but the lower the voltage the less is produced.

Oxygen production is strongly favored over chlorine production because the ocean has far higher concentrations of water molecules than chloride ions (96.66% versus 1.94% by weight under standard ocean salinity of 35 parts per thousand). The ratio between oxygen production and chlorine production can be calculated from the water and chloride concentrations and the voltages applied using the Nernst Equation, but there will always be far more oxygen production than chlorine. However chlorine, as a highly reactive oxidizing agent, can build up in closed systems like aquaria or tanks, and pose problems, for example for fishes in tanks whose gills are highly sensitive to chlorine, or marine mammals whose eyes are affected. In the ocean chlorine is rapidly neutralized by reaction with dissolved organic matter and reduced compounds and elements. We have repeatedly observed that it poses no problem at all for life in the ocean, with fish and corals growing well no farther than a millimeter or two from the anode. Fish swim near the anode, and dissolution of limestone usually takes place only a very short distance away from the anode, removing the acidity produced. But in aquaria with no limestone sediments, the acidity can remain in the water.

The chlorine production side reaction competing with water at the anode also provides another side benefit in that hydrogen ion is not produced by the electron flow, so this acts to make the net reaction at both electrodes one that makes the water more alkaline, and therefore acts to locally reverse ocean acidification from increased atmospheric CO_2. However, as noted before, the effects are small on a global scale, so large-scale electrolysis in the ocean, while LOCALLY reversing ocean acidification, has only a small impact on the ocean's BULK acidity, and only abatement of fossil fuel sources and direct removal of excess atmospheric CO_2 can reverse global ocean acidification.

An interesting variant of this process has been proposed as way to mitigate ocean acidification caused by atmospheric CO_2 buildup by House et al, 2007 and by Rau and Carroll, 2011. They suggest packing the area around the anode with basic minerals, like limestone or igneous rocks high in calcium and magnesium. The acidity at the anode would then increase the dissolution and weathering of these minerals, which serves as a CO_2 sink. They suggest that the water would then turn alkaline, which would promote the dissolution of CO_2 from the atmosphere. However as noted above, the alkalinity generated at the cathode is immediately neutralized by mineral deposition and so would not build up in the water and absorb CO_2 from the air. Furthermore there would be enormous costs for transporting bulky rocks from land to the site of electrolysis, and it is likely that the benefits for reversing ocean acidification or CO_2 buildup would be small.

5. Efficiency and cost of mineral production

The fact that limestone minerals, harder than ordinary concrete, can be grown in the sea in any size and shape, naturally raises the question whether doing so is cost-effective. Hilbertz and Goreau did an experiment in the 1980s at the Discovery Bay Marine Laboratory in Jamaica in which a new battery of known voltage and amp hours was completely discharged through electrodes and the amount of minerals grown on the anode was weighed. The yield was 1.07 Kilograms/Kilowatt hour, very close to the theoretically expected value. A field experiment done in the sea at the Marina Hemingway, near Havana, Cuba measured values of around 0.4-0.5 Kg/KWh (Amat et al., 1994). At this site there were many large steel structures in the water nearby, which attracted stray currents and reduced measured efficiency of mineral production on the cathode.

When one balances the chemical and charge equations, and assuming that all the hydroxyl ions produced by electrolysis of water are neutralized by limestone deposition, one gets 3.7 grams of calcium carbonate per amp hour of electricity.

To calculate the efficiency as yield per watt one must assume a voltage. The Jamaica experiments were done at 1.5 volts, and the Cuban ones at 6 volts. The lower the voltage is (as long as it is above the minimum voltage of 1.23V for electrolysis of water and ignoring junction potentials) the more efficient the process is (Table 1). For standard solar panels at 17 volts, only around 7% of the potential energy is used, and nearly 93% is wasted.

VOLTAGE (VOLTS)	EFFICIENCY (PERCENT)
1.23	100
1.5	82
3	41
6	20.5
12	10.25
17	7.24

Table 1.

Using 6 volts we get a limestone yield of 0.62 Kilograms of calcium carbonate per Kilowatt-hour, which is close to what the Cuban researchers found in the field despite stray current losses!

For high charge rates producing brucite, one produces half as many molecules of brucite for the same charge, because only one hydroxyl ion is needed for each calcium carbonate molecule, but two are needed for each brucite. As brucite molecules weigh 68% as much as limestone, the efficiency in weight produced per kilowatt should be one third that of limestone.

In addition for every two molecules of calcium carbonate (or one molecule of brucite) produced one also produces one molecule of hydrogen gas, which can be used as a fuel in fuel cells. And one would also be producing oxygen and chlorine at the other terminal in a ratio that depends on the voltage and can be calculated from the Nernst Equations.

The energy efficiency of production is inversely related to the voltage above the minimal value for seawater electrolysis because higher voltages produce electrons with much more energy than is needed to break down water, so the excess is wasted as heat. We have never felt or measured significant increases in temperature, so the effect seems to be very small in practice. This decrease of efficiency at higher voltages is equally true of efficiency of hydrogen production using photovoltaic panels. This fact was completely missed in a major review of the subject (Blankenship et al. 2011), which consequently greatly overestimated efficiency of the photovoltaic hydrogen production process. The previous generation of 17 volt photovoltaic panels cause nearly 93% of the potential energy to be wasted when applied to electrolysis for hydrogen production. Such 17 volt panels are now no longer being manufactured, while the new panels, with 24, 48, 60 volts or higher will be even more inefficient for Biorock materials or hydrogen production end uses, so it is clear that efficient use of power requires voltages matched to the minimum end use requirements.

If we assume that the yield is 1 Kg/KWh and that electricity costs from $.03 to $.30 per KWh, the electrical cost of the materials produced ranges from $.03-.30/Kg. This would be highly competitive with cement in many places where transport of cement affects the local cost, especially in small islands surrounded by the sea where cement is expensive because of transport costs.

The materials that are produced, if grown slowly, have a load bearing strength of around 80 MegaPascals, about 3 times stronger than concrete from ordinary Portland Cement, and can be grown in any size or shape. Wolf Hilbertz's original vision was to grow prefabricated construction materials, like roofs, walls, arches, blocks, etc. in the sea and then use them on land for construction. The most effective use would be in what architects call "shells", structures that are thin with regard to their other dimensions like domes, and whose strength in large part comes from tensile forces. Unfortunately the construction market wants buildings immediately, and is rarely willing to wait years for the material to be grown slowly and hard, when concrete will set in days. In addition, in the late 1980s our Biorock work switched away from building material applications to focus on coral reef restoration, and we never had a chance to get back into the construction aspects that Wolf had intended. However the principle is still valid, and such structures would be cost effective in many places far from cement plants.

By applying higher current densities, mineral production can be readily switched from calcium carbonate to magnesium hydroxide. While this material is soft, flaky and not useful for load bearing uses, it has many other applications. This material can be cast in molds to form bricks and blocks or other shapes, and we have done so successfully. Brucite can be readily converted into magnesium carbonate cements by absorbing CO_2 and these are even harder than calcium carbonate.

$$Mg(OH)_2 + CO_2 = MgCO_3 + H_2O \quad (8)$$

The manufacture of Biorock cements therefore removes CO_2 from the atmosphere as they set. In contrast cement manufacture, which combusts limestone to make quicklime, releasing CO_2, is a major global source of greenhouse gases, about 5-10% as much as fossil fuel combustion. Therefore Biorock cements can be readily produced on a large scale that are far harder than contemporary cements and help to reduce global warming instead of causing it like conventional cements do.

Use of Biorock cements can therefore help undo the global warming that cement manufacture contributes to, and hence are truly "green" cements as long as sustainable energy resources like solar, wind, wave, biomass, or tidal current energy are used to make the electricity for its manufacture. We have used all of these energy sources, and currently work with top pioneering groups in the development of all of these energy technologies for growing Biorock materials.

Magnesium carbonate cements are far harder than either calcium carbonate or concrete, and were widely used by the Romans. Roman ruins in Italy built of limestone or marble blocks cemented with magnesium carbonate cements reveal that the limestone is dissolving with acid rain, while the cements are much more resistant. The cements stick out while the building blocks are caved in from dissolution by rain, the opposite effect of bricks whose mortar is crumbling. Using Biorock technology it is now possible to produce such cements in any desired quantity from seawater and hypersaline lakes and lagoons.

6. Biological responses

The biological effects of the electrical fields produced by seawater electrolysis result in astonishing increases in the settlement, growth, survival, and resistance to environmental stress of almost all marine organisms (Hilbertz & Goreau, 1996). Commercial divers on cathodically protected oil rigs, who spend much of their time replacing sacrificial anodes, have to spend a great deal of time scraping off the prolific growth of corals, oysters, and other marine organisms, which are far less abundant on rusting oil rigs that lack cathodic protection. Wolf Hilbertz's first Biorock structures, built at Grand Isle, Louisiana, near the mouth of the Mississippi River, were completely overgrown with oysters that spontaneously settled on them and grew very rapidly. The first experiments with corals done at the Discovery Bay Marine Laboratory in Jamaica in the late 1980s used small pieces of corals attached to Biorock structures. They grew at record rates, up to nearly a centimeter a week (Goreau & Hilbertz, 2005). The results were so dramatic that after 1987 we immediately focused all of our efforts on coral reef restoration, as this is the most sensitive of all ecosystems to increases in temperature, sedimentation, and pollution.

Subsequent work by us and our students on hundreds of Biorock projects across the Atlantic, Pacific, Indian Ocean, and Southeast Asia showed that corals, oysters, seagrasses, saltmarsh, and apparently almost all marine organisms, had much higher settlement, growth, survival, and resistance to environmental stresses, including high temperatures, than genetically identical controls in the same habitat. For example coral growth rates are typically 2-6 times faster than controls, depending on species and conditions, and survival of corals from severe high temperature bleaching stress were 16-50 times higher (Goreau & Hilbertz, 2005). The data are presented in a 2012 book in press, Innovative Methods of Ecosystem Restoration, so the details will not be repeated here. Abstracts of most of the papers in that volume were presented at the Symposium on Innovative Methods of Marine Ecosystem Restoration at the 2011 World Conference on Ecological Restoration, Merida, Yucatan, Mexico, and can be found at: http://www.globalcoral.org/world_conference_on_ecological_r.htm

Initially it was thought that the benefits were due to the higher pH around the Biorock structure, but direct pH measurements showed that the hydroxyl ions generated at the cathode were immediately neutralized by mineral deposition, and very little pH change could be measured in the water even very close by, unless the limestone was broken off to expose the bare shiny metal surface. In addition it was noted that organisms without limestone skeletons also had extraordinary growth rates, that the organisms on the structure had much faster budding and branching, brighter colors, and that there was greater coral settlement and growth in the areas AROUND the structures, not just on them. These benefits were observed to disappear when the power was turned off and growth decreased to levels similar to the controls, but immediately resumed when the power was turned back on. For much more detailed data on biological benefits of electrical fields and their interpretation, please see the forthcoming book, Innovative Methods of Ecosystem Restoration.

Applications of electrolysis to biological phenomena precede even the invention of the battery. In 1791 Luigi Galvani published his book on "Animal Electricity", based on experiments with static electrical discharges that caused the limbs of dead frogs to twitch. This lead to the discovery of electrical propagation in nerves, and a long series of experiments on the effects of electrical currents on limb healing and regeneration in frogs and salamanders, followed by work on electrical stimulation of brains (Becker & Selden, 1985). Much of this work used high voltages or alternating current, and so are fundamentally different phenomena than the low voltage direct currents used in our work. The use of electroshock therapy gave the entire a field a bad name, and a reputation for quackery, so that the legitimate scientific applications of low voltage electrical fields were ridiculed and neglected, to the detriment of scientific understanding (Oschman, 2000, 2003). The "snake oil" or "Frankenstein" reputation has unfairly tarnished the serious science of biological/electrical interactions due to the bogus claims of charlatans or deluded people.

While full explanation of the effects of electrical fields at the biophysical and biochemical level requires further work, the empirical results show enormous benefits for biological health when they are in the right range. Electrical fields that are too low will have little impact, and those that are too high might well have negative impacts, with maximum benefits at some intermediate value. It is long known that organisms maintain a voltage gradient across their cell membrane of around a tenth of a volt, and that they must expend energy to maintain this gradient by enzymatically pumping cations and anions. The resulting voltage gradient drives flows of electrons and protons that are tapped by enzymes to form the high energy biochemical metabolites that serve as the cell's "energy currency", driving synthesis of compounds whose formation would otherwise be thermodynamically prohibited.

Thus electrical gradients of the right magnitude effectively provide living cells with available biological energy at lower cost, leaving them with extra energy for growth, reproduction, healing, and resisting environmental stress. Optimizing these benefits will take much further work on mechanisms. When these are fully explored, the benefits in terms of higher growth and better health will certainly prove revolutionary in many fields of biology, and result in more productive forms of mariculture, aquaculture, and agriculture.

7. Applications

To date electrolysis has largely been used to develop chemical processes in closed systems with controlled chemical composition. The only large-scale environmental application, corrosion control, is operated at the lowest possible level in order to prevent or minimize the applications described in this article. The work we have done since the mid 1970s opens the door to large-scale environmental applications of many novel kinds. These include:

1. protecting coral reefs against global warming, sedimentation, and pollution
2. restoring coral reefs where they have died or been degraded
3. restoring oyster reefs where they have died or been degraded
4. restoring fish habitat
5. restoring shellfish habitat
6. restoring seagrasses
7. restoring saltmarshes
8. mariculture
9. shore protection from erosion and global sea level rise
10. construction materials
11. hydrogen production
12. agricultural applications

Biorock applications involve low voltage and low current densities, and so do not use much electricity, in fact they usually cannot be felt even when one short circuits the system by grabbing the anode and cathode simultaneously with bare hands, since the electrons flow through much more conductive seawater. Using Biorock technology, coral reefs can be grown in front of hotels, which grow the beaches back using about as much electricity as the beach lights, or one or two air conditioners. This is a negligible amount of electricity for places that may be running hundreds of air conditioners at a time, and so the benefits far outweigh the costs. Biorock structures cost a small fraction of the cost of concrete or rock structures with the same dimensions. Reinforced concrete construction first assembles a framework of reinforcing bar, which is a negligible portion of the total structure cost. The concrete poured around it, and the labor, cost many times more than the steel. Biorock construction assembles a steel framework, but instead of purchasing concrete simply wires it to a power source and grows the material over the steel.

Since steel is the cheapest and most available construction material, Biorock costs are largely dependent on the price of electricity. Since most electricity is produced from fossil fuels like coal, oil, and natural gas, it is the largest source of greenhouse gases causing the global warming that is now the major killer of corals worldwide. For this reason we work very closely with the pioneers in sustainable energy systems, in particular wave, tidal, wind, and solar power, so that untapped renewable local energy sources can be used that do not generate CO_2. We are especially focused on use of the development of new wave energy generators that work in waves of less than 10 cm amplitude, which will allow energy to be made along almost any coastline most of the time.

Generation of electricity on-site from renewable energy also avoids power losses in transmission, and will allow much larger structures to be grown with less energy. This will open the possibilities of very large environmental electrolysis projects to save entire coastlines from the effects of global sea level rise and restore their collapsing coral reefs, oyster reefs, and fisheries, while at the same time promoting the development of sustainable energy sources that do not produce produce CO_2 and cause global warming and sea level rise.

Biorock reefs grown in front of severely eroding beaches, with trees and buildings collapsing into the sea, have grown back up to 15 meters (50 feet) of new beach back in a few years, by reducing wave impacts a the shoreline. Therefore they will have major applications as global sea level rise accelerates in the future. Artificial islands can be grown that keep pace with sea level rise, if Biorock technology was used on a large scale.

Author details

Thomas J. Goreau
President, Global Coral Reef Alliance, USA

Acknowledgement

This paper is dedicated to the memory of the late Wolf Hilbertz (1938 – 2007), the inventor of the Biorock Process and an innovator in new applications of electrolysis. This paper is based on 25 years of close work with him in developing the Biorock technology that emerged from his pioneering vision.

8. References

Enrique Amat Balbosa, Juan Carlos Prada, Frank Moore Wedderborn, & Eumelia Reyes Reyes, 1994, Estudio preliminar de la acrecion marina, Revista Arquitectura y Urbanismo, Vol XV, No. 243:1-16

Robert O. Becker & Gary Selden, 1985, The Body Electric: Electromagnetism and the Foundation of Life, William Morrow, New York

Robert A. Berner, 1971, Principles of Chemical Sedimentology, McGraw Hill, New York

Robert A. Berner, J. T. Westrich, R. Graber, J. Smith, & C. S. Martens, 1978, Inhibition of aragonite precipitation from supersaturated seawater: a laboratory and field study, American Journal of Science, 278:816-837

Robert E. Blankenship, David M. Tiede, James Barber, Gary W. Brudvig, Graham Fleming, Marina Ghirardi, M. R. Gunner, Wolfgang Junge, David M. Kramer, Anastasios Melis, Thomas A. Moore, Christopher C. Moser, Daniel C. Nocera, Arthur J. Nozik, Donald R. Ort, William W. Parson, Roger C. Prince, & Richard T. Sayre, 2011, Comparing photosynthetic and photovoltaic efficiencies and recognizing the potential for improvement, Science, 332:805-809

N. A. Buster, C. W. Holmes, T. J. Goreau, & W. Hilbertz, 2006, Crystal habits of the Magnesium Hydroxide mineral Brucite within Coral Skeletons, American Geophysical Union Annual Meeting, Abstract and Poster

T. J. Goreau & W. Hilbertz, 2005, Marine ecosystem restoration: costs and benefits for coral reefs, WORLD RESOURCE REVIEW, 17: 375-409

Wolf H. Hilbertz, 1979, Electrodeposition of minerals in sea water: Experiments and applications, IEEE Journal on Oceanic Engineering, 4:1-19

Wolf H. Hilbertz, 1992, Solar-generated building material from seawater as a sink for carbon, Ambio, 21, 126–129

W. H. Hilbertz & T. J. Goreau, 1996, Method of enhancing the growth of aquatic organisms, and structures created thereby, United States Patent Number 5,543,034, U. S. PATENT OFFICE (14pp.).

Kurt Zenz House, Christopher H. House, Daniel P. Schrag, & Michael J. Aziz, 2007, Electrochemical acceleration of chemical weathering as an energetically feasible approach to mitigating anthropogenic climate change, Environmental Science and Technology, 41:8464-8470

John D. Milliman, 1993, Production and accumulation of calcium carbonate in the ocean: Budget of a nonsteady state, Global Biogeochemical Cycles, 7:927-957

James L. Oschman, 2000, Energy Medicine: The Scientific Basis, Churchill Livingstone, Edinburgh

James L. Oschman, 2003, Energy Medicine in Therapeutics and Human Performance, Butterworth Heinemann, Edinburgh

Greg H. Rau & Susan A. Carroll, 2011, Electrochemical enhancement of carbonate and silicate weathering for CO_2 mitigation, Goldschmidt Conference Abstracts, Mineralogical Magazine, p. 1698

John R. Ware, Stephen V. Smith, & Marjorie L. Reaka-Kudla, 1991, Coral reefs: Sources or sinks of atmospheric CO_2?, Coral Reefs, 11:127-130

World Conference on Ecological Restoration, 2011, Abstracts: http://www.globalcoral.org/world_conference_on_ecological_r.htm